Springer Optimization and Its Applications

Volume 185

Aims and Scope

Optimization has continued to expand in all directions at an astonishing rate. New algorithmic and theoretical techniques are continually developing and the diffusion into other disciplines is proceeding at a rapid pace, with a spot light on machine learning, artificial intelligence, and quantum computing. Our knowledge of all aspects of the field has grown even more profound. At the same time, one of the most striking trends in optimization is the constantly increasing emphasis on the interdisciplinary nature of the field. Optimization has been a basic tool in areas not limited to applied mathematics, engineering, medicine, economics, computer science, operations research, and other sciences.

The series **Springer Optimization and Its Applications (SOIA)** aims to publish state-of-the-art expository works (monographs, contributed volumes, textbooks, handbooks) that focus on theory, methods, and applications of optimization. Topics covered include, but are not limited to, nonlinear optimization, combinatorial optimization, continuous optimization, stochastic optimization, Bayesian optimization, optimal control, discrete optimization, multi-objective optimization, and more. New to the series portfolio include Works at the intersection of optimization and machine learning, artificial intelligence, and quantum computing.

Volumes from this series are indexed by Web of Science, zbMATH, Mathematical Reviews, and SCOPUS.

More information about this series at http://www.springer.com/series/7393

Dionysis D. Bochtis • Simon Pearson
Maria Lampridi • Vasso Marinoudi
Panos M. Pardalos
Editors

Information and Communication Technologies for Agriculture—Theme IV: Actions

Springer

Editors
Dionysis D. Bochtis
Institute for Bio-economy
and Agri-technology (iBO)
Centre for Research & Technology
Hellas (CERTH)
Thessaloniki, Greece

Simon Pearson
Lincoln Institute for Agri-Food
Technology
University of Lincoln
Lincoln, UK

Maria Lampridi
Institute for Bio-economy
and Agri-technology (iBO)
Centre for Research & Technology
Hellas(CERTH)
Thessaloniki, Greece

Vasso Marinoudi
R&D Department
Engineers for Business A/S
Thessaloniki, Greece

Panos M. Pardalos
Department of Industrial
and Systems Engineering
University of Florida
Gainesville, FL, USA

ISSN 1931-6828 ISSN 1931-6836 (electronic)
Springer Optimization and Its Applications
ISBN 978-3-030-84155-3 ISBN 978-3-030-84156-0 (eBook)
https://doi.org/10.1007/978-3-030-84156-0

This Springer imprint is published by the registered company Springer Nature Switzerland AG
The registered company address is: Gewerbestrasse 11, 6330 Cham, Switzerland

Preface

The Fourth Industrial Revolution in the primary sector is captured by the term Agriculture 4.0, which represents the evolution of digital technologies in four aspects. The first aspect is related to the use of data, computing power, and connectivity aiming at the storage and systematization of the information collected leading to the rapid evolution of Big Data, Open Data, Internet of Things (IoT), Machine-to-Machine, and Cloud Computing. Decision Support Systems (DSS) are enabled through the use of IoT processes that integrate data deriving from various sensors and convert them to decisions and actions. These processes include the successful connection between different sensors, the collection of data as well as their processing in order to supply a DSS with all the required information. The simultaneous consideration of environmental, climatic, and cultural factors results in products of increased quality and in input reduction (savings on pesticides, fertilizers, energy, etc.) improving also the sustainability of agricultural systems. The second aspect is the introduction of "Machine Learning," which facilitates decision-making as well as machine performance from the constant collection and analysis of data. The third aspect concerns human–machine interaction through interface processes. Lastly, the fourth aspect involves all the innovative technology that is used for cost and energy rationalization through performance optimization including Robotics, Machine-to-Machine interactions, etc., facilitating the transition from the digital to real.

The above are the directions of Agriculture 4.0 which incorporate the transition to a new era of action in the agricultural sector. Farmers want to improve the sustainability of their production in all of its pillars (Environment, Economy, and Society) by investing on their business. Conventional agricultural practices were gradually improved by the introduction of precision agriculture while the further improvement of technology has led to Agriculture 4.0 which utilizes interconnected technologies that result to yield and sustainability performance improvement and to the increase of production and processing quality. Basic aim is the systematization and automation of agricultural processes leading to increased production. However, the present times call for immediate actions with respect to the transition of agriculture to sustainable agriculture. To that end, certain challenges must be addressed.

The Food and Agriculture Organization has identified these challenges that concern all three pillars of sustainability. Basic challenge is the increase of agricultural productivity in a sustainable manner in order to meet the increasing demand. However, towards that direction a sustainable natural resource base must be ensured. Sustainable agriculture should also address climate change and the intensification of natural hazards, preventing also transboundary pests and diseases. From the socioeconomic perspective, sustainable agriculture must aim towards extreme poverty and inequality elimination, ending hunger and all forms of malnutrition. The basic stimulants of human migration should also be addressed by improving income-earning opportunities in rural areas. Moreover, resilience towards crises, disasters, and conflicts should be increased while food systems must be made more efficient and inclusive. All the above call for coherent and effective national and international governance.

Many countries around the world adopt strategies towards sustainable agriculture. For example, the Italian Government based the development of rural areas on four critical domains. The first domain concerns the safeguarding of agricultural systems in terms of economic efficiency, profitability, sustainability, and stability with respect to crop, livestock, and forestry activities. The second focuses on the supply of environmental services together with the mitigation of climate change along with the preservation of natural resources and the protection of biodiversity. The third domain includes the continuous and undisturbed production of safe and high-quality food while the fourth involves the development of relationship between agriculture and local communities focusing on the improvement of the quality of life in agricultural areas.

The first chapter of the book **"Towards Sustainable Agriculture: Challenges from the Transition to the New Digital Era"** describes the challenges faced by modern agriculture in safeguarding food security and the need for transition towards the new digital era. A variety of innovative technologies that aim at addressing the challenges of modern agriculture in a sustainable manner are presented including precision agriculture and alternative farming while the importance of farmers' well-being is also addressed through an overview of ergonomics in agriculture. Additionally, the socioeconomic issues that arise from the adoption of technological innovations are discussed along with the most important factors that influence their effective integration.

Based on the experience, so far, from the application of cutting-edge technologies in agriculture, there is strong evidence that they can contribute to improving the sustainability of agricultural systems. However, for successful integration, the issues that arise when farmers adopt innovative systems need to be addressed. These technologies are constantly evolving, adapting to the growing demands of users, making it difficult for unfamiliar users to adapt immediately. Also, the cost of these technologies is another deterrent to their widespread application, especially in the case of small agricultural systems, as their complete economic and technical analysis is not yet sufficient. Taking the previously mentioned into account, the adoption rate may vary among farmers while there is still no consensus with respect to the attributes that favor or discourage the adoption of ICT technologies in agriculture.

Nevertheless, a positive relationship is observed between adoption and wealth, experience, education, participation in cooperatives, access to information, proximity, and size of farms as well as credit accessibility. It is undoubtful that ICT use requires basic computer skills from the farmers' side, while the role of specialists (agronomists, researchers, etc.) must be redefined as fundamental for successful guidance and training.

Nevertheless, farmers remain hesitant in adopting innovative technologies due to the still uncertainty of the poorly presented sustainability benefits. The delayed penetration of digital technologies in agriculture may impede the development of the sector in a regional and international level. Thus, and as a first step before investing in ICT technologies, farmers need to understand the potential benefits of this adoption in terms of improving the sustainability of agricultural processes. To this end, the chapter **"Sustainability in a Digital Farming Era: A Cyber-Physical Analysis Approach for Drone Applications in Agriculture 4.0,"** attempts to assess the integration of unmanned aerial vehicles (UAVs) in agriculture. The use of Unmanned Aerial Systems for the facilitation of precision farming operations constitutes a promising innovation as it facilitates the increase of production abiding by sustainability principles. Indicative application of drones in agriculture may include, among others, agrochemical spraying, livestock tracking, remote sensing of crop and soil health, field maturity, harvest readiness as well as crop insurance claims assessment. Moreover, the execution of monitoring tasks offers the most important benefits of drones' use. These tasks may include weed detection, assessment of nitrogen treatments on crops, biomass monitoring, water stress identification, and field mapping. Considering the severity of the consequences of water scarcity on crop health, the chapter focuses on the facilitation of drone use for sustainable farming processes, focusing on monitoring water stress providing information for precision irrigation. Two research questions are posed by the authors that are examined via a multi-method approach. The first concerns the benefits and challenges associated with the application of UAVs in farming operations. The second regards the applicability of "digital twins" in precision farming operations for ensuring water stewardship.

For the first research question, a critical literature review was performed in order to document the benefits and the drawbacks of UAV application in agriculture, while for the second a framework for the analysis of "digital twins" is proposed. More specifically, this methodological framework, focusing on UAVs, investigates the differences between the cyber space analysis and the physical space testing of digital technology systems. For that purpose, the water stress of individual trees was monitored with the help of an emulation modeling tool that captures a UAV navigating across a theoretical orchard. Using the model developed, two real-world pilot case studies were examined in field. The UAVs use carried sensors for water status monitoring of individual plants, feeding with information for precision irrigation. The chapter contributes to the field by promoting an operationalization perspective of digital technologies for sustainable agriculture. Simulations of real-world conditions along with pilot implementations can assist towards the realization of cyber-physical interfaces that lead to a more effective evaluation, facilitating the integration of drones in agricultural activities.

In the chapter, **"Digital Technologies in the Context of Energy: Focus on the Developing World Agriculture"** the relationship between agriculture, rural electrification, and digital technologies is investigated towards poverty alleviation, especially in rural areas of the developing world. The United Nations have recognized extreme poverty as one of the major challenges that world faces and rural development is acknowledged as an important contributor for the eradication of poverty in these areas. Agriculture, and the evolution of the processes and the technologies used can play a vital role, being the single most important economic sector in poor rural regions. To that end, the potential of investing on agricultural electrification is examined, as it can strongly affect the quality and quantity of agricultural production. The increasing of production leads to an increase in income, triggering further economic activities, promoting overall development. Digital technologies are very important in the realization of rural electrification as they can minimize costs of establishment, addressing also technical and non-technical issues that emerge in the operation stage. Consequently, their application should be pursued in order to rip most of the benefits in the future.

Achieving agricultural sustainability requires the holistic assessment of production systems in the entire value-chain which is the aim of the chapter **"A Circular Precision Farming System Towards the Optimization of Dairy Value-Chains."** The improper use of resources in agriculture (e.g., water, fertilizers, agrochemicals) that is usually related to conventional agricultural practices contributes to the majority of the adverse agricultural environmental impacts. The 10% of the total greenhouse gas emissions in the EU-28 are attributed to agriculture. Enteric fermentation from ruminants, manure decomposition as well as soil nitrification and denitrification are considered as the main sources. In particular, agriculture contributes to over 90% of the ammonia emissions with the number increasing rapidly from 2012. These emissions come at a great financial cost for EU as, for example, the impact of nitrogen pollution's environmental impact which is estimated between a total 70 and 320 billion euros per year. Considering the above, conventional farming cannot be characterized as cost-effective while it contributes to environmental degradation.

Farmers are very often unaware of the best practices that should be followed with respect to the optimal management of livestock and the associated crops. The value-chains created in that case are considered as non-optimal, while the lack of effective data handling and processing leads to inferior production systems. For this reason, it is necessary to convert the data into knowledge that can be used in decision-making. In order to bridge that gap, multidisciplinary expertise is essential to monitor and interpret interrelated agricultural indicators such as soil readiness, nutrient efficiency, product quality, and animal welfare. Digital transformation and the penetration of ICT technologies in agricultural supply chains have led to the development of a plethora of tools that mostly focus on partial solutions. However, a holistic approach is needed in order to provide to farmers with useful and actionable advice, towards efficient, consistent, and optimal results. Moreover, the simultaneous consideration of circular economy principles in the development of ICT technologies related to agriculture triggers the evolution of Circular Precision Farming Systems.

Such systems integrate all the elements that synthesize a farming system, offering the users a unified solution. The holistic approach of farm management results in input reduction via efficient resource management, improving the sustainability performance of production systems.

The chapter examines the realization of a Circular Precision Farming System. Its requirements and basic elements are presented for two agricultural production sectors: dairy farming and crop production. In such complex value-chains, as the ones created with livestock farming and the crops grown for feed, decision support systems are essential in guiding stakeholders towards appropriate actions in the entire supply chain. Tangible benefits include profit maximization, improved environmental footprint, and risk minimization.

Health and safety of workers is one of the most important indicators of social agricultural sustainability. The sector employs a very large number of workers around the world, while the advent of technological advances has considerably increased their quality of living. Nevertheless, the occupations of the sector can be still listed among the most dangerous. Agricultural workers operate under extreme conditions and working environments, while the nature of the tasks they perform is physically challenging. Depending on the working conditions, there is a plethora of health problems faced by farmers around the world, including respiratory, cardiovascular, and skin disorders, reproductive impairments, various types of cancers, heat and illnesses related to the use of agrochemicals as well as noise-induced hearing loss. Fatigue is also one of the most important factors causing health problems to farmers. The nature of the work they perform requires repetitive movements (e.g., bending, kneeling, and lifting) and also abnormal posture of the body leads to a high frequency of musculoskeletal problems. Furthermore, new sources of danger have emerged due to the mechanization of the sector, as for example whole body and hand-arm transmitted vibration, that induce new syndromes that require further examination.

Taking the above into consideration, the protection of agricultural workers requires actions for the reduction of the danger in the performed tasks or their replacement using engineering and administrative controls along with the ultimate use of protective equipment. The chapter **"An Analysis of Safety and Health Issues in Agriculture Towards Work Automation"** presents an overview of all the hazards and health problems related to the execution of agricultural tasks, further to the risk assessment and control measures towards their mitigation. Besides the development of effective management policies, the importance of the consideration of human factors and ergonomics in the solutions given is emphasized.

Safeguarding agricultural sustainability through the adoption of ICT technologies brings to the surface the need for integrated management of production systems. More specifically, in order to sustainably plan agricultural tasks, a thorough examination of agri-food systems is required. At the same time, the lack of awareness in the assessment of territorial values is emerging due to the increased farmland consumption. Considering the above, smart agriculture as a driver of sustainable agriculture should also be aimed towards spatial planning for its successful development.

The size of production units and agricultural holdings as well as their structure, in combination with land availability, the health of soils as well as the relevant availability of resources constitute the main factors with respect to the investment in innovative technologies. Smaller agricultural units imply smaller availability of financial resources for investments. Respectively, the reduction of production capital, with the loss of land or water due to urbanization, further diminishes the potential investments on sustainable practices, deteriorating the effective assessment leading to agricultural crisis. The implementation of sustainable spatial planning, in conjunction with the introduction of new generation smart farming, can contribute to the reduction of inputs through the effective management of cultivations.

The above, in the long term, can lead to increased food security as well as land protection not only through the preservation of non-renewable resources but also through the timely forecasting and treatment of adverse weather events, thus safeguarding the preservation of the landscape. It is therefore easily understood that innovative technologies provide opportunities for the effective spatial planning. The availability and interconnection of data can expand the development options considering also sustainability parameters. However, for the successful implementation and effective absorption of the above, cooperation and participation of all the relevant stakeholders, such as farmers and policy makers, is necessary.

Contributing to the above, the chapter **"Smart Farming as a Game-Changer for Regional-Spatial Planning"** aims at investigating the way that smart farming (and its evolution) can be affected by strategic planning and policy options while emphasis is given on the expression of the economic, social, and ecological policies of a region on its general spatial planning. The cooperation between farmers and policy makers should be promoted, aiming at safeguarding sustainable development through the improvement of food security. The so-called Strategic Farms, which are agricultural holdings that integrate ICT technologies in the cultivation process, can play a fundamental part towards that direction. The wider cooperation between stakeholders of different regions can effectively promote the adoption of the new concepts in the emerging strategies.

The 2030 Agenda for Sustainable Development and the 2050 EU's vision are enabled through contribution of regional/spatial planning towards the promotion of smart farming. Therefore, in the first chapter of the second part of the book, the effect of strategic development that focuses only on urbanization, on the developing trend of smart farming was examined along with the potential obstacles created for farmers. During the recent decades, industrial and commercial land uses (along with extended residential areas and construction sites) prevailed that of agricultural holdings. Nevertheless, the farmers' investments on innovations that improve the sustainability and quality of their products need to be appreciated during spatial planning by offering farmers of increased available land.

For the effective integration and development of digital agriculture, there are a number of primary issues, which are related to primary production and must be taken into account. These issues include the conservation of land, water, plant and animal genetic resources, biodiversity and ecosystems, the prevention of soil sealing, the confinement in change of available agricultural and forest and other semi-natural areas as well as the limiting of the transformation to urban artificial land. Addressing these issues is of utmost importance for smallholder farmers, being the foundations of spatial/regional planning, at a variety of levels. The study taking the example of Italy identified the gap between development strategies at global level and regional/spatial planning which impedes the efforts of agriculture towards achieving SDGs.

In order to address these challenges, a decision-making framework is proposed which considers a complex system for urban and regional planning. Basic aim is to incorporate smart farming into regional/spatial planning by identifying "Strategic Farms" as rural sustainable development factors. Preserving food security in soils used for agricultural production leads to further benefits related not only to the conservation of soil and water but also the protection from natural phenomena, thus enhancing the possibilities of enjoying the landscape. That is why the connection of agri-food systems with the corresponding soil systems reveals the way in which strategic spatial planning affects the availability of financial resources and consequently the willingness of producers to invest in innovative practices. Given the above, it is understood that in order to promote the penetration of the 4.0 revolution in agriculture, there must be progress at local, regional, and global level. Innovative technologies support the holistic management of agriculture and food systems. The vertical management of the agricultural industry without the simultaneous evaluation of land uses leads to irreversible results as in this way agriculture achieves its economic but not its environmental goals.

In summary, the first part of the chapter tackles with the future of the development of smart farming examining the most important concerns and opportunities along with potential strategy possibilities for planners and policy makers in Europe. Firstly, the Agenda 2030 key challenges for smart farming are presented which include the increasing of available food along with the achievement of higher quality standards with respect to safety, environment, welfare, energy, and climate change. Subsequently, and in order to guide farmers towards the digital evolution, the consequences of spatial planning on agri-food systems are examined. Finally, the inconsistencies between global development regional/spatial planning strategies are examined, as they form the framework conditions for the development of smart farming. Since farmers, through innovation, aim at improving the environmental and social performance of their production, their intention should be considered thoroughly during spatial planning, while possible approaches towards that direction should be made available to policy makers and planners.

Adding on the importance of regional/spatial planning, the chapter **"Agriculture in Latin America: Recent Advances and Food Demands by 2050"** investigated the potential of agriculture in Latin America. Having a quarter of the world's arable land and a third of the world's freshwater resources, Latin America and the Caribbean can become major contributors in achieving food security in the planet. To the above it should be added that these regions are responsible for 15% of the global export of agricultural products, being the world's largest net food exporting region.

More specifically, the chapter summarizes the current state of agriculture in Latin America, along with the most recent developments for the increase of production via advanced processes that simultaneously aim at mitigating the adverse climate and environmental impacts of agriculture. These advancements aim at maintaining food security for the continuously growing population, meeting the relevant demand projections for the year 2050. The issues elaborated in the chapter include the impact of these regions to climate change, the policies that are used to increase the production of food, the area suitability for agriculture as well as the projections on food demand for 2050, taking into account that these areas' population growth rates rank third below the Asian and African regions.

The chapter **"The Development Opportunities of Agri-Food Farms with Digital Transformation,"** having as a case scenario Italian agri-food farming, aims to provide a picture of the perception that agricultural operators have about the opportunities and limitations related to the adoption of intelligent agribusiness as a part of the digital transformation of agriculture. Authors analyze digital transformation to identify new approaches and opportunities in the agri-food sector based on integrating participatory planning and a novel approach based on suitable tools to acquire and process qualitative and quantitative information concerning the possible alternative scenarios of digital transformation. Results show that the Italian agri-food sector has begun to understand that digital innovation is a strategic lever able to guarantee greater competitiveness to the entire supply chain, from production in the field to food distribution.

However, it becomes clear that simply introducing technologies is not enough to generate results. Digital transformation requires social, economic, and policy systems to provide the basic conditions and enablers for digital transformation. It is worth mentioning here the "law of disruption" stating that although technology changes exponentially, economic and social systems change progressively.

Towards that direction the chapter **"Precision Agriculture's Economic Benefits in Greece: An Exploratory Statistical Analysis"** attempts to examine innovations in agriculture in Greece with the view of their economic benefits. Basic aim of the work is to group Greek regions with respect to the crop type, size of arable land, the process innovations used, and the eventual economic benefit. Three groups were identified. The first group includes the regions of Eastern Macedonia and Thessaly, where the largest agricultural plains are located (fields greater than 40 hectares). This group mostly consists of no horticulture crops, improved sowing, and improved plant disease prevention. Nevertheless, there are no improvements in the use of fertilizers, irrigation, and harvesting while no advancements are reported in labor productivity and the quality of the final products. Even though the latter are not as important as the former, the results indicate that the economic benefits of those regions derive from the availability of land rather than the implied application of innovative technologies such as variable rate planting/seeding (VRP/VRS) and UAVs.

The second cluster concerns the regions of Epirus, Peloponnese, and Crete which contains mostly no arable crops, arable lands below 20 hectares as well as permanent crops. No improvements in sowing are indicated by the research, but a series of less intense characteristics were identified including improved fertilizer use and irrigation. This group is mainly characterized by permanent crops and small arable land; however, only a small number of innovative technologies have been used recently (e.g., variable rate nutrient application - VRNA and variable rate irrigation - VRI) with no tangible economic benefits. The third and last group involves farmers in Western and Central Greece and mostly includes horticulture cultivations. This group applies improved harvesting, ploughing, and agricultural machinery use and demonstrates improved labor productivity, increased income, and quality of final products. In addition, precision agriculture (PA), machine guidance (MG), and UAV are also applied, making this group the only one with tangible economic benefits compared to the other two.

Taking the convention that a business can be characterized as innovative in case it applies at least one innovation in the last 3 years, it is conducted that Greek agriculture is innovative; however, the economic benefits are not calculable yet. More specifically even though all groups displayed innovation on the applied processes, only the third demonstrated economic benefits. The differences between the examined clusters are observed in the crop type and the number of innovative processes. The first group mostly includes arable and the second permanent crops while the third mostly horticulture crops. Simultaneously, in the first two groups two types of innovation are observed while in the third group, which includes the most benefited farmers of Western and Central Greece, there are three. Considering the above, the authors conclude that horticulture crops are considered as more productive, requiring less arable land than the other crops while greenhouses provide with safety against extreme weather conditions which is a major concern for farmers. Additionally, for economic benefits to start increasing, a minimum of three types of innovations should be applied in the farm.

The acceptance of the digital transformation of agriculture requires the successful dissemination of these technologies to the stakeholders through the establishment of interaction methods among users. The development of user-friendly interfaces can benefit the progress of smart agriculture. Simultaneously, the integration of innovative technologies into existing environments with the users that is already familiar can further assist to the diffusion of digital transformation. The chapter **"AI-Based Chatbot System Integration to a Social Media Platform for Controlling IoT Devices in Smart Agriculture Facilities"** introduces an easy-to-use, efficient, and safe framework for the operation of IoT agricultural devices in natural language dialogs, via the development of an intelligent Conversational Agent (chatbot) using Artificial Intelligence (AI). As a communication user interface, an instant messaging application of a popular social media platform is used. The users are offered with context-aware services with respect to monitoring and controlling agricultural facilities through question-answer sessions. Due to its technological readiness and features as well as its high penetration to mobile users, the messenger application of "Facebook" was chosen for the implementation. According to the conclusions of the research, the use of an intelligent conversational agent via a popular social media platform contributes to the maximum penetration of IoT technologies in the agricultural sector, in the most effective and user-friendly manner.

Familiarity with state-of-the-art technologies mainly concerns older farmers who need training to understand them and incorporate them into agricultural processes. Younger farmers are already familiar with ICT technologies in their everyday life, thus the use and the development of ICT technologies has to be integrated into their education programs in order to create effective and flexible learning environments. Flexibility in education is becoming in recent times imperative as certain external factors, as the COVID-19 pandemic, have created extra burden on institutions that struggle to safeguard the continuity of education. These new needs have increased the interest in online education even though its strategic importance in addressing global need of education has been stressed for decades. However, online education has fundamental differences from conventional teaching, and consequently new pedagogical approaches are required especially since until now little attention is paid during the development of online educational material.

A number of institutions worldwide have integrated online programs for resident and distance learning. However, mostly budgetary constraints impede the progress of the adoption of online learning platforms for many educational institutions. Online teaching and learning are considered as an opportunity to promote creativity, critical thinking, and entrepreneurship to students, modernizing conventional education. This is the reason why it is preferred mostly by higher education institutions. Nevertheless, the challenge is to discover the ways to efficiently introduce online teaching in educational programs. The above also apply for agricultural, biological, and engineering educational programs with the added challenge of obtaining the necessary resources (financial and infrastructure) for such courses considering their applicational nature.

Addressing the above issues, the chapter **"IT in Education: Developing an Online Course"** investigates the integration of ICT technologies in education. Towards that direction the elements of the learning environment for online teaching are presented along with the required outcomes and objectives that constitute the desired student behavior. For the enrichment of the above, the basics of learning theory are presented followed by the best practices that should be followed in delivering online courses. Moreover, the differences of instructional and curriculum design are presented, and also the guidelines for designing an online course. The chapter also provides with information for the flipped classroom concept and also the relevant tools that can be used for online teaching.

The chapter **"Assisting DIY Agricultural Robots Towards Their First Real-World Missions"** deals with the issue of the continuity of robotic skill sets obtained through high school and university level education practices that leads to a noticeable gap is between educational and commercial agricultural robotic solutions. This gap should be closed in order to promote future engineering careers, as it can be inferred by observing successful attempts that bridged education and future employment by moving educational effort to mass production. Considering the above, the chapter investigates the potential of transforming university level DIY robotic solutions to marketable products to be able to perform real-world missions. To perform considerable agricultural operations, these vehicles should at least have adequate accuracy and power. The recent advancements in the electronics industry has triggered the development of a large number of devices with a plethora of attributes at a reduced cost. The benefits of this increase in accessibility and affordability are ripped from both students that can develop marketable products in the context of their education and also from entrepreneurs that can design tailored products at lower prices. A project-based learning model (PBL) approach is built upon a previous work on creating/upgrading electric vehicles that can undertake light-duty agricultural tasks. The approach design was based on cost-effectiveness and avoiding complicated processes, focusing on the use of everyday, simple material and electronic components. In the context of the work, two applications are presented. The first concerns a vehicle that performs all-terrain soil-specific measurements and the other a robotic sprayer for fertilizers, pesticides, and herbicides. The basis of both platforms includes Arduino uno boards and raspberry pi units for more complex scenarios along with the use of navigation units (e.g., navio2 and pixy2 cameras). The vehicles were controlled through various methods (e.g., smart phones) while basic automatic control functions were employed. Simple Artificial Intelligence (AI) modules were also integrated in the testing process and both visual and textual programming interfaces were utilized for designing the platforms' logic. Wi-Fi was used for the majority or remote operation scenarios; however, for longer controlling distances, LoRa interfaces were also deployed. Lastly, the use of small solar units in order to increase the autonomy and efficiency of the platforms was also examined.

With respect to actions taken inside educational institutions for the evaluation of technological innovations, the chapter titled **"Evaluation of Spray Coverage and Other Spraying Characteristics from Ground and Aerial Sprayers (Drones—UAVs) Used in a High-Density Planting Olive Groves"** attempts to compare the performance of the most common ground sprayer types against a spraying drone in a high-density olive grove located in Perrotis College of the American Farm School of Thessaloniki, Greece. The work aims at addressing best-practice issues faced by farmers with respect to the accurate estimation of coverage percent, uniformity, drifting, etc. of spraying materials. The development of Unmanned Areas Systems offers new potential to farmers; however, these systems haven't yet been adequately evaluated with respect to their performance. In order to evaluate the spraying coverage and certain other characteristics Water Sensitive Papers and scanning software were used. The results highlighted the potential of drones with respect to material savings and efficiency increase, while important differences were observed between the different types of sprayers. Among the most important limitations of the use of unmanned systems is the deficiencies in the EU legislation.

The rapid progress of digital strategies has revolutionized conventional marketing tactics, thus in order to attract new potential customers, the approaches used within logistics sites must be re-evaluated. For efficient decision-making, developers, marketers, and designers need to take into account the complicated and interconnected behavioral characteristics of the users. Towards that direction in the chapter entitled **"Predictive Model for Estimating the Impact of Technical Issues on Consumers' Interaction in Agri-Logistics Websites,"** an identification approach of the various correlations existing between the variables which affect the efficiency of the digital marketing strategy is presented. Based on existing literature, the presented work sets as a hypothesis that the existing correlations between different web-variables have a direct impact on the efficiency of an agri-logistic digital marketing strategy. This hypothesis is considered with a view to prognosticate the most efficient digital marketing strategies that can be employed by agri-logistic websites and, as a second step, to enable the long-term forecast of digital marketing within the agri-logistic sector. A three-stage methodology is presented starting with the extraction of numerous web analytics from different world-leading agri-logistics websites followed by a statistical analysis for the examination of possible intercorrelations between the web analytics metrics, and finally a Fuzzy Cognitive Map (FCM) approach was implemented to build a predictive model as the basis for a process and agent-based simulation model for the evaluation of the consumers' interaction in agri-logistics digital marketing.

In conclusion, the fourth book of the series *Information and Communication Technologies for Agriculture* under the theme **Actions** investigates the implementation of cutting-edge technologies on real-world applications. From the compilation of the chapters presented, it becomes apparent that the penetration of ICT in agriculture can result in several benefits related to the sustainability of the sector. However, to yield the maximum benefits successful management is required. It must also be highlighted the importance of proper education in the adoption of innovative technologies starting from the adaptation of educational systems to the new era and moving to the familiarization of farmers to the new technologies.

Thessaloniki, Greece — Dionysis D. Bochtis
Lincoln, UK — Simon Pearson
Thessaloniki, Greece — Maria Lampridi
Thessaloniki, Greece — Vasso Marinoudi
Gainesville, FL, USA — Panos M. Pardalos

Contents

Part I
Agriculture Digital Transformation and Sustainability

Towards Sustainable Agriculture: Challenges from the Transition to the New Digital Era

Maria Lampridi, Vasso Marinoudi, Lefteris Benos, Simon Pearson, Dionysis D. Bochtis, and Panos M. Pardalos

1 Main Challenges Placing Pressure on Modern Agriculture

"*Food security exists when all people, at all times, have physical and economic access to sufficient, safe and nutritious food that meets their dietary needs and food preferences for an active and healthy life*" [1]. This is a concise widely accepted definition that elucidates the food security, stated after the report of the World Food Summit in 1996. In particular, the objective of the Sustainable Development Goal 2 (usually abbreviated as SDG 2) is to eliminate hunger and malnutrition worldwide by 2030 by guaranteeing that all people, and especially children, have enough food

M. Lampridi (✉) · L. Benos
Institute for Bio-Economy and Agri-Technology (iBO), Centre for Research and Technology Hellas (CERTH), Thessaloniki, Greece
e-mail: m.lampridi@certh.gr

V. Marinoudi
Lincoln Institute for Agri-Food Technology (LIAT), University of Lincoln, BrayfordWay, Brayford Pool, Lincoln, UK

Engineers for Business S.A. (EfB), Thessaloniki, Greece

S. Pearson
Lincoln Institute for Agri-Food Technology (LIAT), University of Lincoln, BrayfordWay, Brayford Pool, Lincoln, UK

D. D. Bochtis
Institute for Bio-Economy and Agri-Technology (iBO), Centre for Research and Technology Hellas (CERTH), Thessaloniki, Greece

farmB Digital Agriculture P.C., Thessaloniki, Greece

P. M. Pardalos
Department of Industrial and Systems Engineering, University of Florida, Gainesville, FL, USA

D. D. Bochtis et al. (eds.), *Information and Communication Technologies for Agriculture—Theme IV: Actions*, Springer Optimization and Its Applications 185,
https://doi.org/10.1007/978-3-030-84156-0_1

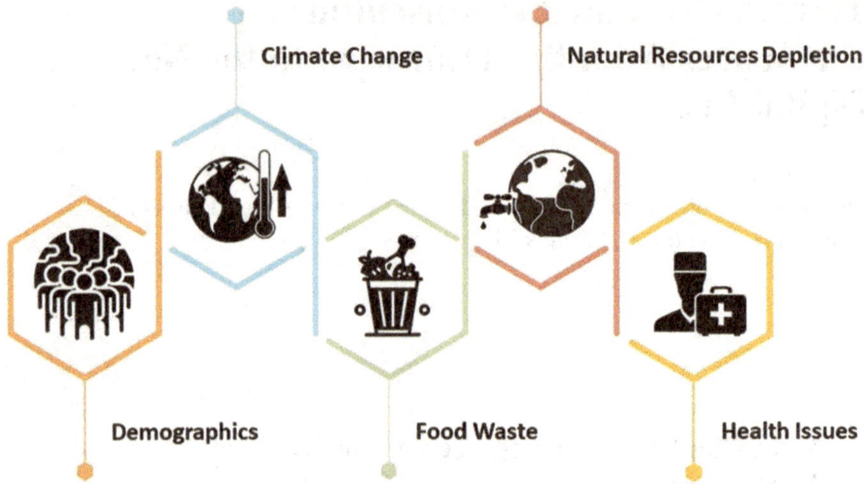

Fig. 1 Summary of the main challenges placing pressure on modern agriculture

every day. Generally speaking, a stable environment within every country constitutes a key factor for the accomplishment of the desired sustainable food security. The bulk of the population which is undernourished neither can afford to buy the necessary products for their livelihood nor have the ability to produce their food. The backbone of the effort for meeting the ever-increasing needs is arguably agriculture, a sector whose development thousands of years ago transformed the style of human life. However, modern agriculture has to overcome some very crucial challenges which put a lot of pressure on it. These challenges are summarised in Fig. 1, while a brief description of them follows.

First of all, the global call for food is expected to drastically be intensified owing to the imminent explosion of the population in earth. This increase in food production has to be on the order of 60–70% by 2050 [2, 3]. In addition, malnutrition, as well as poverty and hunger, is among the primary reasons of the increasing rate of the migration leaving rural areas for urban ones, with the global urban population proportion having risen to 50% in 2010 (from 28.3% that was in 1950) [4, 5]. Moreover, the diet is being altered as a consequence of the growing urban population with an increasing demand in animal protein [6]. This transition to a "Western" diet, which is also associated with high intake of added sugar, refined fats and low intake of vegetables and fruits, has introduced additional challenges for environmental sustainability and health [7]. Raising livestock corresponds to 25% of the amount of water used in agriculture, which, in turn, accounts for approximately 18% of greenhouse gas emissions caused by humans [8]. As far as the health impact is concerned, this dietary shift remarkably increases the incidence of coronary heart disease, type II diabetes and other chronic diseases that lessen the life expectancies. Alternatively, widely adopted healthier diets could reduce the gas emissions, water use, eutrophication, land clearing coming from agriculture and contribute to the prevention of diet-related diseases [9].

One additional issue to be addressed is whether the natural resources are sufficient. Interestingly, a large amount of global farmlands has been degraded [10], namely 10–20% of lands around the world [11]. The land degradation can be characterised as a process of loss or lessening of the biological productivity, with the final state being that of desertification [12]. The loss of the ability of soil to be recovered is attributed principally to the unsustainable practices like intensive tillage. A cause of degradation of farmlands is also the deforestation, with about 80% of the global deforestation being associated with agricultural concerns [13]. Although deforestation for making farmland itself does not necessarily cause soil degradation, it becomes a risk factor indirectly through eroding water resources followed, many times, by poor management. The final conclusion is noteworthy as even though the irrigation systems have enhanced their efficiency, the ever-increasing population make the scarcity and security of water a real concern. Besides, soil erosion can be provoked by vegetation overcutting as well as inappropriately crop rotations instead of balanced ones, unsustainable fallow periods and livestock overgrazing. In order to maintain crop yields or restore them, farmers utilise fertilisers. However, this intervention can result in soil-nutrient imbalance.

In 2012, the necessity for imperative actions to reverse the land degradation was stressed by the United Nations Conference on Sustainable Development [14]. This international awareness resulted in a political document that involves practical measures in order to implement a sustainable development for achieving a "*land-degradation neutral world*". The zero-net land degradation concept/requirement stands for not giving up those lands which have lost their productivity, while recognises that action has to be taken both locally and at international level by implementing integrated strategies for assuring food security. According to Chasek et al. [10], the land degradation neutrality concept relies mainly on two prerequisites, which are: (a) The realisation that fully prevention of land degradation is too ambitious and (b) The land degradation rate can be decreased.

Another important aspect that has to be considered is that the world's biodiversity is currently under threat by virtue of direct impact of climate change [15, 16]. This issue is driven because of the wild natural habitats conversion to agricultural fields and the subsequent soil organic matter loss [17]. In addition, it is well-known that the diversified and relatively small farms demonstrate higher productivity per area as compared with the large monocultures. This phenomenon is known as "inverse farm size-productivity relationship" or as "paradox of scale" [18, 19]. For the sake of accomplishing productive and resilient smallholder systems, policies regarding food security should accentuate a rise in agroecological capacity. This entails sustainable, eco-efficient and environmentally friendly approaches for managing diversified cropland [20, 21]. Moreover, pesticides should be avoided, as much as possible. At the same time, soil fertility strategies should be integrated by intensifying production in tandem with preserving functional biodiversity, thus, staying away from potential environmental risks that smallholders face with [22]. The decisions pertaining to climate adaptation investments differ in scale ranging from distribution of funding among the worldwide biodiversity hotspots to identifying optimum investment in the connectivity of continental habitats, to initiatives for keeping the biodiversity in a specific region [23].

To make matters worse, agriculture is greatly affected by climate change via a bidirectional cause-effect relationship. The term "climate change" encompasses the global warming and the consequential large-scale changes of weather patterns. On the one hand, the agricultural practices produce considerable amounts of greenhouses gas emissions, which are related to intensive farming systems according to the Intergovernmental Panel on Climate Change [24], that definitely affect climate. The agricultural sector has the higher share of emissions concerning methane and nitrous oxides. The increase of the concentration of the greenhouse gases in atmosphere causes subsequent increase in temperature. On the other hand, high values of CO_2 concentration, for example, is associated with drought stress resilience [25], since under these circumstances plants require less stomata to be open and subsequently lose less water via the process of transpiration. This can lead to the increase of water use efficiency. Nonetheless, in some places on earth, especially in the tropics, increased values of CO_2 concentrations probably could not compensate to the climatic change and crop yield is anticipated to be reduced [26]. Moreover, the temperature rise along with alterations in the regime of precipitation has not only negative consequences on the quality, volume and stability of the agricultural production, but also on the environment in which the agricultural practices take place [27]. As in the above CO_2 concentration example, although the crop growth may be enhanced due to higher temperatures, when the temperature values exceed a critical threshold, crop yields can decline [28]. Climate change is responsible for important unfavourable impacts on plenty of aspects such as food security, human health and water resources [29]. Side effects of the aforementioned change of climate are the increase in the precipitation variability, the existence of floods and droughts that turn out to decrease crop yields. Additionally, the existing problems that are related to the environment and have already been mentioned, like soil degradation and groundwater depletion, are expected to be escalated on account of the climate change. A plethora of studies exist in the recent relative literature on climate impacts and possible adaptation strategies, including [30–34], indicating that this multidisciplinary area is of great scientific concern. The degree at which the food security will be affected by the climate change depends on the Gross Domestic Product (GDP) of a nation, adaptability and its geographical location. Predictions indicate that since developing countries have limited resources so as to adapt to the changes, they are going to suffer more due to extreme weather phenomena and temperature variability [33].

An additional challenge, that modern agriculture faces, is the food loss and waste [35, 36]. According to FAO [37], food loss refers to the food reduction or quality in the chain as a result of the actions and decisions of food suppliers. In contrast, food waste is related to the reduction of the quality or quantity of food as a consequence of the actions and decisions of consumers, food service providers and retailers. Food can be wasted throughout the supply chain because: (a) The fresh product deviates from the optimal colour, size and shape, (b) Large amounts of edible food are often discarded from both eating establishments and household kitchens and (c) Products which are close to the "best-before" date are often thrown away by consumers and retailers. Approximately 33–50% of all the foods that are produced on earth is never consumed, whereas at about 800 million people are hungry [38]. A both remarkable and alarming statistical evidence frequently mentioned in literature is that if we consider the food waste as a nation, this nation would account for the third most reported emitter of gases associated with greenhouse phenomenon, after the countries of China and the United States [39]. In fact, food waste constitutes a massive market inadequacy whose kind is not presented in other industries. It is noteworthy highlighting that, overall, it is not only the fact that natural resources are misspent, but also that the lost food creates methane during its decomposition that is very harmful. Eliminating both food waste and loss is a key challenge in favour of establishing a "zero hunger" world as well as safeguarding sustainable production and consumption patterns with positive effects on both livelihoods and climate change.

There are also a lot of illnesses that have been identified via community-based epidemiological studies in the whole world. In particular, it has been recognised that there is an urgent need to prevent farmers from pesticide-related illnesses, hearing loss, cancer cases and respiratory diseases [40, 41]. Concerning the pesticide-related illnesses, the exposure to them may be via inhalation, ingestion or contact with workers' skin. The duration of the exposure, the kind of pesticide and each farmer's health status are determining factors regarding the health outcome. Pesticides can be accumulated within body fat, metabolised and excreted [42]. The numerous adverse effects in human health include dermatological, carcinogenic, neurological, reproductive, gastrointestinal and endocrine effects [43]. Furthermore, there has been an important increase of the farmers' exposure to toxic gases, bioaerosols, organic and inorganic dusts which are related to a number of respiratory diseases including bronchitis, non-immunogenic bronchospasm and sinusitis. The most widely studied respiratory disease comes from the organic dusts exposures, such as animal feeding operations and grain processing, while inorganic dusts exposure is originated from soil components and is a hazard especially at dry climate regions, like California for instance [44]. Finally, epidemiologic studies have related farming practices (such as phenoxyacetic acid herbicides and pesticides) with cancer cases. In brief, agricultural activities have been reported to be associated with leukemia, multiple myeloma, non-Hodgkins lymphoma, melanoma, prostate and soft tissue sarcoma, to mention but a few [45]. Nevertheless, more research is necessary as a means to establish causality.

Last but not least, musculoskeletal disorders (usually abbreviated as MSDs) are the most widespread of all non-fatal disorders in agriculture, particularly those which take place during labour-intensive practices [46, 47]. MSDs cover a wide range of injuries and pains including back, neck and knee pain as well as upper and lower limb disorders. Agricultural workers are at high risk of developing MSDs, because of their interaction with machines, the required physical demands and the repetitive nature of activities. In the Netherlands for example, approximately 61% of the leave claims were by reason of musculoskeletal injuries [48], while in France MSDs correspond to the 93% of the evidence for compensable work-related agricultural diseases [49]. According to the recent review study in [50], central risk factors include repetitive kneeling, lifting of heavy cargo, trunk flexion and working under awkward postures of trunk and wrist. Interestingly, low back pain is identified as the most prevalent symptom, although hip and knee osteoarthritis is also highly affecting the ability of farmers to labour and to maintain a livelihood [51]. Finally, machineries with driving seats appear to be related also to painful low back syndromes, while handheld machines are strongly associated with upper extremities disorders. As concluded by Benos et al. [52], who conducted a survey on ergonomics in agriculture on the subject of mechanised operations, the main reasons of this kind of pains are the whole-body and hand-arm transmitted vibration, respectively.

2 Sustainable Ways to Meet the Challenges

2.1 Need for Sustainable Precision Agriculture

Taking into consideration the above discussion regarding the main challenges that have to be addressed, it can be deduced that there is an urgent need for increasing the degree of effectiveness of the agricultural practices with simultaneous reduce of the environmental burden. As a consequence, these two prerequisites drive the evolution of agriculture towards precision agriculture and smart farming systems in order to establish sustainability and a safe environment.

Agricultural sustainability should evaluate not only environmental issues, but also social and economic challenges associated with the agricultural practices [53]. Generally, sustainability assessment and implementation are very challenging tasks involving a plethora of different variables to be considered. According to the study of Lampridi et al. [54] on agricultural sustainability, the above variables refer to agricultural practices, economic viability, technological level, stakeholders, type of machinery, storage, transportation, agrochemicals, fertilisers, pesticides and crop, location of cultivation and climatic conditions. By investigating the relative literature of the last decade, they concluded that between 2016 and 2019 the interest on agricultural sustainability had been increased. The most common methodologies take account of frameworks, indexes and indicator-based tools. Furthermore, the participation of stakeholders was proved to be fundamental in defining the sustainability level, while the impact of input management and resource usage was mostly studied. Normally, there is no consensus about the agricultural sustainability assessment standardisation. A lot of methodologies have been developed [55]. Among those methodologies, life cycle appraisal is widely used [56], while also several indicator-based techniques exist considering various methods that use different approaches with respect to the agricultural sustainability definition, objective and intended users [57].

Pesticides, fertilisers and water are no longer applied uniformly across the entire rural regions. Instead, precision agriculture uses the minimal needed quantities and targets them towards very specific sites. Hence, its purpose is to improve the agricultural activities by guaranteeing maximal productivity with reliable measurements for the sake of providing an extensive overview of the processes taking place in the farmlands. Precision agriculture appeared in the mid-eighties. Today, the concept has been upgraded and adapted to various crops, practices and countries depending on cropping system, soil conditions and countries, while it is constantly evolving. The contemporary precision agriculture includes also full and partial automatic guidance of agricultural vehicles [58, 59], traceability of the products within supply chains [60, 61], software concerning the agricultural production systems management [62] and on-farm research, to mention but a few. The activities aim at optimising natural resources usage, energy consumption and chemicals pertaining to plant growth as well as pest and weed control. To put it simply, precision agriculture tries to implement the following coincident requirements; the right handling in the right place and time by exploiting the available technologies at that time [63, 64].

Smart farming lies on four pillars, each one of central importance: (a) Optimal management of the natural resources, (b) Landscape and ecosystem conservation, (c) Adequate services and (d) New technologies for farmers as a means to apply the essential changes [65]. In fact, the agricultural sector is undergoing the so-called "fourth revolution" or "agriculture 4.0" [66, 67] motivated by the exponentially increased application of Information and Communication Technology (ICT) [68]. ICT is an essential prerequisite for the precision agriculture implementation and is supported by policy-makers on a global scale.

Although technological innovation has been gradually implemented throughout the previous decades, current technological outbreak has the potential to reinforce agriculture for enhancing the food value chain and addressing the mounting needs of consumers. Undoubtedly, the responsible innovation concept must underpin the agriculture 4.0. In other words, agriculture 4.0 should make sure that novel technologies, which are developed to optimise both productivity and eco-efficiency, will also meet human needs and be socially responsible. The concept of responsible smart farming is not well-established in agriculture [67]. The study of Eastwood et al. [69] for responsible innovation regarding smart dairying gives a valuable initial point for developing a framework. In this fashion, it should be mentioned that the three ways that agriculture 4.0 follows to meet the challenges are, apart from ICT, non-conventional farming techniques and novel technologies [70]. While the last two categories are very important, they are beyond the scope of this book. Nevertheless, a synoptic description follows for the sake of completeness. As far as ICT is concerned, it contains a large variety of technologies that are described in detail in other chapters of this book series. An illustrative graph of these state-of-the-art technologies that are going to contribute to meet the challenges that put pressure on modern agriculture can be seen in Fig. 2.

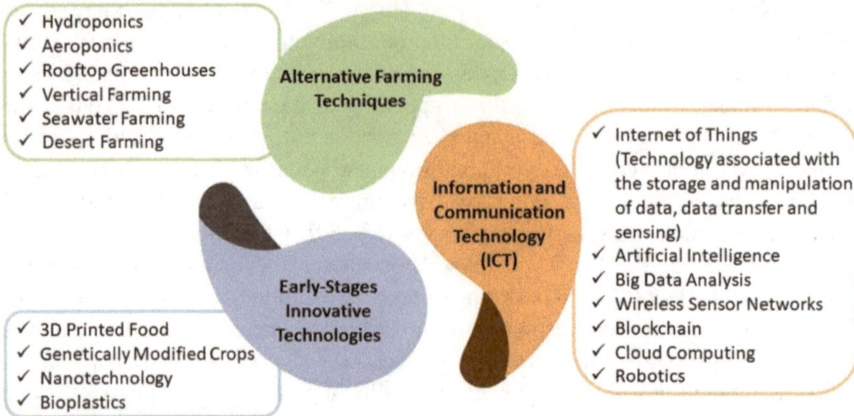

Fig. 2 Different technologies and techniques to meet the challenges

2.2 *Alternative Farming Techniques*

Regarding conventional farming techniques, adverse factors of soils, such as inappropriate soil, soil compaction or degradation, poor drainage and the presence of nematodes and microorganisms causing diseases, prevent the effective growth of crops [71]. However, technological advancements offer the basic elements to overcome these limitations. In particular, the greenhouse industry constantly develops new technologies to adapt to the requirements of the new market [72]. For that reason, soilless crops techniques have been developed such as hydroponics [73]. This kind of horticulture utilises mineral nutrient solutions in aqueous solvents for growing plants without soil by exposing their roots to the nutritious liquid [74]. An inert medium, like gravel and perlite, can support the roots. The nutrients can come, for example, from chemical fertilisers, artificial nutrient solutions and fish excrement. Since these farming systems do not prerequisite fertile land for the sake of being effective, they need less space and water when compared to the conventional systems. Current technologies have contributed to the diversification of water sources regarding irrigation in sites where these sources are the major limiting issue, including the use of desalinated seawater, rainwater or reused water [75–77]. Also, they can support vertical farming production, that obviously increases the crop production per area of land, while as they are practiced under a controlled environment, they can support non-stop production all year round [78]. Vertical farming is the novel practice pertaining to the growing of crops within vertically stacked layers. It is related to the large scale urban farming of vegetables, fruits and grains inside a building which has been manufactured in such a way that can accommodate specific kinds of crops by using hydroponics [79]. Remarkably, the recently emerged COVID-19 pandemic and the precaution measures that were taken to "flatten the curve" (such as lockdown, quarantine, border shutdowns, prohibition of key staple foods exports, etc.) particularly tested the supply chains resilience and displayed the fragility of cities to unanticipated food crises [80]. Urban agriculture with emerging practices (such as hydroponics, rooftop greenhouses, aquaponics, aeroponic and vertical farming) might be a viable solution to these kinds of challenges in conjunction with offering a chance to realise urban organic waste treatment, water recycling and resource circularity [81, 82].

Taking into account the increasing needs for food, the world should turn also to other techniques like seawater and desert farming. Contrary to the problem of desertification that was mentioned above, the potential to transform deserts into arable farmlands may constitute a global answer to climate change and world hunger [83]. As agriculture lies on water supply, farming at arid regions is definitely a challenge but, at the same time, a reality. This alternative approach integrates different technologies to desalinate water for the required irrigation, operate these greenhouses and produce electricity from solar power. Automated water recovery is used to revegetate and improve the soil conditions of the surrounding lands. These systems allow crops growing throughout the entire year with considerable yields and by halves water use [84]. As far as seawater farming is concerned, it addresses the undesirable soil conditions as well as the lack of proper water for farming operations in coastal regions. In these systems, saltwater is used rather than freshwater to enrich the soils and support a variety of rural activities. Indicative projects regarding desert farming are running in Australia and Saharan countries [85], while seawater farming is taking place in several countries including Eritrea, Mexico, the United Arab Emigrate and Australia [86].

2.3 Early-Stages Innovative Technologies

Interestingly, since 1996 some farmers have adopted genetically modified crops, as they can increase the production via creating plants with higher grain yield within a short period of time. The main benefit of the genetic engineering methods is the chance to overwhelm sexual incompatibilities existing between species barriers and plants allowing for introducing of genes originated from other plants, fungi and bacteria [87]. The adoption of genetically modified crops can have plenty of benefits, such as reduced amounts of chemical pesticides, drought-resistant and disease-resistant plants which need fewer fertilisers and water, increased production of more nutritious food with lower cost and prolonged shelf life [88, 89]. However, this practice has triggered controversies, even though strict regulatory processes have been implemented worldwide [90]. Other innovations include also 3D printed food as a means to produce food according to individuals' health status, dietary habits and taste preferences [91] as well as strategies exploiting nanotechnology to enhance the crop protection and nutrition [92]. Finally, enterprises are urged to use food containers, which are biodegradable or can be recycled, instead of disposable plastic packaging bags and containers for food. For this purpose bioplastics are increasingly being used which, however, have not managed to deliver the same packaging capability as plastics [93]. All these technologies are still in their infantile stages, but could constitute, under certain circumstances, game-changers for the coming years.

2.4 Ergonomics Issues

The fourth industrial revolution promotes a human-centred attitude at any working environment. Thus, as it was highlighted above, one of the major challenges that appears during agricultural operations is to assure workers' health and safety through safety management strategies. The importance of these strategies is intensified considering the number of people working in farms worldwide and also the epidemic proportions of health problems and MSDs, in tandem with the resulting economic costs. In order to meet the aim of occupational health and safety, ergonomics (also referred to as human factors) is employed. The term "Ergonomics" originates from the Greek words "ergo" (meaning work) and "nomos" (meaning law). The essentially synonymous phrase "human factors" has been adopted to stress the physical or cognitive attribute of a person or social behaviour adequate to humans which can affect the operation of the technological systems. In plain language, ergonomics and human factors is all about fitting the assignment to worker, via drawing attention to worker-working environment interface by preventing workers from dangerous activities based on physiological and psychological principles. The subject of ergonomics is to improve comfort and safety, decrease human error, increase productivity with a particular emphasis on the interaction between the worker and the thing of interest. The thing of interest may be, for instance, a tool, equipment, sensors, computers or working ecosystem.

Even small changes can result in large differences in reported discomforts [94]. A simple ergonomic intervention in agriculture was indicated by [95]. They showed that using a long-handled hoe considerably reduced the trunk flexion during weeding by, in parallel, increasing the production as compared to other manual weeding practises. Also, in the same study, the usage of pneumatically powered shears for manual cutting proved that even farmers with partial disability could come back to their job. However, ergonomics is a versatile scientific discipline which includes the determination of the risk factors provoking MSDs, the root causes, along with the development, application and assessment of the ergonomic interventions. As highlighted by Benos et al. [96], a state-of-the-art ergonomic analysis (Fig. 3) involves self-reports along with computational and experimental measurements that have a great potential to offer useful input for the risk assessment analysis. Risk assessment encompasses the determination of the crucial factors that can provoke MSDs, evaluation of the risks related to that hazard and examination of the appropriate techniques to eliminate it. Finally, the harmful postures and activities are identified, while safe practices are given via educating farmers through simple guides. Also, systematic international strategies for increasing the risk factors

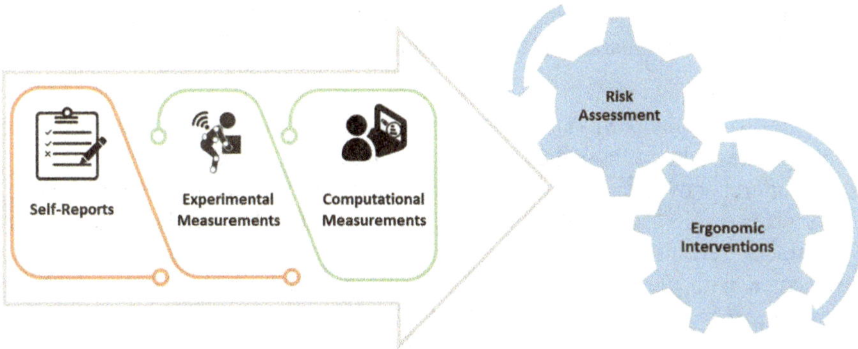

Fig. 3 Overview of the self-reports as well as experimental and numerical measurements for deriving data needed for a risk assessment which, in turn, can lead to ergonomic interventions so as to reduce the chance of the farmer getting injured

awareness connected with MSDs are a long-term goal by re-setting the practical limits like the maximal curried cargo and working hours.

Self-reports include special questionnaires (such as the Nordic questionnaire [97]) and personal interviews in order to determine the most affected body regions. As a matter of fact, the recent studies of Benos et al. [50, 52], that review the relative literature on ergonomics in agriculture over the last decade (2010–2020), found that self-reports were used in the majority of the investigations. In essence, questionnaires try to relate agricultural practices and body postures with MSDs. Furthermore, the questionnaires were related to the exposure to vibrations which are transmitted towards the upper extremities and the whole body because of the contact between parts of the human body with those of the machines, like the handles and seats. Some questionnaires collect also data from the terrain, characteristics of the machines as well as duration and speed of driving. Ways of analysing data, such as logistic regression analysis [98] and other statistical methods, are applied so as to determine the main risk factors such as age, gender, wrong postures, working hours, experience and rest breaks. Self-reports, by themselves, are not able to provide clinical diagnosis and, thus, careful filtering and interpretation are needed. One possible limitation is that the latest musculoskeletal discomforts have a greater chance to be kept in mind than an older pain. Besides, the filling out conditions and the environmental surroundings can impact the results.

Characterisation of the relationship between non-neutral postures and MSDs necessitates direct measurement methods. These methods are extensively used to provide unbiased and precise information content in contrast to purely observation-based methods and self-reports [99]. Driven from the progress in ICT, experimental measurements in agricultural activities usually include Inertial Measurement Units (IMUs) and tri-axial accelerometers and gyroscopes [100–103], electromyography (EGM) [104–106], electrocardiograph [107], electrogoniometers [108], handgrip dynamometers [109], shock absorbers [110], pressure sensors [111], piezoelectric sensors [112], optical markers [113] and cameras [114, 115].

In short:

- Electromyography (EMG) is utilised for the purpose of quantifying the activity of the muscles. Identifying when the muscles are active and to what extent during agricultural activities, turns out to be valuable for physicians for comprehending the mechanisms of musculoskeletal injuries and, accordingly, giving the proper treatment by following rehabilitation protocols. Data originated from EMG do not align with the real values of muscle forces at all times and, thus, careful assessment is required [116]. In [107], EMG, electrocardiography (a chart of electric voltage versus time regarding the electrical heart activity by implementing special electrodes on the human skin) and motion data were concurrently gathered with a multichannel telemetry system in a survey for evaluating the use of an on-body assist suit for farming activities.
- Electrogoniometers are electronic devices which can convert the angle at the joints to a voltage. The voltage can be constantly measured, rendering these devices ideal for quick and accurate capturing the dynamic movements that take place during the agricultural operations.
- Accelerometers and gyroscopes are small-sized portable measurement instruments frequently utilised during in field-based studies as a means to assess the exposure to non-neutral postures. Regarding tri-axial accelerometers and gyroscopes, they give concurrent measurements in three directions and they are widely used for evaluating the exposure that is experienced by a structure. These structures include the seat of tractors and quad bikes or the handlings of the grass trimmers and power tillers, to name but a few [52].
- Inertial Measurement Units (IMUs) are electronic devices, which are also small portable devices and combine information collected via several electromechanical sensors including accelerometers, gyroscopes and magnetometers. They are also used to estimate the spatial object's orientation by using sensor fusion algorithms. IMUs are superior to individual sensors, since the advantages of each electromechanical device help compensating the restrictions of the others. For instance, accelerometer-based orientation measurements arising from the gravity acceleration can be applied to correct the so-called "drift" error which affects the estimates coming from gyroscope [99].
- Handgrip dynamometers are devices that measure the maximal hand isometric strength. They are usually utilised for estimating the athletes' handgrip strength in sports like tennis and during rehabilitation. In agriculture, they have been used to identify the effects of dissimilar kind of pruning shears [117] and adjust an equipment according to the hand-grip strength of each individual [109].
- Suspension systems are used to reduce the vibration which is transmitted from the machine (like tractors) to the drivers. Active and passive suspension systems in tractors have been investigated by many researchers [102, 112]. Suspension systems can be mechanical, hydraulic, pneumatic or a combination of the above.

- During driving an agricultural machine with a seat, large values of pressure under the human thighs and buttocks may provoke damage to the nervous system of these areas resulting in the discomfort of the operator. Thus, measuring of pressure constitutes another important factor. In [111], a carpet with a system of sensors (eight pressure mapping pads) was used which was put on the tractor seat. This barometric mapping allowed for a comparison between three different seats concerning the comfort evaluation.
- Piezoelectric sensors are devices that exploit the piezoelectric effect so as to measure, for example, variations of pressure, force and acceleration though generating voltage. In agriculture, piezoelectric sensors have been used for measuring vibration during operating tractors under several surfaces and operating conditions like terrain roughness and tractor speed [112, 118].
- Optical markers have been used in numerous motion capture systems, as they can be precisely detected. A motion capture system usually uses a set of cameras to find the markers' locations [119]. Gait analysis is the most common application in clinical medicine, while in agriculture a few studies have been conducted. Hudson et al. [113] used special markers to measure the angle of ankle, knee and hip during weeding.

Finally, the real data derived through the above direct measurement methods can be served as input in computational tools, like finite element software, with the intention of calculating the stresses within bones and soft tissues (e.g., meniscus and cartilage) and evaluate numerical what-if scenarios. For the purpose of having realistic estimates, numerous significant aspects must be taken into account, including precise geometrical representation of the body structures, appropriate material models for modelling their mechanical behaviour and application of reasonable boundary and loading conditions [120]. Biomechanical simulations appear to be a useful tool for examining MSDs, related to either misuse (as for instance soft tissue overload during agricultural activities) or with joint diseases (like knee and hip osteoarthritis) and proposing ergonomic interventions to farmers.

3 Socio-economic Challenges on Technologies Adoption by Farmers

3.1 The Social Dimension of Sustainability

Sustainable intensification and ICT systems are closely related, as new technologies constitute a key factor on achieving sustainable intensification [121]. This concept is of great importance in the framework of sustainable agriculture. Sustainable intensification can be defined as the process through which the productivity can be increased, however, without being in detriment to the environment, while contributing to environmental and societal benefits wherever possible [122]. It is also related to the trade-off between economic and ecological performance [123]. In

Fig. 4 Main components of sustainable intensification

plain words, as can be illustrated in Fig. 4, the concept of sustainable intensification encompasses three pillars, namely (a) People, (b) Environment and (c) Profitability.

Unfortunately, sustainable intensification has failed to address equally all these components, as insufficient emphasis has been given on social sustainability. Current research on this dimension of sustainability investigates a wide range of fields by applying different conceptual methodologies. These fields contain political, management and business studies as well as social learning among rural societies. The social sustainability is approached via a theoretical lens by addressing the real meaning of the term and its significance as well as examining the state of the art, or suggesting ways on how to assess this dimension of sustainability with the use of indicators or methods based on supply and value chain [124, 125]. In spite of the progress regarding the integration of social aspects in sustainability, there is no cohesion on their completely understanding. The social system of agriculture includes the actors and their interactions accompanied by social roles. According to Janker et al. [124], these interactions entail the institutionalised interactions (relationships and contracts) and the institutional embedding (in the form of traditions, norms and legal system). In fact, there are no clear boundaries of the social system in agriculture. Although the interactions and the actions are directly associated with the production activities, the former are affected by other systems, namely the preferences of the consumers, the expectations of the society, regulations and market requirements [124]. The central agricultural social system's components are all the stakeholders which are involved indirectly or directly with the farming processes. All the stakeholders, including those in and out of the farm sphere, interact between each other. As long as all the roles of the actors are sufficiently satisfied, their needs are met and, consequently, the social system's sustainability is assured. But, considering a rapidly changed society, how can we know what future generations will need? The answer is not straightforward, as we cannot identify all the future needs. The only thing that we can try to establish is a social system which can be functioning today, while it has an adaptive capacity to be adjusted for the future requirements.

As Rose et al. [67] highlighted, by neglecting the social dimension of sustainability, the challenges originated from this mistreatment can create areas of potential controversy by simultaneously enhancing the fears for food insecurity and resistance to the adoption of new technologies. These challenges are described next.

3.2 Areas of Potential Controversy Enhancing the Resistance to New Technologies

The agricultural sector has been responsible for a number of controversies, such as the usage of genetic modification and chemicals (e.g., DDT and neonicotinoids (a sort of insecticides)) [67]. Besides, we have witnessed how former revolutions in technology and machinery provoked mass unemployment in the agricultural sector. As usually said "*Technology is a two-edged sword*" meaning that it has both positive and negative consequences depending on the way we use it. The possible negative effects of ICT in agriculture are now seriously being taken into consideration. For example, a large amount of data is anticipated to be collected via the extensive use of new technologies. Nevertheless, ownership pertaining to this data as well as how it is going to be stored and exploited still remains a worry. Another concern is that the data, which will be produced by commercial machines and software, could be used to give uncontrollably and free of charge valuable decision-making information to companies. Unless a widely accepted legal framework is developed, a lack of trust may occur. However, considering a lack of trust, concern about possible private enterprises profiting, negative impacts on agricultural employment, the way of farming and the way with which the food is produced may enhance the resistance to new technologies [126]. Another example of worry is the reported scepticism by communities for the use of drones for taking images of their private lives and work [126]. Finally, there is also a public concern on the safety existing in farms because of the simultaneous presence of autonomous farming vehicles or the accidents that can be caused during human-robot interaction [96].

The technologies used for precision agriculture can be roughly divided into two types on the basis of the interaction level and the required operator's learning investment. These are the "embodied knowledge technologies" and the "information intensive technologies". The former do not need additional skills, while the latter require decision-making skills and additional investment in the context of knowledge, software or service support regarding data analysis for instance [127]. The extensive use of ICT may lead to the experiential knowledge marginalisation and generate a disconnection between the farms and the farmers. This could result in losing work satisfaction and worsen the reported high incidence of problems related to mental health in agriculture [126]. Alterations to farming practices can also lead, especially the owners of small farms, to leave their farms in favour of urbanisation.

At the same time, agriculture 4.0 is anticipated to create high-skilled jobs. Nonetheless, these jobs will definitely not suit the majority of the existing agricultural workers who have already experienced marginalisation. Actually, it is not only seasonal workers, graders and sorters that are fearful of losing their jobs in a digitalised agricultural environment. Even better-paid workers, such as farm advisors, are anxious when watching machines making evidence-based decisions for a number of farming operations without human contribution [128]. In the framework of the analysis of Autor et al. [129], in general, the tasks can be divided into routine (manual or cognitive) and non-routine (manual or cognitive) ones. The former are the tasks which can be executed according to explicit directions or programmed instructions, whereas non-routine tasks involve complex and problem-solving operations and need flexibility and other skills. The increasingly incorporation of ICT systems, including AI, seems to have the potential to give robots cognition and awareness, thus, rendering them capable to carry out not only routine and manual tasks (as happened in the past), but non-routine and cognitive tasks as well.

When the level of automation increases throughout a process, there are increased needs with respect to both skills and education. In contrast, low-skilled labour is still required for carrying out the activities taking place within routine tasks. Consequently, there exists a limited space for the middle-skilled workers. Thus, the so-called "job polarisation" phenomenon emerges. This phenomenon stands for the parallel growth of both low-skill and high-skill jobs to the detriment of middle-skill jobs. Taking into account that, in principle, high-skill jobs and low-skill jobs correspond to high- and low-wage occupations, a decreased share of middle-skill jobs in the wage distribution is expected (wage polarisation) [130]. As pointed out by Marinoudi et al. [131], the effects of automation on the agricultural sector are strongly associated with labour substitution and complementarity. On the one hand, substitution refers mainly to low-skill labour, which, in principle, concerns routine tasks' replacement. On the other hand, complementarity is related mostly to high-skill labour, which involves cooperation during cognitive tasks.

3.3 Factors Affecting the Adoption of New Technologies in Agriculture

Adoption of new technologies in agriculture has been investigated by rural sociology, especially by the area of the food and agriculture sociology, which mainly focuses on farm production economics. Thus, profitability coming from the technologies was primary studied as a factor for adoption. With the increasing awareness of the environmental problems, also sustainable technologies for protecting the environment have been considered, particularly in developed countries. Adaptation

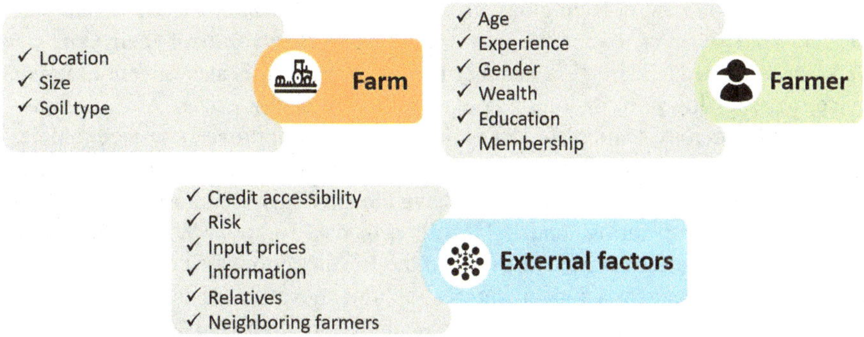

Fig. 5 Factors affecting the adoption of new technologies in agriculture

and diffusion of new technologies in agricultural sector have been examined in a number of studies including [127, 132–134]. Figure 5 summarises the most prevalent factors influencing the adoption by the farmers, which can be simply divided into factors related to the characteristics of the farm and farmers and as well as external motivations or deterrents.

Farm Characteristics

The likelihood of adopting ICT systems is connected with the characteristics of the farm, such as its size, location and soil type. In general, larger farms can incorporate easier the technologies owing to their capacity to absorb risks. Moreover, the larger the farm, the faster the critical information is reached, since farms with larger areas allocate more resources for information. This fact renders large-scale farmers early adopters [134]. Small farms, on the other hand, require more incentives because of scarceness of resources. However, some researchers suggest that there exists no clear connection between technologies adoption and farm size [133]. Studies, such as [135], demonstrate that also the farm location affects the decisions for investment to new technologies. In short, closeness to important market sites and information sources offers more access to market outlets and, thus, ICT adoption turns out to be easier. Furthermore, soil characteristics of the farm influence the adoption, as ICT can provide important information for the soil, thus, leading to better management practices that have the potential to overcome problems associated with natural disasters and climatic conditions.

Farmers Characteristics

Another important factor is, certainly, the age of the farmers, because it affects their attitude on innovative technologies. If the age of the farmer is above 60, as mentioned in [134], the likelihood of adoption declines, although its level of acceptance relies also on the kind of technology. In contrast, some other studies show that older farmers tend to adopt easier the new technologies as a result of their extensive experience [136, 137]. Being a member in an organisation is also of central importance for accessing information on new technologies and practices. Also, as household wealth increases, more tendency is expected to approve new technologies. While inventing on innovation is risky for less wealthy farmers, wealthier farmers (usually with larger areas of farmlands) have the alternative to allocate less regions for the uncertain operations and link profit with risk aversion. Without no doubt, education of the farmer plays a key role on the ICT adoption, as it tends to improve the creative thinking and judgement for taking innovative decisions. Moreover, proper education is related to better accessibility to improved technologies and information acquisition, which may reduce the adoption costs rendering the time for adoption shorter. Finally, the gender factor on technologies adoption is questionable and it seems to be connected with the fact that farmers are often influenced by other farmers of the same sex [137].

External Motivations or Deterrents

The lack of information is another element in technology adoption. Imperfect information enhances farmers' uncertainty, while information diffusion contributes to easier approval. In addition, risk constitutes a usually addressed issue regarding technology acceptance, since farmers who cope with bad climatic conditions often adopt ICT to reduce the level of risk. Also, past prices of input and output along with future expectations of them influence the investment decisions regarding new equipment. Dinar and Yaron [138] suggested that the rise in input and output prices has a positive impact on using modern technologies. As mentioned above, financial assets can shape the adoption. In essence, credit accessibility results in dissimilar adoption rates [134]. Ordinarily, credit access boosts farmers to approve ICT. Credit availability can affect not only investments, but the crop choices as well, since limited credit accessibility results in liquidity problems influencing the crop choices. In the absence of credit, relatives and friends can serve as a financial source which, in turn, may encourage or discourage the implementation of technologies. Also, the level of adoption pertaining to new technologies of neighbouring farms can be a factor, who take on the role of information source. Under social pressure, the farmer may choose to act just like their neighbour and either accept new practices or disapprove them.

4 Conclusions

In summary, there is no doubt that the innovative technologies, such as Artificial Intelligence (AI), the Internet of Things (IoT), Cloud Computing and robotics, have a great potential to shift farming towards the fourth revolution. As a consequence, these emergent technologies can play a key role towards establishing a sustainable agriculture. However, the technologies adoption by farmers constitutes a challenging issue including agriculture-related businesses and policy-makers. In fact, farmers have to engage with a wide-ranging agricultural practices and ICT systems to cope with the continuously changing demands of consumers. What is more, these technologies are considerably evolving, while information on the costs of incorporating ICT in agricultural activities is often not available. Accordingly, decisions for technologies approval are usually made within an uncertain climate via "trial and error" approaches with a subsequent doubtful economic profit.

Considering also that the extent of adoption as well as its rate varies substantially among farmers, even of the same region, arguments can be raised concerning the number of farmers which will afford agriculture 4.0 in the near future. Although there is no clear consensus on which factors affect positively and which ones discourage the ICT implementation, it seems to be a positive relationship between adoption and wealth, experience, education, membership in an organisation, information, neighbouring farms, size of the farmland and credit accessibility. Overall, farmers need some fundamental knowledge on computers and skills so that they can use ICT as well as Web via which they will be able to find out beneficial information about product prices, for instance, or communicate with farmers from other sites to share ideas and experiences. A key aspect is also the role of agronomists and researchers on advising the farmers on various topics. In fact, one of the major ICT advantages is that it contributes to the better communication in favour of not only human relationships, but also the agricultural development and national economy.

References

1. Rome Declaration and Plan of Action
2. Tzounis A, Katsoulas N, Bartzanas T, Kittas C (2017) Internet of Things in agriculture, recent advances and future challenges. Biosyst. Eng. 164:31–48
3. Saiz-Rubio V, Rovira-Más F (2020) From smart farming towards agriculture 5.0: A review on crop data management. Agronomy
4. Wang H, He Q, Liu X, et al (2012) Global urbanization research from 1991 to 2009: A systematic research review. Landsc. Urban Plan. 104:299–309
5. World Bank Open Data | Data
6. Boland MJ, Rae AN, Vereijken JM, et al (2013) The future supply of animal-derived protein for human consumption. Trends Food Sci. Technol. 29:62–73
7. Conrad Z, Niles MT, Neher DA, et al (2018) Relationship between food waste, diet quality, and environmental sustainability. PLoS One 13. https://doi.org/10.1371/journal.pone.0195405
8. Tilman D, Clark M (2014) Global diets link environmental sustainability and human health. Nature 515:518–522. https://doi.org/10.1038/nature13959
9. Ruini LF, Ciati R, Pratesi CA, et al (2015) Working toward Healthy and Sustainable Diets: The "Double Pyramid Model" Developed by the Barilla Center for Food and Nutrition to Raise Awareness about the Environmental and Nutritional Impact of Foods. Front Nutr 2. https://doi.org/10.3389/fnut.2015.00009

10. Chasek P, Safriel U, Shikongo S, Fuhrman VF (2015) Operationalizing Zero Net Land Degradation: The next stage in international efforts to combat desertification? J Arid Environ 112:5–13. https://doi.org/10.1016/j.jaridenv.2014.05.020
11. Safriel UN (2007) The assessment of global trends in land degradation. Environ Sci Eng (Subseries) Environ Sci 1–38. https://doi.org/10.1007/978-3-540-72438-4_1
12. Vogt J V., Safriel U, Von Maltitz G, et al (2011) Monitoring and assessment of land degradation and desertification: Towards new conceptual and integrated approaches. L. Degrad. Dev. 22:150–165
13. De Clercq, M.; Vats, A.; Biel A (2018) Agriculture 4.0 – Farming Technology to Meet Food and Climate Goals
14. United Nations Conference on Sustainable Development, Rio+20.:. Sustainable Development Knowledge Platform
15. Thomas CD, Cameron A, Green RE, et al (2004) Extinction risk from climate change. Nature 427:145–148. https://doi.org/10.1038/nature02121
16. Dawson TP, Jackson ST, House JI, et al (2011) Beyond predictions: Biodiversity conservation in a changing climate. Science (80-.). 332:53–58
17. Gardi C, Jeffery S, Saltelli A (2013) An estimate of potential threats levels to soil biodiversity in EU. Glob Chang Biol 19:1538–1548. https://doi.org/10.1111/gcb.12159
18. Barrett CB, Bellemare MF, Hou JY (2010) Reconsidering Conventional Explanations of the Inverse Productivity-Size Relationship. World Dev 38:88–97
19. Woodhouse P (2010) Beyond industrial agriculture? Some questions about farm size, productivity and sustainability. J Agrar Chang 10:437–453. https://doi.org/10.1111/j.1471-0366.2010.00278.x
20. Ratnadass A, Fernandes P, Avelino J, Habib R (2012) Plant species diversity for sustainable management of crop pests and diseases in agroecosystems: A review. Agron. Sustain. Dev. 32:273–303
21. Keating BA, Carberry PS, Bindraban PS, et al (2010) Eco-efficient agriculture: Concepts, Challenges, And opportunities. Crop Sci 50:S-109-S-119. https://doi.org/10.2135/cropsci2009.10.0594
22. Tscharntke T, Clough Y, Wanger T, et al (2012) Global food security, biodiversity conservation and the future of agricultural intensification. Biol Conserv 151:53–59
23. Wintle BA, Bekessy SA, Keith DA, et al (2011) Ecological--economic optimization of biodiversity conservation under climate change. Nat Clim Chang 1. https://doi.org/10.1038/NCLIMATE1227
24. Revised 1996 IPCC Guidelines for National Greenhouse Gas Inventories—IPCC
25. Jin Z, Ainsworth EA, Leakey ADB, Lobell DB (2018) Increasing drought and diminishing benefits of elevated carbon dioxide for soybean yields across the US Midwest. Glob Chang Biol 24:e522–e533. https://doi.org/10.1111/gcb.13946
26. Pugh TAM, Müller C, Elliott J, et al (2016) Climate analogues suggest limited potential for intensification of production on current croplands under climate change. Nat Commun 7:1–8. https://doi.org/10.1038/ncomms12608
27. Agovino M, Casaccia M, Ciommi M, et al (2019) Agriculture, climate change and sustainability: The case of EU-28. Ecol Indic 105:525–543. https://doi.org/10.1016/j.ecolind.2018.04.064
28. 2015 - The State of Food Insecurity in the World 2015 | World Food Programme
29. Magadza CHD (2000) Climate change impacts and human settlements in Africa: Prospects for adaptation. Environ Monit Assess 61:193–205. https://doi.org/10.1023/A:1006355210516
30. Fonta WM, Kedir AM, Bossa AY, et al (2018) A Ricardian valuation of the impact of climate change on Nigerian cocoa production: Insight for adaptation policy. Int J Clim Chang Strateg Manag 10:689–710. https://doi.org/10.1108/IJCCSM-05-2016-0074
31. Loboguerrero A, Campbell B, Cooper P, et al (2019) Food and Earth Systems: Priorities for Climate Change Adaptation and Mitigation for Agriculture and Food Systems. Sustainability 11:1372. https://doi.org/10.3390/su11051372

32. Abd-Elmabod SK, Muñoz-Rojas M, Jordán A, et al (2020) Climate change impacts on agricultural suitability and yield reduction in a Mediterranean region. Geoderma 374:114453. https://doi.org/10.1016/j.geoderma.2020.114453
33. Anderson R, Bayer PE, Edwards D (2020) Climate change and the need for agricultural adaptation. Curr. Opin. Plant Biol. 56:197–202
34. Liu M, Xu X, Jiang Y, et al (2020) Responses of crop growth and water productivity to climate change and agricultural water-saving in arid region. Sci Total Environ 703:134621. https://doi.org/10.1016/j.scitotenv.2019.134621
35. Hodges RJ, Buzby JC, Bennett B (2011) Postharvest losses and waste in developed and less developed countries: Opportunities to improve resource use. J Agric Sci 149:37–45. https://doi.org/10.1017/S0021859610000936
36. Salihoglu G, Salihoglu NK, Ucaroglu S, Banar M (2018) Food loss and waste management in Turkey. Bioresour. Technol. 248:88–99
37. Food Loss and Food Waste | FAO | Food and Agriculture Organization of the United Nations
38. Noack AL, Pouw NRM (2015) A blind spot in food and nutrition security: where culture and social change shape the local food plate. Agric Human Values 32:169–182. https://doi.org/10.1007/s10460-014-9538-y
39. Food waste is responsible for 6% of global greenhouse gas emissions - Our World in Data
40. Arcury TA, Quandt SA, Russell GB (2002) Pesticide safety among farmworkers: Perceived risk and perceived control as factors reflecting environmental justice. Environ. Health Perspect. 110:233–240
41. Nicolopoulou-Stamati P, Maipas S, Kotampasi C, et al (2016) Chemical Pesticides and Human Health: The Urgent Need for a New Concept in Agriculture. Front Public Heal 4:1. https://doi.org/10.3389/fpubh.2016.00148
42. Alewu B, Nosiri C (2011) Pesticides and Human Health. In: Pesticides in the Modern World—Effects of Pesticides Exposure. InTech. https://www.intechopen.com/chapters/19601; https://doi.org/10.5772/18734
43. Mnif W, Ibn A, Hassine H, et al (2011) Effect of Endocrine Disruptor Pesticides: A Review. OPEN ACCESS Int J Environ Res Public Heal 8:2265–2303. https://doi.org/10.3390/ijerph8062265
44. Nieuwenhuijsen MJ, Noderer KS, Schenker MB, et al (1999) Personal exposure to dust, endotoxin and crystalline silica in California agriculture. Ann Occup Hyg 43:35–42
45. Cerhan JR, Cantor KP, Williamson K, et al (1998) Cancer mortality among Iowa farmers: Recent results, time trends, and lifestyle factors (United States). Cancer Causes Control 9:311–319. https://doi.org/10.1023/A:1008877204830
46. Fathallah FA (2010) Musculoskeletal disorders in labor-intensive agriculture. Appl Ergon 41:738–743. https://doi.org/10.1016/j.apergo.2010.03.003
47. McCurdy SA, Samuels SJ, Carroll DJ, et al (2003) Agricultural injury in California migrant Hispanic farm workers. Am J Ind Med 44:225–235. https://doi.org/10.1002/ajim.10272
48. Hartman E, Oude Vrielink HHE, Huirne RBM, Metz JHM (2003) Sick leave analysis among self-employed Dutch farmers. Occup Med (Chic Ill) 53:461–468. https://doi.org/10.1093/occmed/kqg089
49. Bernard C, Tourne M (2007) [Musculoskeletal disorders in agriculture]. Rev Prat 57:45–50
50. Benos L, Tsaopoulos D, Bochtis D (2020) A Review on Ergonomics in Agriculture. Part I: Manual Operations. Appl Sci 10:1905. https://doi.org/10.3390/app10061905
51. Kirkhorn S, Greenlee RT, Reeser JC (2003) The Epidemiology of Agriculture-related Osteoarthritis and its Impact on Occupational Disability. Wis Med J 102:38–44
52. Benos L, Tsaopoulos D, Bochtis D (2020) A Review on Ergonomics in Agriculture. Part II: Mechanized Operations. Appl Sci 10:3484. https://doi.org/10.3390/app10103484
53. Van Pham L, Smith C (2014) Drivers of agricultural sustainability in developing countries: A review. Environ. Syst. Decis. 34:326–341
54. Lampridi MG, Sørensen CG, Bochtis DD (2019) Agricultural Sustainability: A Review of Concepts and Methods. Sustainability 11:27. https://doi.org/10.3390/su11185120

55. De Olde EM, Oudshoorn FW, Sørensen CAG, et al (2016) Assessing sustainability at farm-level: Lessons learned from a comparison of tools in practice. Ecol Indic 66:391–404. https://doi.org/10.1016/j.ecolind.2016.01.047
56. Yan MJ, Humphreys J, Holden NM (2011) An evaluation of life cycle assessment of European milk production. J Environ Manage 92:372–379. https://doi.org/10.1016/j.jenvman.2010.10.025
57. Binder CR, Feola G, Steinberger JK (2010) Considering the normative, systemic and procedural dimensions in indicator-based sustainability assessments in agriculture. Environ. Impact Assess. Rev. 30:71–81
58. Zhou K, Jensen AL, Bochtis D, et al (2020) Metric map generation for autonomous field operations. Agronomy 10. https://doi.org/10.3390/agronomy10010083
59. Jensen MAF, Bochtis D, Sorensen CG, et al (2012) In-field and inter-field path planning for agricultural transport units. Comput Ind Eng 63:1054–1061
60. Gonzalez-Amarillo CA, Corrales-Munoz JC, Mendoza-Moreno MA, et al (2018) An IoT-Based traceability system for greenhouse seedling crops. IEEE Access 6:67528–67535. https://doi.org/10.1109/ACCESS.2018.2877293
61. Qian J, Fan B, Wu X, et al (2017) Comprehensive and quantifiable granularity: A novel model to measure agro-food traceability. Food Control 74:98–106. https://doi.org/10.1016/j.foodcont.2016.11.034
62. Bochtis D, Sørensen CAG, Kateris D (2018) Operations management in agriculture. Academic Press. https://doi.org/10.1016/C2015-0-06290-6. ISBN: 978-0-12-809786-1
63. Gebbers R, Adamchuk VI (2010) Precision agriculture and food security. Science (80-.). 327:828–831
64. Robert PC (2002) Precision agriculture: A challenge for crop nutrition management. Plant Soil 247:143–149
65. Zecca F (2019) The Use of Internet of Things for the Sustainability of the Agricultural Sector: The Case of Climate Smart Agriculture. Int J Civ Eng Technol 10:494–501
66. Walter A, Finger R, Huber R, Buchmann N (2017) Opinion: Smart farming is key to developing sustainable agriculture. Proc Natl Acad Sci 114:6148–6150. https://doi.org/10.1073/pnas.1707462114
67. Rose DC, Chilvers J (2018) Agriculture 4.0: Broadening Responsible Innovation in an Era of Smart Farming. Front Sustain Food Syst. https://doi.org/10.3389/fsufs.2018.00087
68. Sørensen CAG, Kateris D, Bochtis D (2019) ICT Innovations and Smart Farming. In: Communications in Computer and Information Science
69. Eastwood C, Klerkx L, Ayre M, Dela Rue B (2019) Managing Socio-Ethical Challenges in the Development of Smart Farming: From a Fragmented to a Comprehensive Approach for Responsible Research and Innovation. J Agric Environ Ethics 32:741–768. https://doi.org/10.1007/s10806-017-9704-5
70. Agriculture 4.0 – The Future Of Farming Technology
71. Khan FA (2018) A Review an Hydroponic Greenhouse Cultivation for Sustainable Agriculture. Int J Agric Environ Food Sci 2:59–66. https://doi.org/10.31015/jaefs.18010
72. Fernández JA, Orsini F, Baeza E, et al (2018) Current trends in protected cultivation in Mediterranean climates. Eur. J. Hortic. Sci. 83:294–305
73. Chen P, Zhu G, Kim HJ, et al (2020) Comparative life cycle assessment of aquaponics and hydroponics in the Midwestern United States. J Clean Prod 275:122888. https://doi.org/10.1016/j.jclepro.2020.122888
74. dos Santos JD, Lopes da Silva AL, da Luz Costa J, et al (2013) Development of a vinasse nutritive solution for hydroponics. J Environ Manage 114:8–12. https://doi.org/10.1016/j.jenvman.2012.10.045
75. Shekarchi N, Shahnia F (2019) A comprehensive review of solar-driven desalination technologies for off-grid greenhouses. Int J Energy Res 43:1357–1386. https://doi.org/10.1002/er.4268

76. Aznar-Sánchez JA, Velasco-Muñoz JF, López-Felices B, Román-Sánchez IM (2020) An Analysis of Global Research Trends on Greenhouse Technology: Towards a Sustainable Agriculture. Int J Environ Res Public Health 17:664. https://doi.org/10.3390/ijerph17020664
77. Aznar-Sánchez J, Belmonte-Ureña L, Velasco-Muñoz J, Valera D (2019) Aquifer Sustainability and the Use of Desalinated Seawater for Greenhouse Irrigation in the Campo de Níjar, Southeast Spain. Int J Environ Res Public Health 16:898. https://doi.org/10.3390/ijerph16050898
78. Cifuentes-Torres L, Mendoza-Espinosa LG, Correa-Reyes G, Daesslé LW (2020) Hydroponics with wastewater: a review of trends and opportunities. Water Environ J wej.12617. https://doi.org/10.1111/wej.12617
79. Al-Chalabi M (2015) Vertical farming: Skyscraper sustainability? Sustain Cities Soc 18:74–77. https://doi.org/10.1016/j.scs.2015.06.003
80. Bochtis D, Benos L, Lampridi M, et al (2020) Agricultural workforce crisis in light of the COVID-19 pandemic. Sustain 12. https://doi.org/10.3390/su12198212
81. Pulighe G, Lupia F (2020) Food First: COVID-19 Outbreak and Cities Lockdown a Booster for a Wider Vision on Urban Agriculture. Sustainability 12:5012. https://doi.org/10.3390/su12125012
82. Lal R (2020) Home gardening and urban agriculture for advancing food and nutritional security in response to the COVID-19 pandemic. Food Secur. 1–6
83. Köberl M, Müller H, Ramadan EM, Berg G (2011) Desert Farming Benefits from Microbial Potential in Arid Soils and Promotes Diversity and Plant Health. PLoS One 6:e24452. https://doi.org/10.1371/journal.pone.0024452
84. Innovations in desert and drylands farming | Support to Investment | Food and Agriculture Organization of the United Nations
85. Sahara Forest Project
86. Seawater Farming | Mission 2014: Feeding the World
87. González FG, Rigalli N, Miranda PV, et al (2020) An Interdisciplinary Approach to Study the Performance of Second-generation Genetically Modified Crops in Field Trials: A Case Study With Soybean and Wheat Carrying the Sunflower HaHB4 Transcription Factor. Front Plant Sci 11:178. https://doi.org/10.3389/fpls.2020.00178
88. Klümper W, Qaim M (2014) A meta-analysis of the impacts of genetically modified crops. PLoS One 9. https://doi.org/10.1371/journal.pone.0111629
89. Perry ED, Ciliberto F, Hennessy DA, Moschini GC (2016) Genetically engineered crops and pesticide use in U.S. maize and soybeans. Sci Adv 2:1600850. https://doi.org/10.1126/sciadv.1600850
90. Kamle M, Kumar P, Patra JK, Bajpai VK (2017) Current perspectives on genetically modified crops and detection methods. 3 Biotech 7
91. Sun J, Peng Z, Zhou W, et al (2015) A Review on 3D Printing for Customized Food Fabrication. In: Procedia Manufacturing. Elsevier B.V., pp. 308–319
92. Kah M, Tufenkji N, White JC (2019) Nano-enabled strategies to enhance crop nutrition and protection. Nat Nanotechnol 14:532–540. https://doi.org/10.1038/s41565-019-0439-5
93. Srinivasan B, Kulshreshtha G (2020) Recent developments in food-based bioplastics production. In: Handbook of Environmental Chemistry. Springer Science and Business Media Deutschland GmbH, pp. 107–127
94. Fathallah F, Duraj V (2017) Small Changes Make Big Differences: The role of Ergonomics in Agriculture. Resour Mag 24:12–13
95. Ramahi AA, Fathallah FA (2006) Ergonomic evaluation of manual weeding practice and development of an ergonomic solution. In: Proceedings of the Human Factors and Ergonomics Society. pp. 1421–1425
96. Benos L, Bechar A, Bochtis D (2020) Safety and ergonomics in human-robot interactive agricultural operations. Biosyst Eng 200:55–72. https://doi.org/10.1016/j.biosystemseng.2020.09.009

97. Deakin JM, Stevenson JM, Vail GR, Nelson JM (1994) The use of the Nordic questionnaire in an industrial setting: a case study. Appl Ergon 25:182–185. https://doi.org/10.1016/0003-6870(94)90017-5
98. Kumar P, Chakrabarti D, Patel T, Chowdhuri A (2016) Work-related pains among the workers associated with pineapple peeling in small fruit processing units of North East India. Int J Ind Ergon 53:124–129. https://doi.org/10.1016/j.ergon.2015.11.006
99. Schall MC, Fethke NB, Chen H, et al (2016) Accuracy and repeatability of an inertial measurement unit system for field-based occupational studies. Ergonomics 59:591–602. https://doi.org/10.1080/00140139.2015.1079335
100. Merino G, da Silva L, Mattos D, et al (2019) Ergonomic evaluation of the musculoskeletal risks in a banana harvesting activity through qualitative and quantitative measures, with emphasis on motion capture (Xsens) and EMG. Int J Ind Ergon 69:80–89. https://doi.org/10.1016/j.ergon.2018.10.004
101. Milosavljevic S, McBride DI, Bagheri N, et al (2011) Factors associated with quad bike loss of control events in agriculture. Int J Ind Ergon 41:317–321. https://doi.org/10.1016/j.ergon.2011.02.010
102. Gialamas T, Gravalos I, Kateris D, et al (2016) Vibration analysis on driver's seat of agricultural tractors during tillage tests. Spanish J Agric Res 14. https://doi.org/10.5424/sjar/2016144-9664
103. Kociolek AM, Lang AE, Trask CM, et al (2018) Exploring head and neck vibration exposure from quad bike use in agriculture. Int J Ind Ergon 66:63–69. https://doi.org/10.1016/j.ergon.2018.02.009
104. Ulrey BL, Fathallah FA (2012) Evaluation of a personal device in reducing the risk of low back disorders during stooped work. In: Work. pp. 2381–2383
105. Młotek M, Kuta Ł, Stopa R, Komarnicki P (2015) The Effect of Manual Harvesting of Fruit on the Health of Workers and the Quality of the Obtained Produce. Procedia Manuf 3:1712–1719. https://doi.org/10.1016/j.promfg.2015.07.494
106. Kuta Ł, Stopa R, Szyjewicz D, Komarnicki P (2019) Determination of comfortable position for tractor Driver's hands based on dynamic load. Int J Ind Ergon 74. https://doi.org/10.1016/j.ergon.2019.102866
107. Dewi NS, Komatsuzaki M (2018) On-body personal assist suit for commercial farming: Effect on heart rate, EMG, trunk movements, and user acceptance during digging. Int J Ind Ergon 68:290–296. https://doi.org/10.1016/j.ergon.2018.08.013
108. Ulrey BL, Fathallah FA (2013) Effect of a personal weight transfer device on muscle activities and joint flexions in the stooped posture. J Electromyogr Kinesiol 23(1):195–205. https://doi.org/10.1016/j.jelekin.2012.08.014
109. Dewangan KN, Gogoi G, Owary C, Gorate DU (2010) Isometric muscle strength of male agricultural workers of India and the design of tractor controls. Int J Ind Ergon 40:484–491. https://doi.org/10.1016/j.ergon.2010.05.008
110. Vallone M, Bono F, Quendler E, et al (2016) Risk exposure to vibration and noise in the use of agricultural track-laying tractors. Ann Agric Environ Med 23:591–597. https://doi.org/10.5604/12321966.1226852
111. Romano E, Pirozzi M, Ferri M, et al (2019) The use of pressure mapping to assess the comfort of agricultural machinery seats. Int J Ind Ergon. https://doi.org/10.1016/j.ergon.2019.102835
112. Loutridis S, Gialamas T, Gravalos I, et al (2011) A study on the effect of electronic engine speed regulator on agricultural tractor ride vibration behavior. J Terramechanics 48:139–147. https://doi.org/10.1016/j.jterra.2010.10.002
113. Hudson DS, Copeland JL, Hepburn CG, Doan JB (2014) Stooped Postures Are Modified by Pretask Walking in a Simulated Weed-Pulling Task. J Agromedicine 19:27–34. https://doi.org/10.1080/1059924X.2013.865572
114. Pinzke S, Lavesson L (2018) Ergonomic conditions in manual harvesting in swedish outdoor cultivation. Ann Agric Environ Med 25:481–487. https://doi.org/10.26444/aaem/93334

115. Juntaracena K, Neubert MS, Puntumetakul R (2018) Effects of muddy terrain on lower extremity muscle activity and discomfort during the rice planting process. Int J Ind Ergon 66:187–193. https://doi.org/10.1016/j.ergon.2018.03.009
116. Escamilla R (2009) Electromyographic Activity During Upper Extremity Sports. In: The Athlete's Shoulder. Elsevier Inc., pp. 385–400
117. Çakmak B, Ergül E (2018) Interactions of personal and occupational risk factors on hand grip strength of winter pruners. Int J Ind Ergon 67:192–200. https://doi.org/10.1016/j.ergon.2018.05.002
118. Gomez-Gil J, Javier Gomez-Gil F, Martin-De-Leon R (2014) The Influence of Tractor-Seat Height above the Ground on Lateral Vibrations. Sensors 14. https://doi.org/10.3390/s141019713
119. Lazaridis S, Patikas DA, Bassa E, et al (2018) The acute effects of an intense stretch-shortening cycle fatigue protocol on the neuromechanical parameters of lower limbs in men and prepubescent boys. J Sports Sci 36:131–139. https://doi.org/10.1080/02640414.2017.1287932
120. Benos L, Stanev D, Spyrou L, et al (2020) A review on finite element modelling and simulation of the anterior cruciate ligament reconstruction. Front Bioeng Biotechnol 8:967. https://doi.org/10.3389/FBIOE.2020.00967
121. Dicks L V., Rose DC, Ang F, et al (2019) What agricultural practices are most likely to deliver "sustainable intensification" in the UK? Food Energy Secur 8:e00148. https://doi.org/10.1002/fes3.148
122. Gunton RM, Firbank LG, Inman A, Winter DM (2016) How scalable is sustainable intensification? Nat Plants 2:16065. https://doi.org/10.1038/nplants.2016.65
123. Gadanakis Y, Bennett R, Park J, Areal FJ (2015) Evaluating the Sustainable Intensification of arable farms. J Environ Manage 150:288–298. https://doi.org/10.1016/j.jenvman.2014.10.005
124. Janker J, Mann S, Rist S (2019) Social sustainability in agriculture – A system-based framework. J Rural Stud 65:32–42. https://doi.org/10.1016/j.jrurstud.2018.12.010
125. Wolsko C, Marino E, Doherty TJ, et al (2016) Systems of access: A multidisciplinary strategy for assessing the social dimensions of sustainability. Sustain Sci Pract Policy 12. https://doi.org/10.1080/15487733.2016.11908156
126. Rose DC, Wheeler R, Winter M, et al (2021) Agriculture 4.0: Making it work for people, production, and the planet. Land use policy 100:104933. https://doi.org/10.1016/j.landusepol.2020.104933
127. Barnes AP, Soto I, Eory V, et al (2019) Exploring the adoption of precision agricultural technologies: A cross regional study of EU farmers. Land use policy 80:163–174. https://doi.org/10.1016/j.landusepol.2018.10.004
128. Eastwood C, Ayre M, Nettle R, Dela Rue B (2019) Making sense in the cloud: Farm advisory services in a smart farming future. NJAS - Wageningen J Life Sci 90–91:100298. https://doi.org/10.1016/j.njas.2019.04.004
129. Autor DH, Levy F, Murnane RJ (2003) The Skill Content of Recent Technological Change: An Empirical Exploration. Q J Econ 118:1279–1333. https://doi.org/10.1162/003355303322552801
130. Heyman F (2016) Job polarization, job tasks and the role of firms. Econ Lett 145:246–251. https://doi.org/10.1016/J.ECONLET.2016.06.032
131. Marinoudi V, Sørensen CG, Pearson S, Bochtis D (2019) Robotics and labour in agriculture. A context consideration. Biosyst Eng. https://doi.org/10.1016/j.biosystemseng.2019.06.013
132. Paustian M, Theuvsen L (2017) Adoption of precision agriculture technologies by German crop farmers. Precis Agric 18:701–716. https://doi.org/10.1007/s11119-016-9482-5
133. Liu T, Bruins R, Heberling M (2018) Factors Influencing Farmers' Adoption of Best Management Practices: A Review and Synthesis. Sustainability 10:432. https://doi.org/10.3390/su10020432
134. Sansel Tandogan N, Gedikoglu H (2020) Socio-Economic Dimensions of Adoption of Conservation Practices: What Is Needed to Be Done? In: Organic Agriculture. IntechOpen

135. Kassie M, Jaleta M, Shiferaw B, et al (2013) Adoption of interrelated sustainable agricultural practices in smallholder systems: Evidence from rural Tanzania. Technol Forecast Soc Change 80:525–540. https://doi.org/10.1016/j.techfore.2012.08.007
136. Wollni M, Andersson C (2014) Spatial patterns of organic agriculture adoption: Evidence from Honduras. Ecol Econ 97:120–128. https://doi.org/10.1016/j.ecolecon.2013.11.010
137. Deressa TT, Hassan RM, Ringler C, et al (2009) Determinants of farmers' choice of adaptation methods to climate change in the Nile Basin of Ethiopia. Glob Environ Chang 19:248–255. https://doi.org/10.1016/j.gloenvcha.2009.01.002
138. Dinar A, Yaron D (1992) Adoption and abandonment of irrigation technologies. Agric Econ 6:315–332. https://doi.org/10.1016/0169-5150(92)90008-M

Sustainability in the Digital Farming Era: A Cyber-Physical Analysis Approach for Drone Applications in Agriculture 4.0

Naoum Tsolakis, Dimitrios Bechtsis, Giorgos Vasileiadis, Ioannis Menexes, and Dionysis D. Bochtis

1 Introduction

The food sector provides pioneering grounds for utilizing intelligent automations and robotic systems with a notable example being Ocado's Customer Fulfilment Centre in Andover, England, utilizing 1,300 bots that result in delivery punctuality by 95% and order accuracy by 99% [1]. The scope of utilizing intelligent systems in food supply chains depends upon the particular strategic objectives articulated by the involved stakeholders. On the one end, in operations-focused cases similar to Ocado, robotic automation enables operational efficiency downstream the supply chain to ensure high service-levels and increased responsiveness to market demand [2, 3]. On the other end, at a strategic level, the Food and Agriculture Organization of the United Nations reported the use of unmanned aerial vehicles (UAVs), commonly known as drones, in agricultural production to increase efficiency in upstream operations to further ensure food security and sustainability [4], particularly in emerging economies. At this latter policy-making level, foresight programmes at both national and regional levels envision the sustainable future of agricultural production and further define strategies to deliver this vision [5–8]. Indicatively, the Danish Green Technological Foresight on Environmental Agriculture provided a technology foresight study to support the adoption of technology solutions that could promote environmentally friendly agriculture [5].

N. Tsolakis (✉) · G. Vasileiadis · I. Menexes · D. D. Bochtis
Institute for Bio-Economy and Agri-Technology (iBO), Centre for Research and Technology Hellas (CERTH), Thessaloniki, Greece
e-mail: n.tsolakis@certh.gr

D. Bechtsis
Department of Industrial Engineering and Management, International Hellenic University (IHU), Thessaloniki, Greece

D. D. Bochtis et al. (eds.), *Information and Communication Technologies for Agriculture—Theme IV: Actions*, Springer Optimization and Its Applications 185,
https://doi.org/10.1007/978-3-030-84156-0_2

Unmanned aerial systems posit a promising technological application for enabling precision farming operations and ensuring increased crop productivity in a sustainable manner [9, 10], considering also the nutritional needs of the predicted 9.8 billion global population by 2050 [11]. The global market for drone-based solutions in agriculture is projected to amount US$6.52 billion in 2026 demonstrating a compound annual growth rate of 22.6% [12]. Indicative agricultural activities where drones are utilized include: tracking livestock [13], spraying pesticides [14], remote sensing of crop health [15], evaluating field maturity and harvest readiness [16], and facilitating crop insurance claims [17]). Most importantly, drones enable precision farming operations, like: detecting weed patches [18], exploring the effect of nitrogen treatments on crops [19], monitoring crop biomass [20], identifying water stress in crops [21], and mapping vineyard vigor [22].

Considering the vital role of freshwater resources in agriculture, along with the pressing issue of water scarcity in major food producing countries like India, UAV-enabled remote sensing capabilities could be valuable for farmers to ensure water stewardship and promote environmental sustainability in the sector [23]. Indicatively, results from the "Unmanned Aerial Vehicles—Wireless Sensor Network" scheme applied on over 12,000 ha of farmland in the Republic of China demonstrated that drones can ensure irrigation efficiency with significant water savings by up to 67% [4].

Notwithstanding the documented applicability and benefits of UAVs in agriculture, especially in the light of sustainability, financial viability concerns exist due to the acquisition cost of the required sensors and the supporting infrastructure [24, 25]. To that end, proactive assessment of UAV applications is needed to inform stakeholders' decision-making process and foster the adoption of digital technologies by farmers [6, 7, 26, 27]. Nevertheless, the assessment of digital technology applications in agriculture is challenging due to a range of factors involved at an operational level, while the majority of existing approaches only enables qualitative analysis thus not allowing the quantification of prospective risks and benefits [28].

Farmers need to become aware of the functionality and the tangible gains associated with the adoption of digital technologies, like UAVs, in agricultural field operations prior to investing substantially. To this end, researchers and businesses either develop simulation models or directly implement pilot technologies to engage farmers at a cyber or at a physical space, respectively. However, unreliable results, poor communication and ineffective dissemination of information hinder farmers from developing a genuinely positive attitude towards the adoption of digital technologies [29]. Therefore, the effective transition towards an Agriculture 4.0 era, similarly to Industry 4.0 in the manufacturing sector, could be supported by the development and application of "digital twins" to understand and clearly communicate to involved stakeholders the implications of digital technologies in agriculture [30].

This research explores the utilization of UAVs in agriculture towards ensuring environmental sustainability in farming operations. Specifically, motivated by the evident need to tackle the challenge of water scarcity and ensure farmers' livelihood [31], the objective of this research is to provide a methodological approach for facilitating the anticipated use of drones for sustainable farming operations, particularly in terms of monitoring crops' water stress status and informing precision irrigation activities. In this regard, this research addresses the following Research Questions (RQs):

- RQ#1—What are the benefits and challenges associated with the application of UAVs in farming operations?
- RQ#2—Is the development of "digital twins" valid for UAVs to foresee their applicability in precision farming operations for ensuring water stewardship?

In order to address the enunciated RQs, this study applies a multiple methods approach. Firstly, a critical literature taxonomy was performed to identify and summarize advantages and disadvantages related to the applicability of UAVs in agriculture to tackle RQ#1. In an attempt to answer RQ#2, an integrated methodology to analyze "digital twins" in agriculture was proposed. Especially, the proposed methodology explores the underlining dichotomy between the cyber space analysis and the physical space testing of digital technology systems, particularly focusing on UAVs. To this end, an emulation modelling tool was developed which captures a rotary wing UAV that navigates across a conceptual orchard and monitors the water stress level of individual trees. Thereafter, based on the emulation model, two real-world pilot use cases of actual UAV systems were tested on an agricultural field. The UAVs were equipped with sensors for identifying the water status of each plant in the field to inform the planning of precision irrigation activities. This research contributes to the foresight field by adopting an operationalization view over digital technologies for sustainable agriculture and through proposing an integrated cyber-physical analysis approach for drone applications in agriculture, comprising of both an emulation-based research tool and physical assets.

2 Materials and Methods

The basic terminology, theoretical lens and research approach pertinent to this study are specified in the subsections that follow. The materials and methods were developed with a focus on UAVs, as a digital technology application, for the effective water management in agricultural fields.

2.1 *Basic Terminology*

The extant body of literature documents the use of "digital twins" for integrating information regarding the management of resources to then inform equivalent real-world implementations [32]. Therefore, as the focus of this research is "digital twins" for environmentally sustainable agriculture, it is necessary to define the terms in this context.

"Digital Twins"

"Digital twins" is a relatively nascent concept and the ambiguity characterizing the term is evident as most scientific articles and business reports adopt either an asset-based [33] or a supply chain-centric [34] view over the term. This research adopts a hybrid view over the term "digital twins". In particular, we claim that a "digital twin" should capture virtual emulation models of the working environment and the actual hardware system(s) performing operations in order to: (i) enable the ex-ante evaluation of functionality and operations efficiency at the cyber space; and (ii) inform the design and calibration of the actual operational units to support efficiency at the physical space. At the same time the transmission of sensed data from the physical space could be used to update the cyber space constructs, while the information should be shared across end-to-end network echelons to dynamically adjust operations according to the entire supply chain optimal performance requirements (Fig. 1). This research focuses on the first part of our definition that infers engagement at a cyber-physical interface.

Sustainable Agriculture

"Sustainable agriculture" embraces the triple-helix model of sustainability (i.e., environmental, economic and social pillars) applied to the agro-food system domain. Considering that water management has strategic significance for ensuring food security and sustainability in agriculture, particularly in water scarce regions [6–8, 35], this research adopts the environmental sustainability pillar with a specific focus on freshwater appropriation in orchards investigated from the perspective of plantations' precision irrigation needs.

2.2 *Theoretical Lens*

In principal, this research adopts the lens of Foresight Theory as it aims to provide a research methodology and respective analysis toolset to "*create actionable and domain/context specific information or knowledge about the future*" [36]. In particular, this research is positioned at the level of foresight process and impact, in alignment to Piirainen and Gonzalez [36], considering that we propose a research process that allows stakeholders to proactively evaluate a technology intervention in the context of agriculture.

At a greater extent, considering the multifaceted character of foresight and our focus on the impact of an intervention to tackle environmental sustainability challenges, this research responds to the technologies' roadmap proposed by Borch [5]. Specifically, we introduce emulation and testbeds' application as a "*descriptive and systematic evaluation of the (perceived) consequences of applying a technology*" [5],

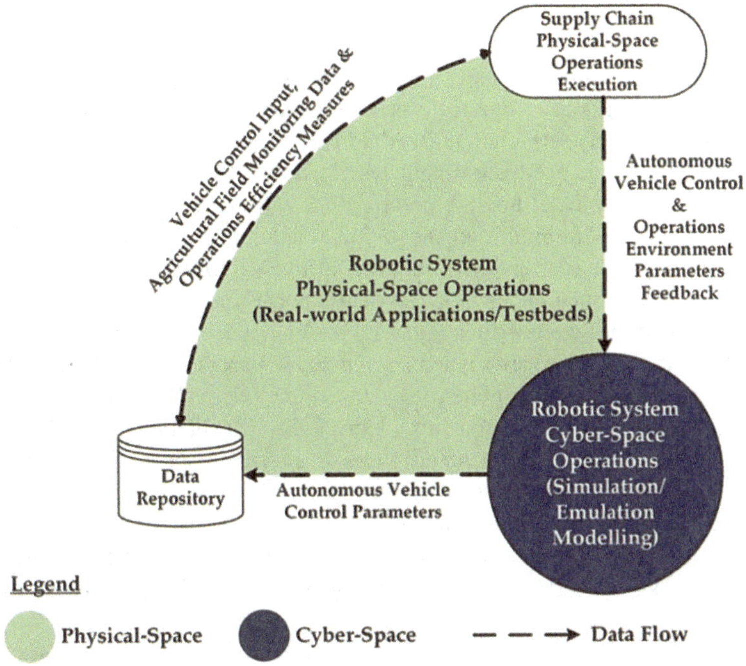

Fig. 1 "Digital Twins" in technology-driver operations

in order to inform farmers with regard to the operationalization of sustainability in agriculture via introducing automated technologies in farming activities. The proposed cyber-physical analysis approach comprising of emulation modelling and real-world technology applications can provide verification of cognitive-wise assertions about a future state of automated agricultural practices [37], hence contributing to the foresight activity.

2.3 *Research Approach*

This research was conducted by deploying a multistage methodological research, as depicted in Fig. 2. Initially, a literature review along with text mining and a critical taxonomy of the retrieved scientific articles were conducted to identify the benefits and challenges associated with drones in agriculture (1st Research Stage), in a robust and systematic manner. Thereafter, following the digital technologies' assessment framework proposed by Tsolakis et al. [38], the stages of emulation modelling (2nd Research Stage) and the real-world implementation of physical UAVs (3rd Research Stage) were investigated. The critical taxonomy along with the emulation model and the real-world pilot implementation of drones are specified in the subsections that follow.

Critical Taxonomy

In order to identify main benefits and challenges regarding the use of drones in agriculture, existing knowledge from peer-reviewed literature was synthesized. In this respect, to identify relevant published articles, we performed structured searches using the terms "unmanned aerial vehicles", "drones" and "intelligent aerial vehicles", in the 'Article Title' field, in combination with the terms "precision agriculture" and "precision farming", in the 'Article Title, Abstract, Keywords', in the Scopus database. The timespan was set from 'All years' to 'Present'. The additional use of the terms "emulation" and/or "Agriculture 4.0" did not generate any results. The reviewed articles were written in the English language. Our review was limited to scientific articles and reviews whereas conference papers were excluded from our analysis. Grey literature and online secondary sources were also retrieved to identify policy and commercial developments in the field. The literature search was not exhaustive as our aim was to identify the main advantages and disadvantages stemming from the use of drones in agriculture.

By November 2nd, 2019, a total of 22 articles studying the use of drones in agriculture was identified for review. The annual allocation of the retrieved articles is presented in Fig. 3. The recent research interest about UAVs in agriculture is evident

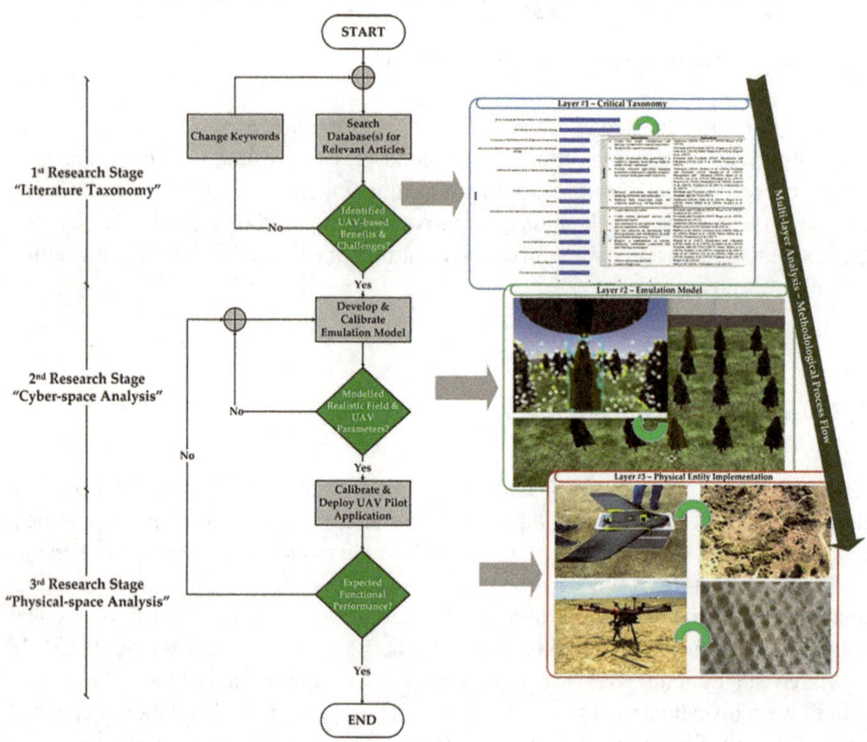

Fig. 2 Multistage methodological analysis flowchart

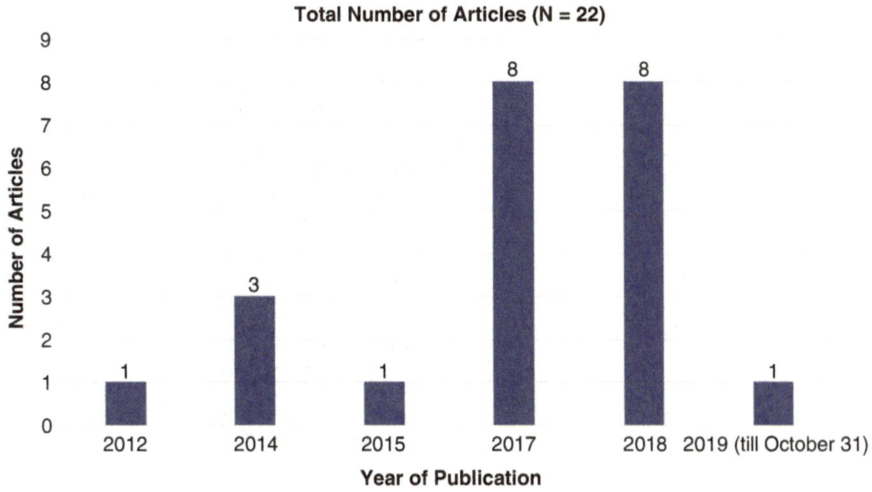

Fig. 3 Published articles by year

as the first related article was published in 2012. The rapid increase in the number of published articles demonstrates the increasing awareness about the application of UAVs in precision agriculture. Likewise, the distribution of the reviewed articles by journal is depicted in Fig. 4. Notably, the distribution of the studies among the scientific journals is quite even, thus indicating the multifaceted research opportunities stemming from the application of drones in farming operations.

In addition, we performed a text mining analysis in the abstracts of the reviewed articles through developing a bespoke programming code in R, an open-source language and environment for statistical computing. Text mining is a technique applied for natural language processing in order to unveil interesting information [39]. Figure 5a illustrates a cloud diagram that depicts the significance of the terms "drone" and "agriculture". Furthermore, Fig. 5b illustrates a circular dendrogram confirming the relevance of intelligent aerial vehicles (marked as "UAV"—unmanned aerial vehicle), including drones, for precision agriculture operations. More specifically, the 'complete-linkage' hierarchical clustering method was applied by calculating the Euclidean distance between term vectors.

Cyber-Space Analysis: Emulation Modelling

In this research, a conceptual orchard was recreated in a three-dimensional environment where an emulated model of an actual quadrotor drone could operate to monitor the water status of individual trees (Fig. 6). This model was created at the Gazebo emulation environment for representing the geomorphological characteristics of the field, thus allowing the spatial modelling across the X-, Y- and Z-axes [40]. At a next step, the emulated UAV could hover, rotate and capture canopy

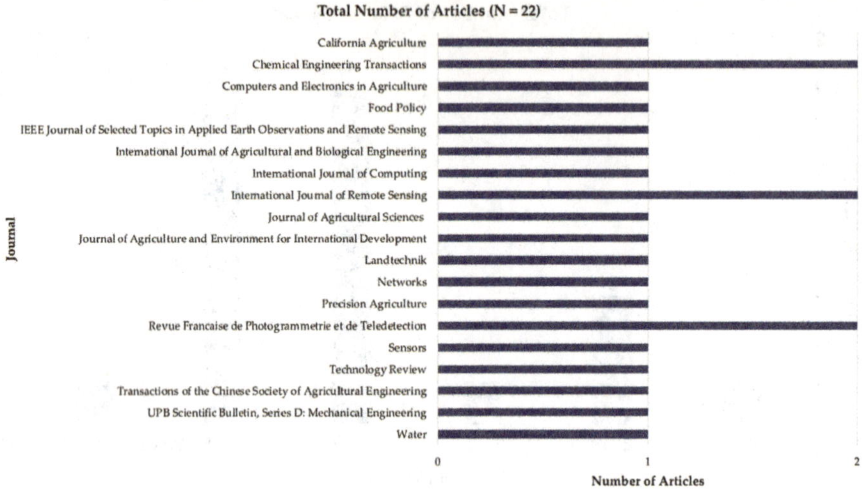

Fig. 4 Published articles by journal

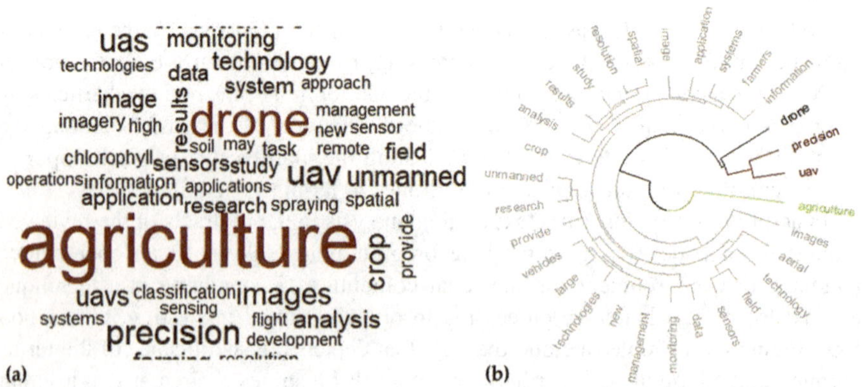

Fig. 5 Relevance of UAVs in precision agriculture demonstrated through: (**a**) a word cloud diagram comprising of 50 words; and (**b**) a circular dendrogram highlighting four clusters of the most associated terms

images of the crops at nearly any optical angle using the Robot Operating System (ROS). For the path tracking of the UAV, ROS used the Simultaneous Localization and Mapping (SLAM) procedure. Furthermore, the emulated Light Detection and Ranging (LiDAR) sensor enabled the UAV to adjust its flying altitude depending on the varying geomorphology and topography of the agricultural field, thus avoiding possible collisions. In particular, the emulated drone is the commercially available rotary wing UAV model DJI S1000.

Fig. 6 Emulation model of an agricultural field environment and a quadrotor drone

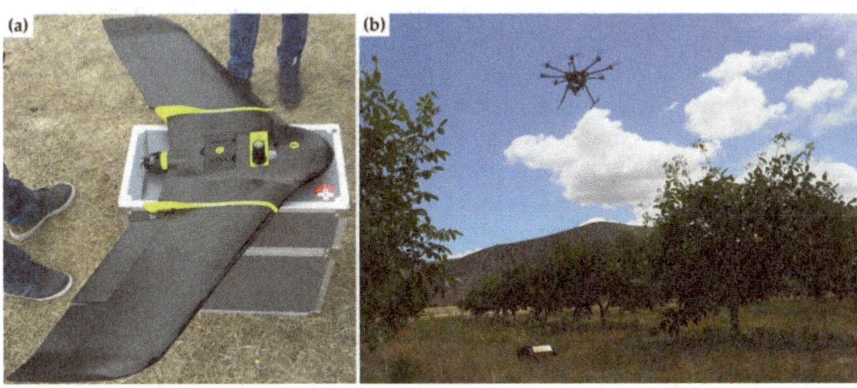

Fig. 7 Pilot implementations of: (a) an actual bespoke fixed wing drone in an agricultural field (model eBee); and (b) an actual commercial rotary wing drone (model DJI S100) along with a Husky vehicle in collaborative field operations

Physical Space Analysis: Pilot System Implementation

A real-world pilot testing for monitoring agricultural fields was performed by using an actual fixed wing drone, the model eBee provided by senseFly (Fig. 7a), along with the commercial rotary wing drone model DJI S1000 equipped with the Ardupilot Pixhawk 2 autopilot hardware (Fig. 7b). The drones are available at the Institute for Bio-Economy and Agri-Technology (iBO), an Institute of the Centre for Research and Technology—Hellas (CERTH).

3 UAVs in Agriculture: Benefits and Challenges

UAVs are one of the major platforms utilized for remote sensing in agriculture, along with satellites and balloons. Owing to their technological development, decreasing cost, increased level of modularity and enhanced flexibility, UAVs are a preferred solution for precision farming applications in both developed and developing countries. Nevertheless, the adoption of UAVs in precision farming applications requires consideration of the associated technical benefits and challenges, depending on the scope of the intended farming application.

Otto et al. [41] reviewed over 200 scientific articles on UAVs and identified agriculture as one of the most promising areas for commercial applications of such technological systems. The authors further identified promising areas for modelling research regarding UAVs. Zhang and Kovacs [42] provided a review of unmanned aerial systems used for environmental monitoring and precision agriculture activities. The authors discussed both the benefits and challenges of UAVs and stressed the necessity for additional research on the field in order to ultimately provide reliable systems which are appreciated and embraced by farmers. Furthermore, Bansod et al. [43] provided a review comparing the benefits and challenges between satellite- and drone-based solutions applied in precision farming operations. The study specifically stressed the challenges associated with the use of UAVs in agriculture across the technical, reliability, privacy rights and safety domains. Shamshiri et al. [9, 10] reviewed automated systems applied in agricultural operations and emphasized the potential of collaborating automated systems, combining multiple field robots and UAVs, in order to collect data, reveal concealed information, and optimize the use of farming inputs.

In a generic agricultural field context, the work presented by Vigneau et al. [44] discussed the capability of UAVs to monitor vegetation indices and, through data analytics and image processing, obtain biochemical and biophysical variables about crops. The authors suggested that the ability to repeat drone flights and collect data over crops' cycle stimulates research and practice interest. Simic Milas et al. [45] reported the use of UAVs for retrieving crops' structural and biochemical parameters to determine their chlorophyll content. More specifically, the authors used an unmanned aerial system to monitor and determine the chlorophyll content of corn agricultural field segments in Michigan, United States of America. Huuskonen and Oksanen [46] introduced a precision farming application for soil sampling to better inform fertilization activities in Southern Finland. The technological solution comprised of a UAV able to scan the selected agricultural field and a set of augmented reality glasses to guide users towards generated soil sampling points.

From a water management perspective, Cancela et al. [47] published a Special Issue on the use of UAVs and satellite systems for water management in agriculture. The Special Issue particularly focused on identifying methodologies for efficiently leveraging such remote sensing technology systems for water management in agriculture. Hogan et al. [48] summarized research applications of small unmanned aerial systems, along with their parameters and limitations. The authors specifically reported the ability to use UAVs for detecting water stress in plants/crops by capturing and analyzing the canopy spectral signature. Anderson [49] discussed the catalytic role of drones in introducing the big data narrative to the precision agriculture domain through examining the case of a vineyard in San Francisco, United States of America. The author of the study supported that the use of UAVs to collect accurate data can reduce water use and lower the chemical load on the environment.

Focusing on technological and technical aspects per se, Barbey et al. [50] compared Pléiades (i.e., a satellite platform) and UAV images retrieved during precision viticulture applications in France. The authors realized that for narrow vine distance rows and small structures, UAVs posit an effective imagery technology for the accurate characterization of vineyards. The work of Ipate et al. [51] presented a guideline for designing a quadrotor drone. Thereafter, the authors deployed the drone to inspect the exterior polyethylene film structure of a greenhouse and examine crops' health. Sarghini and De Vivo [52] also presented a Computational Fluid Dynamics analysis of two different heavy lift multirotor configurations to investigate the resulting aerodynamic effects in the case of spraying pesticides or fertilizers. The authors reported that multirotor UAVs can spray large areas of farmland, around 4,000–6,000 m^2, in about 10 min by achieving savings of about 20–40% in the chemicals' volume and without exposing the operator to health risks. Additionally, Sarghini and De Vivo [53] discussed the merits of intelligent aerial vehicles in agriculture and investigated the technical requirements of multirotor drones for performing agricultural tasks. The authors focused on the mechanical elements of the drone, particularly on the propulsion system, and the resulting payload and flight length capabilities of the system for performing tasks like the application of fertilizers and pesticides. In the work of Lan et al. [54] the challenging issue of obstacles' avoidance in farmlands by UAVs was investigated. The authors compared obstacle avoidance technologies and suggested the use of multisensor fusion on a UAV system to recognize distorting obstacles and enable intelligent autonomous navigation. Liu et al. [55] developed and tested a small-sized and low-cost attitude measurement unit that could be applied to agriculture-focused drones.

From a mainly methodological viewpoint, Lysenko et al. [56] proposed a Robot Plane Vegetation Index, adapted to technological capabilities of UAVs, to monitor the nitrogen nutrition of wheat plants in Ukraine. Additionally, Murugan et al. [57] developed an algorithmic approach to segregate sparse and dense areas in an Indian sugarcane field by leveraging images captured from both a satellite and a drone. The aim of the authors was to ensure precision agriculture monitoring while minimizing the cost of utilizing UAVs in India. Szantoi et al. [58] used images captured through a UAV to map orangutan habitat and agricultural areas in Indonesia. The study concluded that, in contrast to the exclusive use of satellite imagery, UAV-gathered data combined with existing satellite imagery and image classification algorithms provide a cost-effective and high-resolution imagery solution in a variety of land mapping applications. Yamamoto et al. [59] applied a super-resolution image scaling method to process low-resolution images of tomatoes in order to automatically detect and identify plant diseases. The utilized method was intended to be used to low-resolution images captured by UAVs to accelerate phenotyping and vigor diagnosis in agricultural fields.

Finally, Reger et al. [60] discussed and summarized the legislative schemes and regulations regarding the use of UAVs in Germany, the European Union, the United States of America and Japan. The authors suggested that restrictions and gaps in international regulations should be revised and addressed to avoid negative social response to UAV missions in agriculture. In the same context, Freeman and Freeland [61] discussed the regulatory landscape regarding the use of UAVs in agriculture in the United States of America. The authors highlighted the role of regulations in fostering the integration of UAVs in the American airspace to propel their commercial use in agriculture.

Table 1 summarizes the main benefits and challenges associated with the use of UAVs in agriculture and taxonomizes accordingly the retrieved scientific studies. A description for the referenced advantages and disadvantages is also provided to better comprehend the associated views on UAVs in agriculture.

4 "Digital Twins" and UAVs: Monitoring Water Stress in Orchards

In case a crop is in a water stress condition, changes in its leaves occur that generate unique electromagnetic "signatures" [48]. These changes are typically detectable in the visible light spectrum. In addition, changes in the texture of a crops' waxy coating (i.e., cuticle) might be detectable in the invisible infrared light. Therefore, the capability of UAVs to monitor water stress in orchards, inform farmers and support water stewardship depends on both the technical specifications of the system and the quality/calibration of the installed sensory equipment.

Following our proposed methodological approach on the evaluation of UAVs in agriculture via "digital twins", particularly for monitoring the water status of crops, in the subsections that follow we present the emulation model (i.e., cyber space analysis) that was developed as part of this research along with the pilot implementation (i.e., physical space analysis) of actual drone systems in an orchard. Therefore, the proposed methodology allows the creation of cyber-physical interfaces to enable more robust decision-making over the evaluation and adoption of digital technologies in agriculture.

Table 1 UAVs in agriculture: Benefits and challenges

	Description	References
Benefits	• Create less crops' compaction and damage compared to manual inspection	Anderson [49]; Lan et al. [54]; Reger et al. [60]
	• Require relatively low capital expenditure	Bansod et al. [43]; Freeman and Freeland [61]; Hogan et al. [48]; Otto et al. [41]; Sarghini and De Vivo [53]; Simic Milas et al. [45]; Szantoi et al. [58]; Zhang and Kovacs [42]
	• Enable on-demand data gathering, in a repetitive manner, even during night or under cloudy conditions	Freeman and Freeland [61]; Huuskonen and Oksanen [46]; Lan et al. [54]; Vigneau et al. [44]
	• Provide efficient agricultural field mapping resolution, specifically compared to satellite imagery, for narrow fields and small structures	Anderson [49]; Bansod et al. [43]; Barbey et al. [50]; Freeman and Freeland [61]; Hogan et al. [48]; Huuskonen and Oksanen [46]; Cancela et al. [47]; Ipate et al. [51]; Liu et al. [55]; Murugan et al. [57]; Reger et al. [60]; Shamshiri et al. [10]; Szantoi et al. [58]; Vigneau et al. [44]; Yamamoto et al. [59]; Zhang and Kovacs [42]
	• Remove poisoning hazard during spraying fertilizers and pesticides	Freeman and Freeland [61]; Lan et al. [54]; Sarghini and De Vivo [52]
	• Reduce land inspection costs for relatively small fields (e.g., <20 ha)	Anderson [49]; Otto et al. [41]; Reger et al. [60]; Sarghini and De Vivo [53]; Simic Milas et al. [45]; Szantoi et al. [58]; Zhang and Kovacs [42]
Challenges	• Could imperil air-safety	Bansod et al. [43]; Freeman and Freeland [61]; Reger et al. [60]; Sarghini and De Vivo [53]; Szantoi et al. [58]; Zhang and Kovacs [42]
	• Could violate personal privacy and landowner rights	Bansod et al. [43]; Freeman and Freeland [61]; Reger et al. [60]; Szantoi et al. [58]; Zhang and Kovacs [42]
	• Are restricted to low-altitude inspection due to regulatory ceilings	Anderson [49]; Huuskonen and Oksanen [46]; Sarghini and De Vivo [53]; Szantoi et al. [58]; Reger et al. [60]; Zhang and Kovacs [42]
	• Are not effective in identifying field heterogeneities and variabilities in wide agricultural settings (e.g., >20 ha)	Barbey et al. [50]; Otto et al. [41]; Reger et al. [60]; Simic Milas et al. [45]; Yamamoto et al. [59]
	• Require a combination of canopy databases, calibration, correction and data filtering techniques	Hogan et al. [48]; Huuskonen and Oksanen [46]; Ipate et al. [51]; Sarghini and De Vivo [53]; Simic Milas et al. [45]; Szantoi et al. [58]; Vigneau et al. [44]
	• Depend on sensory devices	Bansod et al. [43]; Lan et al. [54]; Liu et al. [55]; Otto et al. [41]; Szantoi et al. [58]; Vigneau et al. [44]

(continued)

Table 1 (continued)

	Description	References
	• Allow maximum payload	Bansod et al. [43]; Reger et al. [60]; Sarghini and De Vivo [52]
	• Allow limited flight time	Bansod et al. [43]; Cancela et al. [47]; Otto et al. [41]; Yamamoto et al. [59]

4.1 Cyber Space: Emulation Modelling

An emulation model can be used to evaluate the performance of a UAV equipped with appropriate sensors. A drone can navigate across an agricultural field or an orchard, detect individual plants, monitor water stress level and detect freshwater requirements of crops, and inform precision irrigation activities. An emulation model can be used to first assess the functional characteristics of a UAV within the environment of operations, assess the performance of the sensors used to scan the crops, map the spatial characteristics of the orchard and autonomously navigate the aerial vehicle in the orchard at an optimal route.

In particular, the developed model consists of emulated constructs of the: (i) orchard layout; (ii) trees within the orchard; (iii) a UAV; and (iv) sensors and cameras equipping the drone. The UAV can then navigate autonomously within the orchard based on the aerial vehicle's routing algorithm embedded in the model and the signals received from the emulated sensors, as depicted in Fig. 8a. The Simultaneous Localization and Mapping procedure along with the perception of the Light Detection And Ranging sensor in the emulated orchard environment are demonstrated in Fig. 8b. The view of the on-board multispectral camera embedded on the UAV is illustrated in Fig. 8c. A camera was emulated to enable plant detection, allow water stress status identification per plant, and ensure vehicle's safety during the autonomous operations in the emulated orchard. Real-time object detection and processing in agricultural environments is exceedingly complex as opposed to typical industrial settings where autonomous robotic systems may be operating [62].

The emulation model further enables the UAV to monitor the water stress level of multiple trees through a single camera (Fig. 9a). The implementation of the flora recognition and the water stress status identification are based on color detection as well as template matching. In this regard, the monitoring per tree is based on continuous sampling of the orchard and tress (when identified), and a corresponding matching of the retrieved signals to the tree reference models stored in the images' library of the emulation model (Fig. 9a). The UAV can then identify and indicate the water status of trees both in cases of water need (e.g., light green trees) and in no water stress situations (e.g., dark green trees), as indicated in Fig. 9b.

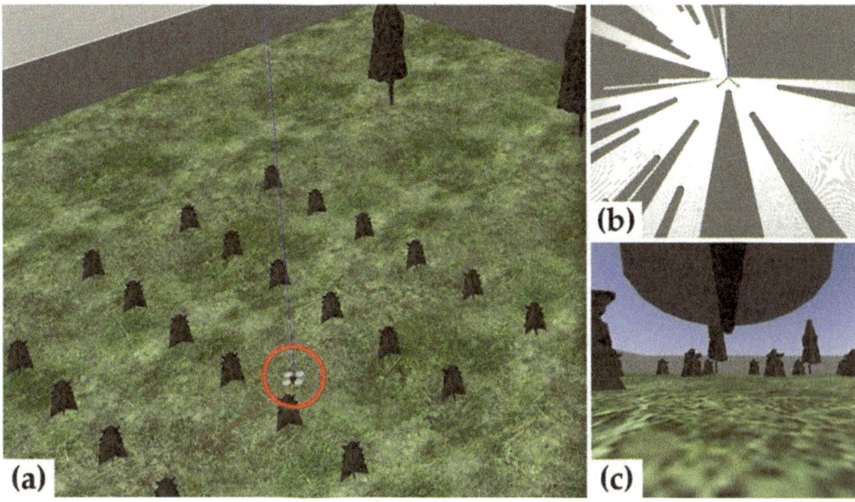

Fig. 8 Emulation model of: (**a**) an orchard environment and a UAV; (**b**) sensors used for navigating the UAV in the orchard; (**c**) an on-board multispectral camera capturing the view of the UAV over the orchard trees

4.2 Physical Space: UAV Systems Deployment

The deployed pilot UAV systems can be used to monitor the status of an agricultural field and identify the water stress level of plants using multispectral cameras. Field irrigation status and water bodies can be identified using band combinations from multispectral and hyperspectral cameras. Hyperspectral imaging has been extensively used for recognizing physiological and structural characteristics in plants and crops. Existing studies suggest the use of machine vision for 3D imaging to enable plant phenotyping (e.g., in potatoes) that could be then used to inform farmers about recommended water application [63]. To that effect, UAVs can effectively monitor the status of agricultural fields and communicate with Farming Information Systems for storing the gathered data.

At the pilot study of the fixed wing unmanned aerial system, the eBee drone with the Sequoia multispectral camera was used for measurements. This type of camera could also be applicable for determining vegetation and other ground features that are captured by the UAV. The band combinations from the multispectral camera are illustrated in Fig. 10.

Finally, the use of the compact digital camera Sony RX100 III with the rotary wing DJI S1000 drone could monitor the status of the irrigation equipment at the agricultural field and possibly control any automated valves for the execution of precision irrigation activities (Fig. 11). The precision irrigation activities could be controlled accordingly to improve freshwater management, depending on various environmental conditions.

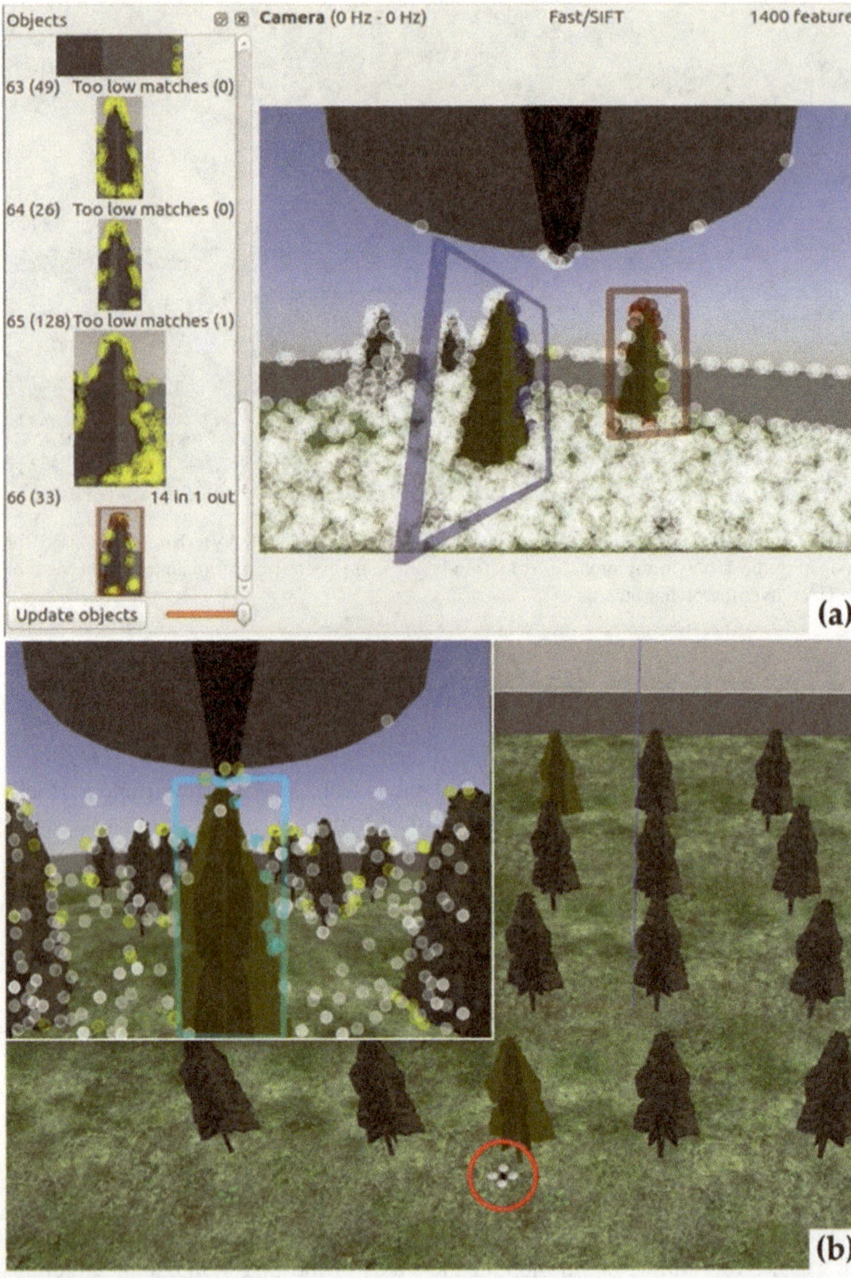

Fig. 9 Functionality of the emulation model includes: (**a**) monitoring of multiple trees through matching input signals to the data library; (**b**) identifying and indicating the water status of trees both in cases of water need (e.g., light green trees) and in no water stress situations (e.g., dark green trees)

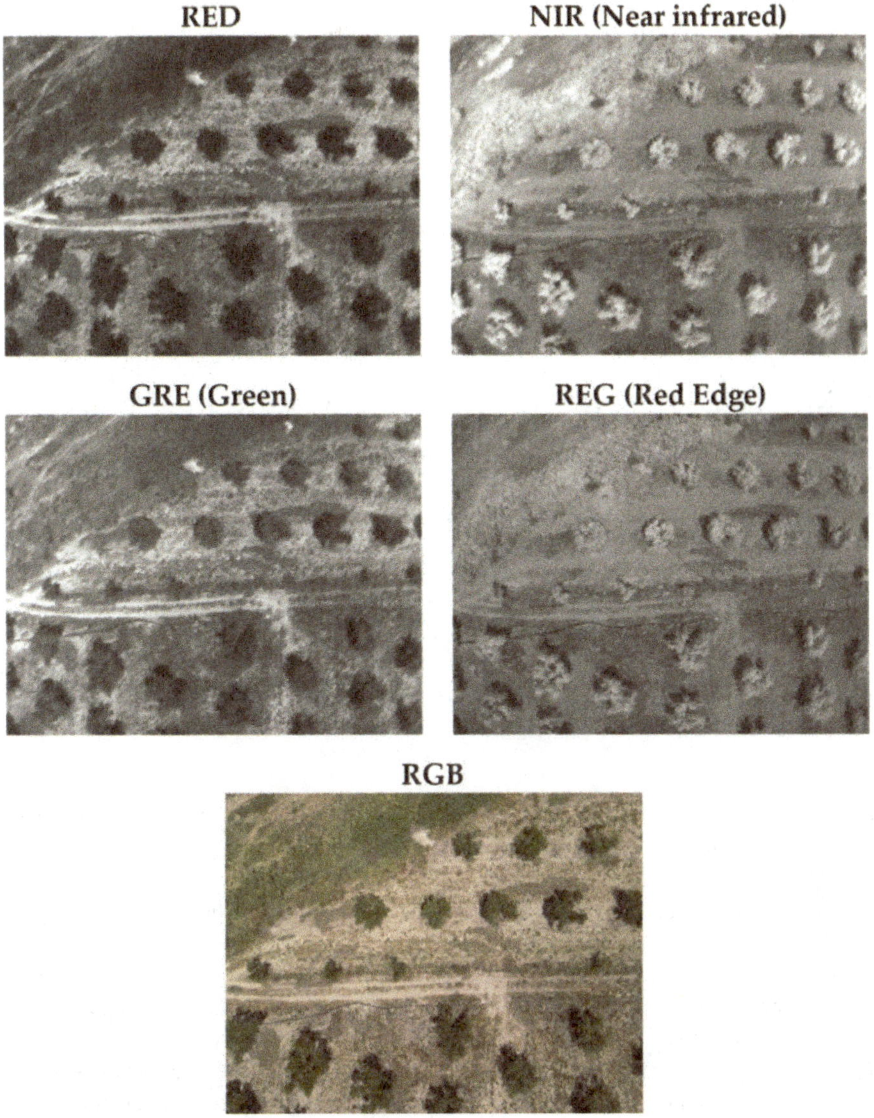

Fig. 10 Band combinations from the multispectral camera

5 Conclusions

Our multistage cyber-physical analysis approach is used to address the articulated research queries. In particular, out critical taxonomy helped respond to RQ#1 with the main advantage of UAVs in agriculture being the efficient mapping of agricultural fields. Furthermore, the greatest identified challenge refers to the need for

Fig. 11 Testing of a commercial rotary wing UAV in a neophyte orchard for irrigation system monitoring

equipment calibration along with data filtering techniques to streamline with the structure of existing canopy-related databases to facilitate the field mapping operations. Regarding RQ#2, an emulation model, that could comprise a "digital twin" for the ex-ante assessment of UAVs during precision farming operations, was developed to inform orchard-related irrigation decisions for water stewardship. The detection of water stress can be performed by multispectral cameras that capture near-infrared light canopy reflections. Emulation models could be also used to comparatively assess the impact of alternative digital technology options in operations. Furthermore, the pilot implementation demonstrates that drones can be used to assess water stress across large farms at a high accuracy level to then plan precision farming (e.g., irrigation) operations. However, sensors need to be first calibrated and databases of canopy spectral signatures have to be developed to reliably detect crops under water stress.

Agriculture 4.0 can be realized by investigating the interplay and synergistic operation of automated vehicles enabled by data exchange in the cyber-physical space. Indicatively, recent advances focus on the joint implementation of drones with augmented reality (e.g., wearable technologies, smart glasses) to assist farmers in gathering data and inform precision farming operations [46, 64]. Regarding the assessment of drone systems' efficiency for water stewardship, the identification of performance indicators could inform about the appropriate sensors to install on the drones. Representative indicators are the 'Normalized Difference Vegetation Index' and the 'Land Surface Wetness Index' which are used to specify crop vigor and crop water status, respectively.

5.1 *Theory Contributions*

The literature on the digitalization of agriculture is inchoate as the extant research efforts myopically focus on the technical and functional aspects of innovative technologies and overlook the related Operations Management and sustainability-wise implications [65]. Additionally, foresight is in principal an instrument for both the executive and legislative branches of governmental authorities aiming at informing policy designs and implementation [66]. To this effect, technology foresight analysis exercises over intelligent aerial vehicles could be argued that are often decoupled from the quantification of subsequent environmental pressures at a granular level of operations.

This research attempts to contribute to the foresight field by adopting an operationalization view over digital technologies for sustainable agriculture via proposing a multistage methodological analysis approach comprising of: (i) academic literature review and critical taxonomy; (ii) emulation modelling; and (iii) testbed application. The adoption of this approach and the engagement in the different levels of analysis could help interrogate UAVs' operational aspects with regard to monitoring water stress levels of individual plants in orchards to then inform the planning of precision irrigation activities. In particular, emulation modelling of real-world agricultural settings and UAVs, along with the pilot implementation of the emulated vehicles, could allow the creation of cyber-physical interfaces to enable more robust performance evaluation and foster the adoption of drones in agriculture. At a greater extent, the proposed "digital twin" analysis perspective of the operational environment (i.e., orchard), in conjunction with the applied digital technologies (i.e., drone and sensors), methodologically contributes to the field of robotic science [67, 68].

5.2 *Practice Implications*

The real-world operational context and the tangible sustainability benefits attained via the adoption of digital technologies in agriculture are often uncertain or ill-defined thus often creating uncertainty and ambiguity to farmers [29, 69]. To this end, the adoption rate of innovative technologies in agricultural operations stagnates, hence possibly impacting the sustainability performance of the sector both regionally (e.g., exploitation of local natural resources and activities' impact on the surrounding ecosystem) and internationally (e.g., virtual flows of natural resources such as freshwater). In addition, scholars in social sciences are concerned with regard to the impact of digital agricultural technologies to rural communities via highlighting the possible exploitation and marginalization of farmers by corporations and landowners [70, 71].

In the light of the abovementioned concerns, this research promotes the adoption of UAVs for freshwater stewardship in farming operations by: (i) identifying and summarizing advantages and disadvantages related to the utilization of UAVs in agriculture; and (ii) examining "digital twins" in agriculture by developing a cyber-physical analysis approach for UAVs that can help farmers to become aware about the functionality, operational characteristics and sustainability merits of physical drone counterparts. In this regard, farmers can have access to low-cost ex-ante, yet informative, assessments of the functional capabilities and performance of alternative UAV applications they foresee for their operations. In addition, farmers can use a "digital twin" of the agricultural field to articulate alternative foresight scenarios with regard to the dipole "drone application—appropriation of freshwater resources" (i.e., groundwater or surface water reserves) and plan their crop rotations accordingly. This need is particularly prominent in water scarce regions like the State of Punjab in India or South East England in the UK. Concerning the water sustainability scope, the emulation model could also allow the operational assessment of alternative intelligent vehicles and sensory equipment which are commercially available [72].

5.3 Limitations

In conducting this research, some technical limitations exist which provide stimulating grounds for exploring future research avenues. Firstly, the literature review considered only one database (i.e., Scopus), hence it was not possible to identify particular UAV-related benefits and challenges that could be covered in other databases. Secondly, the water stress level of each individual tree in the emulation model was programmed to be binary (i.e., water stress and no water stress). The inclusion of an algorithm for simulating the water requirements of particular plants could enable the emulation model to project the long-term irrigation requirements in an agricultural holding. Furthermore, the emulation tool could incorporate weather data to account for the flight capability and functional stability of a UAV system. Thirdly, the applied multistage methodological analysis approach could be expanded to include further analysis modules to enable a more scientific evidence-based decision-making process over the adoption of digital technologies for achieving particular sustainability goals.

5.4 *Future Research*

Agriculture 4.0 technologies have proven benefits, predominantly to agricultural small and medium-sized enterprises, in terms of [73]: (i) increased yields; (ii) reduced costs; (iii) greater profits; (iv) informed decisions; and (v) sustainability. Nevertheless, the adoption of digital technology applications in farms is still circumscribed as the underlining opportunity of evidence-based knowledge in farming is not recognized, yet. In this regard, considering future research directions, we are planning to enrich the applied approach with further analysis stages based on an active engagement with farmers and digital technology solution providers to motivate managerial beliefs that dictate adoption decisions on smart agriculture.

Moreover, future research efforts should expand the view of "digital twins" from the unit of operations echelon (i.e., orchard) to an agro-food supply network system level in order to assess the end-to-end sustainability impact of digital technologies [74]. In this regard, we will be able to make contribution to the Operations Management field by investigating the impact of digital technologies on inventory control, responsiveness and resilience across end-to-end agri-food supply networks. The synergistic action between automated vehicles and drones, or humans and drones, unfolds further research opportunities.

Acknowledgments This research has received funding from the General Secretariat for Research and Technology (GSRT) under reference no. 2386, Project Title: "Human-Robot Synergetic Logistics for High Value Crops" (project acronym: SYNERGIE).

References

1. Ocado (2019). *Changing the Way the World Shops – Ocado Group plc: Annual Report and Accounts*. Hertfordshire, United Kingdom: Ocado Group.
2. Iakovou, E., Bochtis, D., Vlachos, D., Aidonis, D. (eds) (2015). Supply Chain Management for Sustainable Food Networks. Wiley, John Wiley & Sons, Ltd., UK. https://doi.org/10.1002/9781118937495
3. Rodias, E., Berruto, R., Busato, P., Bochtis, D., Sørensen, C.G., Zhou, K. (2017). Energy savings from optimised in-field route planning for agricultural machinery. Sustainability, 9, 1956. https://doi.org/10.3390/su9111956
4. FAO, ITU (2018). *E-agriculture in Action: Drones for Agriculture*. Bangkok, Thailand: Food and Agriculture Organization of the United Nations and International Telecommunication Union.
5. Borch, K. (2007). Emerging technologies in favour of sustainable agriculture. *Futures*, 39, 1045–1066.
6. Lampridi, M.G., Sørensen, C.G., Bochtis, D. (2019a). Agricultural sustainability: A review of concepts and methods. *Sustainability*, 11, 5120.
7. Lampridi, M.G., Kateris, D., Vasileiadis, G., Marinoudi, V., Pearson, S., Sørensen, C.G., Balafoutis, A., Bochtis, D. (2019b). A Case-Based Economic Assessment of Robotics Employment in Precision Arable Farming. Agronomy, 9, 175. https://doi.org/10.3390/agronomy9040175

8. Lampridi, M., Kateris, D., Sørensen, C.G., Bochtis, D. (2020). Energy footprint of mechanized agricultural operations. *Energies*, 13, 769.
9. Shamshiri, R.R., Hameed, I.A., Balasundram, S.K., Ahmad, D., Weltzien, C., Yamin, M. (2018a). Fundamental research on unmanned aerial vehicles to support precision agriculture in oil palm plantations. In: Zhou, J., Zhang, B. (Eds.). *Agricultural Robots – Fundamentals and Applications*. Rijeka, Croatia: IntechOpen, 91–116.
10. Shamshiri, R.R., Weltzien, C., Hameed, I.A., Yule, I.J., Grift, T.E., Balasundram, S.K., Pitonakova, L., Ahmad, D., Chowdhary, G. (2018b). Research and development in agricultural robotics: A perspective of digital farming. *International Journal of Agricultural and Biological Engineering*, 11, 1–14.
11. King, A. (2017). Technology: The future of agriculture. *Nature*, 544, S21-S23.
12. Reports and Data (2019). *Agriculture Drone Market By Component (Hardware and Software & Services), By Integrant (Controller System, Propulsion System, Camera System, Frames), By Application, And Segment Forecasts, 2016–2026*. New York City, NY, United States of America: Reports and Data.
13. Vayssade, J.-A., Arquet, R., Bonneau, M. (2019). Automatic activity tracking of goats using drone camera. *Computers and Electronics in Agriculture*, 162, 767–772.
14. Faiçal, B.S., Costa, F.G., Pessin, G., Ueyama, J., Freitas, H., Colombo, A., Fini, P.H., Villas, L., Osório, F.S., Vargas, P.A., Braun, T. (2014). The use of unmanned aerial vehicles and wireless sensor networks for spraying pesticides. *Journal of Systems Architecture*, 60, 393–404.
15. Candiago, S., Remondino, F., De Giglio, M., Dubbini, M., Gattelli, M. (2015). Evaluating multispectral images and vegetation indices for precision farming applications from UAV images. *Remote Sensing*, 7, 4026–4047.
16. Usha, K., Singh, B. (2013). Potential applications of remote sensing in horticulture – A review. *Scientia Horticulturae*, 153, 71–83.
17. Michels, M., von Hobe, C.-F., Musshoff, O. (2020). A trans-theoretical model for the adoption of drones by large-scale German farmers. *Journal of Rural Studies*, 75, 80–88.
18. Hardin, P.J., Jackson, M.W., Anderson, V.J., Johnson, R. (2007). Detecting squarrose knapweed (Centaurea virgata Lam. Ssp. squarrosa Gugl.) using a remotely piloted vehicle: A Utah case study. GISci. *Remote Sensing*, 44, 203–219.
19. Hunt, E.R., Cavigelli, M., Daughtry, C.S.T., Mcmurtrey, J.E., Walthall, C.L. (2005). Evaluation of digital photography from model aircraft for remote sensing of crop biomass and nitrogen status. *Precision Agriculture*, 6, 359–378.
20. Swain, K.C., Thomson, S.J., Jayasuriya, H. (2010). Adoption of an unmanned helicopter for low-altitude remote sensing to estimate yield and total biomass of a rice crop. *American Society of Agricultural and Biological Engineers*, 53, 21–27.
21. Berni, J.A.J., Zarco-Tejada, P.J., Suarez, L., Fereres, E. (2009). Thermal and narrowband multispectral remote sensing for vegetation monitoring from an unmanned aerial vehicle. *IEEE Transactions on Geoscience and Remote Sensing*, 47, 722–738.
22. Primicerio, J., Di Gennaro, S.F., Fiorillo, E., Genesio, L., Lugato, E., Matese, A., Vaccari, F.P. (2012). A flexible unmanned aerial vehicle for precision agriculture. *Precision Agriculture*, 13, 517–523.
23. Sharma, D.C. (2004). Technologies for the people: A future in the making. *Futures*, 36, 733–744.
24. Angelopoulou, T., Tziolas, N., Balafoutis, A., Zalidis, G., Bochtis, D. (2019). Remote sensing techniques for soil organic carbon estimation: A review. Remote Sensing, 11, 676. https://doi.org/10.3390/rs11060676
25. Ojha, T., Misra, S., Raghuwanshi, N.S. (2015). Wireless sensor networks for agriculture: The state-of-the-art in practice and future challenges. *Computers and Electronics in Agriculture*, 118, 66–84.
26. Gómez-Limón, J.A., Gómez-Ramos, A., Sanchez Fernandez, G. (2009). Foresight analysis of agricultural sector at regional level. *Futures*, 41, 313–324.

27. Marinoudi, V., Sørensen, C.G., Pearson, S., Bochtis, D. (2019). Robotics and labour in agriculture. A context consideration. *Biosystems Engineering*, 184, 111–121.
28. Rasmussen, B., Borch, K. (2003). Risk and science: Are we moving into the fourth age of risk concerns? *Risk, Decision and Policy*, 8, 1–15.
29. Morris, W., Henley, A., Dowell, D. (2017). Farm diversification, entrepreneurship and technology adoption: Analysis of upland farmers in Wales. *Journal of Rural Studies*, 53, 132–143.
30. Tsolakis, N., Bechtsis, D., Bochtis, D. (2019a). AgRos: A robot operating system based emulation tool for agricultural robotics. *Agronomy*, 9, 403.
31. Trendov, N.M., Varas, S., Zeng, M. (2019). *Digital Technologies in Agriculture and Rural Areas – Status Report*. Rome, Italy: Food and Agriculture Organization of the United Nations.
32. Wang, Y., Wang, S., Yang, B., Zhu, L., Liu, F. (2020). Big data driven hierarchical digital twin predictive remanufacturing paradigm: Architecture, control mechanism, application scenario and benefits. *Journal of Cleaner Production*, 248, 119299.
33. Grieves, M., Vickers, J. (2017). Digital twin: Mitigating unpredictable, undesirable emergent behavior in complex systems. In: Kahlen, F.-J., Flumerfelt, S., Alves, A. (Eds.). *Transdisciplinary Perspectives on Complex Systems*. Springer, Cham, Switzerland. 85–113.
34. Liotine, M. (2020). Chapter 8 – Unlocking digital innovation: Guiding principles for driving digital technology in the supply chain. In: Pagano, A.M., Liotine, M. (Eds.). *Technology in Supply Chain Management and Logistics: Current Practice and Future Applications*. Amsterdam, The Netherlands: Elsevier, 143–167.
35. Entezari, A., Wang, R.Z., Zhao, S., Mahdinia, E., Wang, J.Y., Tu, Y.D., Huang, D.F. (2019). Sustainable agriculture for water-stressed regions by air-water-energy management. *Energy*, 181, 1121–1128.
36. Piirainen, K.A., Gonzalez, R.A. (2015). Theory of and within foresight – "What does a theory of foresight even mean?" *Technological Forecasting and Social Change*, 96, 191–201.
37. Öner, M.A. (2010). On theory building in Foresight and Futures Studies: A discussion note. *Futures*, 42, 1019–1030.
38. Tsolakis, N., Bechtsis, D., Srai, J.S. (2019b). Intelligent autonomous vehicles in digital supply chains: From conceptualisation, to simulation modelling, to real-world operations. *Business Process Management Journal*, 25, 414–437.
39. Feldman, R., Sanger, J. (2007). *The Text Mining Handbook: Advanced Approaches In Analyzing Unstructured Data*. Cambridges, UK: Cambridge University Press.
40. Bechtsis, D., Tsolakis, N., Vlachos, D., Iakovou, E. (2017). Sustainable supply chain management in the digitalisation era: The impact of Automated Guided Vehicles. *Journal of Cleaner Production*, 142, 3970–3984.
41. Otto, A., Agatz, N., Campbell, J., Golden, B., Pesch, E. (2018). Optimization approaches for civil applications of unmanned aerial vehicles (UAVs) or aerial drones: A survey. *Networks: An international Journal*, 72, 411–458.
42. Zhang, C., Kovacs, J.M. (2012). The application of small unmanned aerial systems for precision agriculture: A review. *Precision Agriculture*, 13, 693–712.
43. Bansod, B., Singh, R., Thakur, R., Singhal, G. (2017). A comparision between satellite based and drone based remote sensing technology to achieve sustainable development: A review. *Journal of Agriculture and Environment for International Development*, 111, 383–407.
44. Vigneau, N., Chéron, C., Verger, A., Baret, F. (2017). Aerial imaging by drone: Data mining for precision agriculture. *Revue Francaise de Photogrammetrie et de Teledetection*, January(213–214), 125–132.
45. Simic Milas, A., Romanko, M., Reil, P., Abeysinghe, T., Marambe, A. (2018). The importance of leaf area index in mapping chlorophyll content of corn under different agricultural treatments using UAV images. *International Journal of Remote Sensing*, 39, 5415–5431.
46. Huuskonen, J., Oksanen, T. (2018). Soil sampling with drones and augmented reality in precision agriculture. *Computers and Electronics in Agriculture*, 154, 25–35.
47. Cancela, J.J., González, X.P., Vilanova, M., Mirás-Avalos, J.M. (2019). Water management using drones and satellites in agriculture. *Water*, 11, 874.

48. Hogan, S.D., Kelly, M., Stark, B., Chen, Y.Q. (2017). Unmanned aerial systems for agriculture and natural resources. *California Agriculture*, 71, 5–14.
49. Anderson, C. (2014). Agricultural drones. *Technology Review*, 117, 58–60.
50. Barbey, C., Helbert, J., Jaën, A., Pagot, É., Samalens, J.-C., Valette, L., Germain, C., Guyon, D., Wigneron, J.-P. (2014). Comparison of Pleiades and UAV images for precision viticulture in the earth lab program framework. *Revue Francaise de Photogrammetrie et de Teledetection*, 208, 123–129.
51. Ipate, G., Voicu, G., Dinu, I. (2015). Research on the use of drones in precision agriculture. *UPB Scientific Bulletin, Series D: Mechanical Engineering*, 77, 263–274.
52. Sarghini, F., De Vivo, A. (2017a). Interference analysis of an heavy lift multirotor drone flow field and transported spraying system. *Chemical Engineering Transactions*, 58, 631–636.
53. Sarghini, F., De Vivo, A. (2017b). Analysis of preliminary design requirements of a heavy lift multirotor drone for agricultural use. *Chemical Engineering Transactions*, 58, 625–630.
54. Lan, Y., Wang, L., Zhang, Y. (2018). Application and prospect on obstacle avoidance technology for agricultural UAV. *Transactions of the Chinese Society of Agricultural Engineering*, 34, 104–113.
55. Liu, Y., Noguchi, N., Okamoto, H., Ishii, K. (2018). Development of a small-sized and low-cost attitude measurement unit for agricultural robot application. *Journal of Agricultural Sciences*, 24, 33–41.
56. Lysenko, V., Opryshko, O., Komarchuk, D., Pasichnyk, N., Zaets, N., Dudnyk, A. (2018). Information support of the remote nitrogen monitoring system in agricultural crops. *International Journal of Computing*, 17, 47–54.
57. Murugan, D., Garg, A., Singh, D. (2017). Development of an adaptive approach for precision agriculture monitoring with drone and satellite data. *IEEE Journal of Selected Topics in Applied Earth Observations and Remote Sensing*, 10, 5322–5328.
58. Szantoi, Z., Smith, S.E., Strona, G., Koh, L.P., Wich, S.A. (2017). Mapping orangutan habitat and agricultural areas using Landsat OLI imagery augmented with unmanned aircraft system aerial photography. *International Journal of Remote Sensing*, 38, 2231–2245.
59. Yamamoto, K., Togami, T., Yamaguchi, N. (2017). Super-resolution of plant disease images for the acceleration of image-based phenotyping and vigor diagnosis in agriculture. *Sensors*, 17, 2557.
60. Reger, M., Bauerdick, J., Bernhardt, H. (2018). Drones in agriculture: Current and future legal status in Germany, the EU, the USA and Japan. *Landtechnik*, 73, 62–80.
61. Freeman, P.K., Freeland, R.S. (2014). Politics & technology: U.S. polices restricting unmanned aerial systems in agriculture. *Food Policy*, 49, 302–311.
62. Sadgrove, E.J., Falzon, G., Miron, D., Lamb, D.W. (2018). Real-time object detection in agricultural/remote environments using the multiple-expert colour feature extreme learning machine (MEC-ELM). *Computers in Industry*, 98, 183–191.
63. Smith, L.N., Zhang, W., Hansen, M.F., Hales, I.J., Smith, M.L. (2018). Innovative 3D and 2D machine vision methods for analysis of plants and crops in the field. *Computers in Industry*, 97, 122–131.
64. Benos, L., Bechar, A., Bochtis, D. (2020). Safety and ergonomics in human-robot interactive agricultural operations. Biosyst Engineering, 200, 55–72. https://doi.org/10.1016/j.biosystemseng.2020.09.009
65. Bochtis, D., Sorensen, C.G., Kateris, D. (2018). Operations Management in Agriculture, 1st edition. Academic Press-AP, Elsevier. https://doi.org/10.1016/C2015-0-06290-6
66. Weber, K.M., Gudowsky, N., Aichholzer, G. (2019). Foresight and technology assessment for the Austrian parliament — Finding new ways of debating the future of industry 4.0. *Futures*, 109, 240–251.
67. Bochtis, D., Griepentrog, H.W., Vougioukas, S., Busato, P., Berruto, R., Zhou, K. (2015). Route planning for orchard operations. Computers and Electronics in Agriculture, 113, 51–60. https://doi.org/10.1016/j.compag.2014.12.024

68. Kostavelis, I., Gasteratos, A. (2015). Semantic mapping for mobile robotics tasks: A survey. *Robotics and Autonomous Systems*, 66, 86–103.
69. Moysiadis, V., Tsolakis, N., Katikaridis, D., Sørensen, C.G., Pearson, S., Bochtis, D. (2020). Mobile Robotics in Agricultural Operations: A Narrative Review on Planning Aspects. Applied Science, 10, 3453.
70. Papageorgiou, K., Singh, P.K., Papageorgiou, E., Chudasama, H., Bochtis, D., Stamoulis, G. (2020). Fuzzy Cognitive Map-Based Sustainable Socio-Economic Development Planning for Rural Communities. Sustainability, 12, 305. https://doi.org/10.3390/su12010305
71. Rotz, S., Gravely, E., Mosby, I., Duncan, E., Finnis, E., Horgan, M., LeBlanc, J., Martin, R., Neufeld, H.T., Nixon, A., Pant, L., Shalla, V., Fraser, E. (2019). Automated pastures and the digital divide: How agricultural technologies are shaping labour and rural communities. *Journal of Rural Studies*, 68, 112–122.
72. Tsolakis, N., Aivazidou, E., Srai, J.S. (2019c). Sensor applications in agrifood systems: Current trends and opportunities for water stewardship. *Climate*, 7, 44.
73. Annosi, M.C., Brunetta, F., Monti, A., Nat, F. (2019). Is the trend your friend? An analysis of technology 4.0 investment decisions in agricultural SMEs. *Computers in Industry*, 109, 59–71.
74. Defraeye, T., Tagliavini, G., Wu, W., Prawiranto, K., Schudel, S., Assefa Kerisima, M., Verboven, P., Bühlmann, A. (2019). Digital twins probe into food cooling and biochemical quality changes for reducing losses in refrigerated supply chains. *Resources, Conservation and Recycling*, 149, 778–794.

Digital Technologies in the Context of Energy: Focus on the Developing World Agriculture

George Kyriakarakos, Maria Lampridi, and Dionysis D. Bochtis

1 Introduction

More than 700 million people are still living in extreme poverty. The majority of these people live in rural areas of the developing world. United Nations had acknowledged this reality and Sustainable Development Goal 1 is "No Poverty" aiming at eradicating extreme poverty by 2030. Many studies have been published investigating pathways out of poverty [1]. One of the most recent ones [2] has evaluated a number of cases globally and identified four major pathways towards poverty alleviation: industrialization, rural development, social welfare and oil generated employment. For the poorest countries and rural areas, rural development is one of the most viable and realistic approaches towards poverty alleviation.

Rural development can be defined as the process of improving the quality of life and economic well-being of people living in rural areas, often relatively isolated and sparsely populated areas [3]. Traditionally rural development has been focusing on agriculture and forestry as well as on natural resources extraction. In recent years, though, the focus has changed to tourism, recreation and decentralized manufacturing [4], while it has been extensively acknowledged that social infrastructure (education and health) also plays a very important role [5]. When the focus changes to rural areas of the developing world though, agriculture is still considered to be the key [6].

As a matter of fact, in rural areas, agriculture can be the key as an eliminator of poverty since it is the major supplier of food. Since the 86% of people in rural areas depend on agriculture for their living, the lack of sufficient production of food and

G. Kyriakarakos (✉) · M. Lampridi · D. D. Bochtis
Institute for Bio-economy and Agri-technology (iBO), Centre for Research and Technology Hellas (CERTH), Thessaloniki, Greece
e-mail: g.kyriakarakos@certh.gr

D. D. Bochtis et al. (eds.), *Information and Communication Technologies for Agriculture—Theme IV: Actions*, Springer Optimization and Its Applications 185,
https://doi.org/10.1007/978-3-030-84156-0_3

fiber, that could satisfy human needs around the world, can be strongly related to poverty [7]. It is also indisputable that, access to food is becoming a serious global issue, especially due to the increased population which leads to the exponential increase of demand for food [8, 9]. All the above are framed within the second Sustainable Development Goal which refers to zero hunger. As reported by the World Food Program issued by the United Nations the 13% of the world's population does not have access to adequate amounts of food [10]. Moreover, the Food and Agriculture Organization (FAO) characterizes hunger as the most important cause of disease in the world [11]. According to the United Nations definition, the second goal aims to eliminate hunger while achieving food security and improved nutrition through the promotion of sustainable agriculture [12]. Also, in the context of the wider notion of sustainability, development must be used in order to not only sustain but also improve quality of life.

The adoption of sustainable agriculture is a challenging task especially since its definition still remains an unresolved issue within the research community [13]. The term sustainable agriculture is often confused with the one of agricultural sustainability, however both terms have a different context that is related to the subject of examination [14]. More specifically agricultural sustainability defines the principles and rules that need to be followed in order to achieve sustainability in the agricultural sector [15]. On the other hand, sustainable agriculture describes the agricultural practices and processes that abide by the principles of agricultural sustainability. Agricultural sustainability is audited with indicators [16] that are related to the evaluation and the assessment of agricultural processes with respect to their effect on the environment, the economy and the society. Among the existing categories of indicators, the indicative themes of income, nutrition, quality (both environmental and quality of life), security and resource use demonstrate the need for monitoring the sustainability of rural development.

Monitoring of agricultural sustainability became imperative due to its intensification that followed the need for increased primary production [17]. The eventual intensification of agriculture introduced technology within the natural agricultural processes. As it is also the case with industry, this rapid evolution came at a considerable cost on humans and the environment [18]. Thus, taking into account of the limited capacity of our planet new rules had to be set for its protection and sustainability. Along with all other human activities, agriculture must also adapt to the new standards and requirements and become sustainable [19]. In that case, agricultural sustainable development can be achieved by introducing new technologies mainly aiming at the increase of production while protecting the available resources. The introduction of new technologies is of utmost importance in rural areas that are deprived of basic goods [20]. However, no new technology can be applied without safeguarding adequate amount of energy for its application.

Electricity is probably the most useful type of energy due to is high transformation potential [21]. Hence, providing electricity in rural areas can accelerate their development in a variety of ways [22]. Towards that direction this chapter aims at introducing an overview of rural electrification as an enabler of sustainable development in developing countries. The productive uses of energy are presented along with ways to evaluate access to energy. Then the techno-economic considerations in rural electrification are examined and the basic steps in the deployment of a rural electrification project are presented. Basic aim of rural electrification is the improvement of the quality of life. Nevertheless, this development can facilitate the introduction of digital technologies. For that reason, a detailed overview of digital technologies in the context of energy is also entailed in order to highlight the full potential that is generated.

2 Digital Technologies Overview in the Context of Energy

Digital technologies can be defined as *electronic tools, systems, devices and resources that generate store or process data* [23]. ICT on the other hand can be defined as *an extensional term for information and technology that stresses the role of unified communications and the integration of telecommunications (telephone lines and wireless signals) and computers, as well as necessary enterprise software, middleware, storage, and audiovisual systems, that enable users to access, store, transmit, and manipulate information* [24]. In simple terms digital technologies are more concerned with how the technologies are created and developed and ICT technologies are more concerned with how these technologies are used.

Table 1 presents an extensive list of various digital technologies used in energy systems in general and rural electrification systems in particular. The technological status, as well as use examples are presented. As is expected with disruptive technologies, start-up companies lead the way. The various technologies can impact single or multiple stages of the development of a rural electrification project, as was described in Sect. 3.5.

2.1 Commercially Available Digital Technologies

Artificial Intelligence

Artificial Intelligence or as also called machine intelligence refers to the ability of a device to perceive its own environment, being able to take actions that increase its possibility to achieve a requested outcome [25]. When referring to a computer's capacity to learn how to perform a particular task based on already existing data or observation, artificial intelligence can be characterized as Computational Intelligence. Computational intelligence includes several branches such as Fuzzy logic, Artificial Neural Networks, Evolutionary Computation, Machine Learning and Probabilistic methods [26, 27]. Moving forward to the cooperation of computational intelligent systems, multi-agent as well as expert systems have emerged [28]. Many Artificial Intelligence paradigms have been used in rural electrification with most

Table 1 List of digital technologies

Digital technology	Application in the energy sector	Application in rural electrification	Commercial availability	Near-to market technology	Middle to long term technology
Artificial intelligence Computational intelligence Fuzzy logic, artificial Neural networks Evolutionary computation Machine learning theory Probabilistic methods Multi agent systems Expert systems	√	√	√[1]		
Internet of Things	√	√	√[2]		
Blockchain Permissionless or public Permissioned or private	√	√	√[3]		
Cloud computing	√	√	√[4]		
Edge computing	√	√	√[5]		
Mobile computing	√	√	√		
Internet	√	√	√		
Virtual reality Augmented reality	√		√[6]		
Big data	√	√	√[7]		
eSignature Electronic identity	√	√	√		
Cyber security	√	√	√		
Wireless networks Wireless data networks Mobile telephony networks Wireless local area	√	√	√	√	√

(continued)

Table 1 (continued)

Digital technology	Application in the energy sector	Application in rural electrification	Commercial availability	Near-to market technology	Middle to long term technology
networks Wireless sensor networks Satellite communication networks Geosynchronous satellites Terrestrial microwave networks Wireless power networks Near field networks Wireless power networks Near field networks Far-field networks					
Unmanned vehicles	√	√	√[8]		
3D printing	√	√	√[9]	√	√
Printed electronics	√	√	√[10]	√	√
Ubiquitous computing Sentient computing Ubiquitous commerce Context aware computing	√	√	√[11]	√	√
Fintech	√	√	√	√	√
Alternative finance Peer-to-peer lending Crowdfunding Cryptocurrencies	√	√	√[12]	√	√
Civic technology	√		√	√	√
Quantum computing					√

(continued)

Table 1 (continued)

Digital technology	Application in the energy sector	Application in rural electrification	Commercial availability	Near-to market technology	Middle to long term technology
Molecular electronics					√
Artificial photosynthesis					√
Optical rectenna					√
Cognitive robotics Superintelligence DNA digital data storage Exascale computing Brain–computer interface					√

Sources:
[1]https://www.advancedmicrogridsolutions.com
[2]https://www.dajie.eu
[3]https://lo3energy.com
[4]https://www.sparkmeter.io
[5]https://steama.co
[6]https://www.se.com/ww/en/work/services/field-services/industrial-automation/performance-optimization-services/ecostruxure-augmented-operator-advisor.jsp
[7]https://www.odysseyenergysolutions.com
[8]http://powercorner.com
[9]http://rcamtechnologies.com
[10]https://infinitypv.com
[11]https://www.theminigridgame.org
[12]https://www.ecoligo.investments

applications in the design and sizing of the microgrids, as well as in the energy management system used.

Internet of Things (IoT)

The Internet of Things (IoT) is a scheme of interconnecting devices, each one bearing a unique identifier. The scheme offers the ability of data transfer within this network of individual devices excluding human involvement [29]. IoT is enabled through the use of a variety of networks, sensors, automation and control systems [30]. Many sensors and actuators used in microgrids are deployed on Internet of Things topologies.

Blockchain

A blockchain is the term used to describe a series of information blocks which are interconnected with the use of cryptographic hashes [31]. Except from the individual ID of the previous item, each block holds information in its header, along with data from the current action, which are used to safeguard the consistency within the blockchain with the use of a cryptographic hash [32]. Additionally, authorization and validation mechanisms can be found within each chain for the reading and writing operations of each block, a process that guarantees the security of the transaction [33]. In rural electrification projects blockchain can be used both as a backbone for implementing a pay-as-you-go scheme [34], but also as a tool for financing microgrid investments [35].

Cloud Computing

The term cloud computing refers to the availability of computer systems resources upon request It usually concerns the storage of data as well as computing power that does not require the immediate assessment by the user [36]. Most often the term is used to characterize data centers that are available to a large number of users. Cloud computing contributes towards economies of scale and consistency through the sharing of resources [37]. In the energy sector cloud computing is increasing its use and is essentially the location to effectively store big data and manipulate them in order to reach meaningful conclusions.

Edge Computing

Edge computing is essentially an optimization of cloud computing where data processing is performed at the edge of the network [38]. This can be really important for remote microgrids where internet access through mobile telephony networks is intermittent. In those cases, operations can take place locally and when internet connection becomes available functions like data storage on the cloud can take place.

Mobile Computing

Mobile computing refers to the synergy of humans and computers in which the computer performs normally (allowing data transmission) but does not necessarily maintain a specific position as it can be transported according to the user's requirements [39]. Mobile computing is on the rise and even in rural areas of the developing world many people have smart phones. Many pay-as-you-go models offer payments through mobile phone applications.

Internet

Internet is essentially the backbone of most ICT technologies.

Virtual Reality/Augmented Reality

Virtual or Augmented reality refers to a simulated experience that is designed to resemble real world experiences [40]. Virtual reality is used mainly in training activities and it has been proposed for use in remote areas so as to be able to train local people for basic maintenance tasks. Augmented reality applications can bring virtual reality features in the real environment. Current applications in the energy sector include software that can aid a technician in performing tasks by seeing extra information complementing his natural vision.

Big Data

The term big data is used to describe the computational field that deals with the analysis and the extraction of data that cannot be processed with conventional data analysis methods due to their size or complexity [41]. Storing large data and being able to process them in order to get meaningful conclusions can be a very important tool in logging trends and making the necessary decisions in order to increase energy consumption, increase efficiency, forecast energy production from renewables and overall better managing energy.

e-Signature/Electronic Identity

Electronic identification is a digital way to prove the identity of an individual or an organization in order to access services or benefits that are managed for example by companies, banks, governments etc. [42]. In the case of signing a document, the electronic identity is usually verified with the use of e-signature. Estonia is one of the world champions in the use of electronic identity for energy sector applications. Electronic identities are used in the developing world for the implementation of pay-as-you-go schemes for electricity.

Cyber Security

Cyber Security refers to the safeguarding of computer systems and networks against damage or theft of their tangible and intangible parts (hardware, software or data), along with the disturbance of the services they provide [43]. All digital systems need to be protected. Proper cyber security measures need to be in place in any application including for example the energy management system of a microgrid or the platform that implements the payments for the electricity.

Wireless Networks

A wireless network refers to a computer network that employs wireless data connections between the network nodes that it is comprised of [44]. There is a vast array of wireless data network technologies used extensively in energy systems in general and in microgrids in the developing world in particular. Wireless power transmission technologies are al-ready used in terms of charging very low power sensors in microgrids. Research is ongoing for far-field power transmission technologies.

Unmanned Vehicles

Unmanned vehicles are autonomous vehicles that do not require an operator in order to move. They can be either remotely operated or they can be autonomous and able to perceive their environment and navigate on their own [45]. Unmanned vehicles can be both ground and aerial (UGV's and UAV's respectively) and are used for the automation of a variety of operations that require intensive labour, or considerable amounts of time or resources [46, 47]. In rural electrification projects, unmanned aerial vehicles have been used in microgrid projects in order to provide the input data for the design of grids (pole locations).

3D Printing

3D printing (or additive manufacturing) refers to the manufacturing of three-dimensional objects in which the material is joined under the control of a computer [48]. The ability to be able to print parts in a remote location is invaluable. As the 3D printers' cost is decreasing, more applications are expected to be deployed commercially.

Printing Electronics

Printing electronics is a term used to describe all those methods that are employed for the manufacturing of electrical devices on a variety of different substrates [49, 50]. Printed photovoltaics and batteries are already commercially available. The ability to produce PVs and batteries in the country of application without the need of very high cost equipment can further decrease costs.

Ubiquitous Computing

Ubiquitous computing refers to a computer science and software engineering approach in which computing is available anytime and everywhere and in devices beyond the traditional desktop computing [51]. Ubiquitous computing is supported with several other digital technologies including internet, networks, sensors etc. As a term it is not used commercially many commercial applications can fall under this technology. For example, a smart thermostat is essentially such a technology, as are serious games. A mini-grid game has been developed in order to raise awareness among the final users of how a mini-grid works.

Fintech

FinTech is a very broad term that can encompass any type of innovation related to the financing of products or services. Mobile money for electricity pre-payments is the most common example of use in rural electrification.

Alternative Finance

The term alternative finance refers to all those instruments, processes and channels that are developed out of the traditional financial system. Many start-ups are developing alternative finance solutions for rural electrification. Crowdfunding for financing projects in the developing world is one example of this.

Civic Technology

Civic technology improves government-citizen relationships through the embedment of information and communication technologies in decision-making, service delivery, communications and political processes with the use of software [52]. A digital nation based on civic technology can provide applications in the energy sector as well. The best current example is Estonia. Rwanda is one of the Sub-Saharan Africa countries that has embarked on this route, with most current applications being in the transport sector.

2.2 Under Development Digital Technologies

Quantum Computing

Quantum Computing refers to the employment of quantum-mechanical phenomena (e.g. Superposition and entanglement) for the execution of computing [53]. Actual quantum computers are still in infancy, even though there is a model on sale for 15 mil USD [54]. Processing time on a quantum computer is also sold as a service. As quantum computing becomes affordable and increase its performance against conventional super computers it will allow extremely high computing potential for low cost, expanding other applications as artificial intelligence, cloud computing etc. facilitating and decreasing the cost of these technologies.

Molecular Electronics

Molecular electronics is the study and application of molecular building blocks for the fabrication of electronic components. Still at the research stage, photovoltaics based on molecular electronics could increase efficiencies in the future and decrease costs.

Artificial Photosynthesis

Artificial photosynthesis is a chemical process that replicates the natural process of photosynthesis, a process that converts sunlight, water, and carbon dioxide into carbohydrates and oxygen. Research of this topic includes the design and assembly of devices for the direct production of solar fuels, photoelectrochemistry and its application in fuel cells, and the engineering of enzymes and photoautotrophic microorganisms for microbial biofuel and biohydrogen production from sunlight [55].

Optical Rectenna

An optical rectenna is a rectenna (rectifying antenna) that works with visible or infrared light. While rectennas have long been used for radio waves or microwaves, an optical rectenna would operate the same way but with infrared or visible light, turning it into electricity. Essentially an optical rectenna will improve considerably the efficiency of photovoltaics while decreasing the cost [56].

Cognitive Robotics/Superintelligence/DNA Digital Data Storage/Exascale Computing/Brain-Computer Interface

These technologies are still in their infancy but could in the future have applications in the energy domain as well.

2.3 *Digital Technology-Based Products and Services*

As is observed in Table 1 there is a big number of digital technologies that have found their way to rural electrification solutions. Innovative technologies can bring cost down and help the developing countries to technology leapfrogging. While some of these technologies are products on their own as for example is a printed photovoltaic, other technologies need to be combined together in order to develop a product or service that can meaningfully address the needs of rural electrification. This is showcased below with the example of the Pay-As-You-Go business model.

Revenue collection is one of the biggest challenges faced by rural electrification investors [57]. Various models based on pre-payment have been used with success in Sub-Saharan Africa in order to address this issue with success [58]. A traditional post-paid meter measures the consumed energy. Most often an employee of the electricity company goes and notes down the consumption in a predefined period of time (e.g. monthly, semesterly etc.) and the electricity bills are derived from that. Basic pre-paid meters measure the energy consumed as a post-paid meter but allow the input of tokens or codes in order to allow further electricity consumption. Smart meters measure and communicate payment and electricity consumption information automatically. In rural electrification smart meters are further used to implement energy management and control schemes. In its simplest version a smart meter is equipped with a relay switch that activates and deactivates the provision of electricity. The smart meters communicate with the control platform of the electricity operator either wirelessly or with data cables. This control platform enables the system operator (Solar Home System based or microgrid based) to monitor the microgrid and at the same time the platform can allow the connection and disconnection of a consumer based on the pre-payment status.

In order for a consumer to top-up the electricity account a visit to a shop might be in place. This has many drawbacks, with the most important one that the shop is never open constantly. Mobile money is extensively used in the developing world [59]. This is due to the fact that a big number of the population of these countries doesn't have access to the traditional banking system [60]. In this manner a mobile phone application, or even an sms-based service can be used to pre-pay for electricity. The pre-payment data is forwarded automatically to the system operator, who in turns automatically activates the purchased electricity. As is presented a big number of digital technologies are used in order to implement the pay-as-you-go service. While this combination of technologies requires higher technical and business capacities of the entities implementing it has been proven to have a strong potential for faster rate of adoption than comparable approaches [61].

3 The Case of Rural Electrification

3.1 Overview

Rural electrification can provide major benefits to rural populations, since it can facilitate improvement in health [62], education [63] and finally in the economic development status [64]. Global experiences from ranging from the 30s up to the current times have shown the benefits rural electrification can bring to the rural communities in general and agriculture in particular. A review of the US electrification program that took place between the 30s and the 60s showed that the short-term benefits included new jobs creation and increase of rural farms population, boosting the rural economy. In the long-term it has been observed that the electrification of agricultural activities has produced long-lasting effects to the local communities with the ones that got electrified first to show constantly increased economic growth even after the whole country was electrified [65]. In South Korea the rural electrification program lasted 15 years, increasing rural access to electricity from 12% up to 98%. This contributed extensively to increased income and improvement in the quality of life for the majority of the population [66]. In more recent years, the Chinese rural electrification program was deployed. It focused mainly in agriculture and one of the most important outcomes was a considerable increase of the farmers' income [67].

3.2 *Productive Uses of Energy*

The only way to ensure revenues and make a rural electrification investment viable is to create sustainable economic activity in the area [68]. This translates to powering specific devices and appliances such as water pumping for irrigation, water desalination, refrigeration of agricultural produce, space heating and cooling, incubators for poultry farming, milking machines, rice and maize hullers, polishers, threshers, graters, grain mills, oil presses, tailoring, workshop machinery (e.g., drills, chainsaws, rotary hammers, grinders, jigsaws, routers, etc.) and hairdresser equipment, among others [69]. Agriculture related loads need to be in priority, since experience has shown that these are activities causing from the short-term income increase [70, 71].

3.3 *Measuring Access*

The Sustainable Development Goal 7 of the United Nations is "Ensuring access to affordable, reliable, sustainable and modern energy for all". While the target is precise enough the question arises on what is considered having access. Having access is certainly any type of grid connection. The answer is more trivial for off-grid access. What kind of access does a Solar Home System (SHS) provide or is it considered having access if you have a solar lantern that is able to also charge your mobile phone? In order to respond to this need, the United Nations Global Tracking Network for Sustainable Energy for All has proposed a multi-tier framework, which is a comprehensive approach in measuring access [72]. The main attributes of this framework are:

- Five-tier framework
- Based on six attributes of electricity supply
- With electricity supply improvement, there is an increase in the possible electricity services availability

Table 2 Access Tier attributes

Attributes	Tier 0	Tier 1	Tier 2	Tier 3	Tier 4	Tier 5
Peak available capacity (W)	–	>3	>50	>200	>800	>2000
Duration (h)	–	≥4	≥4	≥8	≥16	≥23
Evening supply (h)	–	≥1	≥2	≥3	≥4	≥4
Reliability	–	–	–	–	Max 14 disruptions per week	Max 3 disruptions per week of total duration <2 h
Quality	–	–	–	–	Voltage problems do not affect the use of desired appliances	
Affordability	–	–	–	Cost of a standard consumption package of 365 kWh year^{-1} <5% of the household income		
Legality	–	–	–	–	Bill paid to the utility, pre-paid card seller or authorized representative	
Health and safety	–	–	–	–	Absence of past accidents and perception of high risk in the future	

Source: [72] Bhatia M., Angelou N., "Beyond Connections—Energy Access Redefined", 2015, ESMAP, World Bank

The Index of access to electricity is defined as: $i = \sum P_T \times T$, where : PT refers to the proportion of households at tier $T \in \{0, 1, \ldots, 5.\}$.

Table 2 presents the main attributes of each tier, while Table 3 presents the main services provided. As is understandable, high energy efficiency devices and appliances need to be used in order to achieve cost effectively each Tier access.

An important remark that can be made is that the solar lamp able to provide phone charging is indeed considered having access. The 1 bn people currently in the world without access to electricity [73] do not have access even to this very basic service.

3.4 *Techno-Economic Considerations in Rural Electrification*

There are three main categories of approaches that can be used for rural electrification [74]:

- Grid Extension

  Grid extension is essentially concerned with developing new high, medium and low voltage lines to reach the points of electricity use.
- Microgrids

  According to the IEEE Standard 2030.7-2017 a microgrid is a group of interconnected loads and distributed energy resources with clearly defined electrical boundaries that acts as a single controllable entity with respect to the grid and can connect and disconnect from the grid to enable it to operate in both grid-

Table 3 Services provided in each Tier

Tier 0	Tier 1	Tier 2	Tier 3	Tier 4	Tier 5
–	Task lighting **AND** Phone charging	General lighting **AND** phone charging **AND** television **AND** fan (if needed)	Tier 2 **AND** Any medium-power appliances	Tier 3 **AND** Any high-power appliances	Tier 2 **AND** Any very high-power appliances

Source: [72] Bhatia M., Angelou N., "Beyond Connections—Energy Access Redefined", 2015, ESMAP, World Bank

connected or island modes. As can be understood, when a microgrid is deployed in a rural area it operates in islanded mode. If in the future the main grid is expanded to reach the area previously served by the microgrid, the microgrid infrastructure can continue to be utilized.

- Solar Home Systems (SHS)

  Solar home systems are stand-alone systems most often consisting of photovoltaic panels (PV) and batteries. They can provide either AC or DC electricity depending on its size. The SHS market is currently booming in Africa [75] and usually these systems are sold in kits which include the appliances/devices that can be powered by it, most often lamps, fans, radios and televisions.

  The related cost for each of the above approaches is estimated between 2000 and 3000 USD for grid extension, between 500 and 1200 USD for microgrids and between 150 and 500 USD for Solar Home Systems [76]. As is understandable, and given the very challenging economic conditions of developing countries, grid extension is many times prohibitive due to the related high cost. As such, off-grid systems are expected to play a very important role in future rural electrification activities [77]. The decision whether to go for SHSs or a microgrid is the result of a techno-economic evaluation of both approaches for any given location, with the population density being one of the most important factors [78].

  A note has to be made here in relation to high power autonomous systems that can be deployed in order to meet a specific load in a given location. An example for this might be a system to power a water pump. Currently, even for those systems, the microgrid topology is used, with no actual grid infrastructure, powering only the single load.

  The International Finance Corporation of the World Bank Group has performed an extensive study of microgrids in operation in Sub-Saharan Africa. The main results of this study are presented in Table 4. It is interesting to see that households pay on average 7 USD per month for their monthly electricity bill, while the average access is Tier 2. The average connection cost is 920 USD. The payback period is more than 7 years on average, while it has to be noted that various forms of financing including grants are considered. Finally, a very

Table 4 Summary of findings of Benchmarking microgrids in Sub-Saharan Africa

Indicator	Value
Monthly average revenue per user	7 USD
Average investment per user	920 USD
Tier 2 average residential consumption	11 kWh m^{-1}
Average generation capacity	34 kW
Average number of connections	~100
A/C vs. D/C	85% vs. 15%
OPEX as a % of revenue	58%
Average capital expenditure (CAPEX) payback period	>7 years
Split of CAPEX spending on distribution vs generation	50–50%
Average Distance from National Grid	23 km

Source: International Finance Corporation WBG. Benchmarking Mini-grid DESCOs 2017 update—Summary of Findings, 2018

important aspect to note is that currently the cost split between distribution and generation is 50–50%. This means that the poles and cables cost as much as the photovoltaics and batteries used in the microgrids. This highlights the importance of population density, as a highly dispersed population in a given area might be better electrified through solar home systems instead of microgrids.

3.5 Deployment of a Rural Electrification Project

A rural electrification project is deployed in three major stages, each consisting of a number of sub-stages (Fig. 1) [79].

1. Project development and pre-installation
2. Design, procurement, installation and commissioning
3. Post commissioning and sustaining the project

Each of these phases is further detailed in the following sections.

Project Development and Pre-installation

(a) *Site selection*

During this step you identify the site where the microgrid will be installed. Many factors can affect this choice like distance from the main grid, economic status of the community, acceptance of the electrification project by the community, renewable energies potential etc.

(b) *Feasibility studies and surveys*

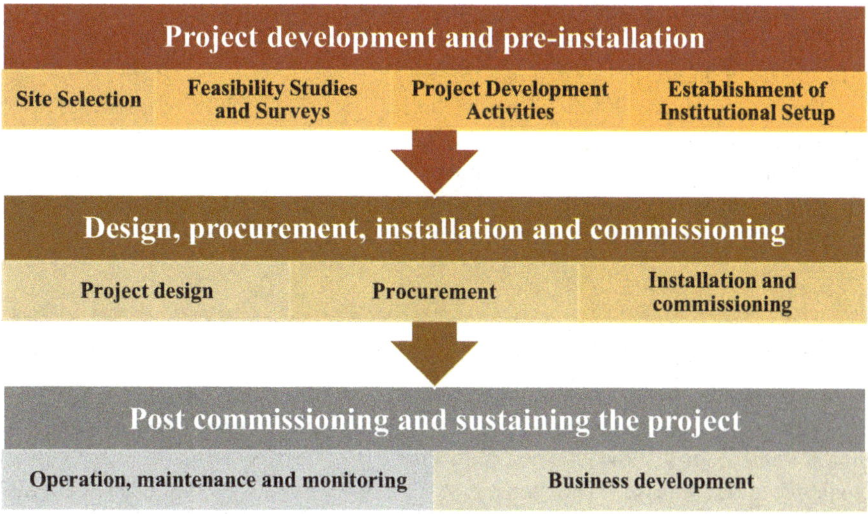

Fig. 1 Stages for the deployment of a rural electrification project

In this step data is collected for the chosen site. Surveys take place in order to develop the load profile and the agricultural productive uses, as well as market related data.

(c) *Project development activities*

The availability and cost of various technologies that can potentially be employed is assessed. The process of obtaining needed licenses and approvals is initiated. Local contracts are also developed in this stage.

(d) *Establishment of the institutional setup*

This step is concerned with the setup of a local governance body. Capacity building activities for both the management structure and the users is initiated.

Design, Procurement, Installation and Commissioning

(a) *Project design*

The final project design is developed. This includes the development of the topology and the detailed list of the various components that are going to be needed.

(b) *Procurement*

A thorough market search is made and the needed components are procured.

(c) *Installation and commissioning*

The microgrid is installed. This includes both the generation and storage as well as the distribution grid. The system is commissioned and delivered to the system operator.

Post Commissioning and Sustaining the Project

(a) *Operation, maintenance and monitoring*
This is concerned with the proper technical operation of the microgrid.
(b) *Business development*
This is concerned with the business model implementation, which needs to provide the proper cash-flows and profit of the investment on a whole, ensuring its economic viability and sustainability.

4 Discussion: Conclusions

This paper has investigated the interrelations of agriculture, rural electrification and digital technologies. Poverty is acknowledged by the UN as the greatest global challenge to be addressed. The rural areas of the developing world are where extreme poverty is observed. Rural development is recognized as the most important approach in poverty alleviation in those areas. Thus, agriculture still remains the main economy sector to be targeted first in the rural areas of the poorest of the poor countries. Agricultural electrification is the investment that makes the most sense in those rural locations, since it can have a very strong impact from the start in terms of the quality and quantity of agricultural production. Rural electrification ensures the availability of energy in remote and developing areas which is crucial in agricultural operations, especially when there is a demand for increased primary production which is the foundation of development.

The increased agricultural production increases the income of the rural communities and this in turn can stimulate other economic activities in the area, fostering overall development. Economic development, offering new employment opportunities, of rural areas can contribute to poverty and hunger elimination adding also to the social benefits. The improvements of economic and social status of developing regions can lead to the adoption of new technologies that can contribute to increased production through the improvement of operations efficiency. New technologies also reduce the environmental impact of agricultural production which is severe taking into account the traditional intensified method of agricultural production.

Towards that direction digital technologies aim at reducing the required agricultural inputs increasing the overall sustainability of agricultural production. Digital technologies are essentially an enabler of rural electrification. They can decrease costs in all stages of a rural electrification project deployment and also address technical and non-technical challenges faced in the operation stage. Their application has to be sought and supported. Research is still strong and further benefits are expected in the future.

References

1. Papageorgiou K, Singh PK, Papageorgiou EI, et al (2020) Participatory modelling for poverty alleviation using fuzzy cognitive maps and OWA learning aggregation. https://doi.org/10.1371/journal.pone.0233984
2. Saleem Z, Donaldson JA (2016) Pathways to poverty reduction. Dev Policy Rev 34:671–90. https://doi.org/10.1111/dpr.12167
3. Moseley M (2003) Rural Development: Principles and Practice. London
4. Ward N, Brown DL (2009) Placing the rural in regional development. Reg Stud 43:1237–44. https://doi.org/10.1080/00343400903234696
5. Rowley TD (1996) Rural Development Research: A Foundation for Policy. Greenwood Press
6. Anríquez G, Stamoulis K (2007) Rural development and poverty reduction: is agriculture still the key?
7. Van Pham L, Smith C (2014) Drivers of agricultural sustainability in developing countries: A review. Environ. Syst. Decis. 34:326–341
8. de Vries M, van Middelaar CE, de Boer IJM (2015) Comparing environmental impacts of beef production systems: A review of life cycle assessments. Livest Sci 178:279–288. https://doi.org/10.1016/j.livsci.2015.06.020
9. Achillas C, Bochtis D (2020) Toward a green, closed-loop, circular bioeconomy: Boosting the performance efficiency of circular business models. Sustain 12:1–6. https://doi.org/10.3390/su122310142
10. World Food Programme (2014) Hunger Statistics. United Nations
11. FAO (2012) FAO POLICY ON GENDER EQUALITY: Attaining Food Security Goals in Agriculture and Rural Development. FAO
12. United Nations (2017) The Sustainable Development Goals Report. United Nations Publ. https://doi.org/10.18356/3405d09f-en
13. De Luca AI, Iofrida N, Leskinen P, et al (2017) Life cycle tools combined with multi-criteria and participatory methods for agricultural sustainability: Insights from a systematic and critical review. Sci. Total Environ. 595:352–370
14. Peano C, Tecco N, Dansero E, et al (2015) Evaluating the sustainability in complex agri-food systems: The SAEMETH framework. Sustain 7:6721–6741. https://doi.org/10.3390/su7066721
15. Lampridi MG, Sørensen CG, Bochtis DD (2019) Agricultural Sustainability: A Review of Concepts and Methods. Sustainability 11:27. https://doi.org/10.3390/su11185120
16. Bockstaller C, Guichard L, Makowski D, et al (2008) Review article Agri-environmental indicators to assess cropping and farming systems. A review. Agron Sustain Dev 28:139–149. https://doi.org/10.1051/agro:2007052
17. Struik PC, Kuyper TW (2017) Sustainable intensification in agriculture: the richer shade of green. A review. Agron. Sustain. Dev.
18. Lampridi M, Kateris D, Sørensen CG, Bochtis D (2020) Energy footprint of mechanized agricultural operations. Energies 13:1–15. https://doi.org/10.3390/en13030769
19. Giovannucci D, Scherr SJ, Nierenberg D, et al (2012) Food and Agriculture: The Future of Sustainability. SSRN Electron J. https://doi.org/10.2139/ssrn.2054838
20. Papageorgiou K, Singh PK, Papageorgiou E, et al (2020) Fuzzy cognitive map-based sustainable socio-economic development planning for rural communities. Sustain 12. https://doi.org/10.3390/su12010305
21. Erlinghagen S, Markard J (2012) Smart grids and the transformation of the electricity sector: ICT firms as potential catalysts for sectoral change. Energy Policy. https://doi.org/10.1016/j.enpol.2012.09.045
22. Kyriakarakos G, Balafoutis AT, Bochtis D (2020) Proposing a paradigm shift in rural electrification investments in Sub-Saharan Africa through Agriculture. Sustain. https://doi.org/10.3390/SU12083096

23. Dibaba W (2017) THE ROLE OF EFFECTIVE INTEGRATION OF ICT IN EDUCATION, ESPECIALLY IN PRIMARY AND SECONDARY EDUCATION OF REMOTE SETTINGS. Int J Adv Res Comput Sci 8:10–3. https://doi.org/10.26483/ijarcs.v8i9.4923
24. Wikipedia contributors (2019) Information and communications technology. In: Wikipedia
25. Ligeza A (1995) Artificial Intelligence: A Modern Approach. Neurocomputing. https://doi.org/10.1016/0925-2312(95)90020-9
26. Anagnostis A, Asiminari G, Papageorgiou E, Bochtis D (2020) A Convolutional Neural Networks Based Method for Anthracnose Infected Walnut Tree Leaves Identification. Appl Sci 10:469. https://doi.org/10.3390/app10020469
27. Papageorgiou K, Carvalho G, Papageorgiou EI, et al (2020) Decision-making process for photovoltaic solar energy sector development using fuzzy cognitive map technique. Energies. https://doi.org/10.3390/en13061427
28. McArthur SDJ, Davidson EM, Catterson VM, et al (2007) Multi-agent systems for power engineering applications - Part II: Technologies, standards, and tools for building multi-agent systems. IEEE Trans Power Syst. https://doi.org/10.1109/TPWRS.2007.908472
29. Gregorio F, González G, Schmidt C, Cousseau J (2020) Internet of Things. In: Signals and Communication Technology
30. Katikaridis D, Bechtsis D, Menexes I, et al (2017) A software tool for efficient agricultural logistics. In: CEUR Workshop Proceedings
31. Silva RF, Mostaço GM, Xavier F, et al (2020) Proposal of an architecture for data integration at agricultural supply chains, considering the implementation of IoT, NoSQL blockchain technologies. In: Proceedings of the 12th EFITA International Conference. Rhodes, 27–29 June 2019, pp 90–95
32. Zheng Z, Xie S, Dai H, et al (2017) An Overview of Blockchain Technology: Architecture, Consensus, and Future Trends. In: Proceedings - 2017 IEEE 6th International Congress on Big Data, BigData Congress 2017
33. Tian F (2016) An agri-food supply chain traceability system for China based on RFID & blockchain technology. In: 2016 13th International Conference on Service Systems and Service Management, ICSSSM 2016
34. Kyriakarakos G, Papadakis G (2018) Microgrids for productive uses of energy in the developing world and blockchain: A promising future. Appl. Sci.
35. Goranovic A, Meisel M, Fotiadis L, et al (2017) Blockchain applications in microgrids: An overview of current projects and concepts. In: Proceedings IECON 2017 - 43rd Annual Conference of the IEEE Industrial Electronics Society
36. Wu C, Chen Z, Wang D, et al (2020) A cloud-based in-field fleet coordination system for multiple operations. Energies 13:1–15. https://doi.org/10.3390/en13040775
37. Etro F (2014) The economics of cloud computing. In: Cloud Technology: Concepts, Methodologies, Tools, and Applications
38. Huh JH, Seo YS (2019) Understanding Edge Computing: Engineering Evolution with Artificial Intelligence. IEEE Access. https://doi.org/10.1109/ACCESS.2019.2945338
39. Zaphiris P, Kurniawan S (2006) Human computer interaction research in web design and evaluation
40. Van Krevelen DWF, Poelman R (2010) A Survey of Augmented Reality Technologies, Applications and Limitations. Int J Virtual Real. https://doi.org/10.20870/ijvr.2010.9.2.2767
41. Kaisler S, Armour F, Espinosa JA, Money W (2013) Big data: Issues and challenges moving forward. In: Proceedings of the Annual Hawaii International Conference on System Sciences
42. Chen B, Tan C, Zou X (2017) Cloud service platform of electronic identity in cyberspace. Cluster Comput. https://doi.org/10.1007/s10586-017-0731-9
43. Wang W, Lu Z (2013) Cyber security in the Smart Grid: Survey and challenges. Comput. Networks
44. Royer EM, Toh CK (1999) Review of current routing protocols for ad hoc mobile wireless networks. IEEE Pers Commun. https://doi.org/10.1109/98.760423

45. Moysiadis V, Tsolakis N, Katikaridis D, et al (2020) Mobile Robotics in Agricultural Operations : A Narrative Review on Planning Aspects. Appl Sci
46. Lampridi MG, Kateris D, Vasileiadis G, et al (2019) A Case-Based Economic Assessment of Robotics Employment in Precision Arable Farming. Agronomy 9:175. https://doi.org/10.3390/agronomy9040175
47. Marinoudi V, Sørensen CG, Pearson S, Bochtis D (2019) Robotics and labour in agriculture. A context consideration. Biosyst Eng. https://doi.org/10.1016/j.biosystemseng.2019.06.013
48. Murr LE (2016) Frontiers of 3D Printing/Additive Manufacturing: from Human Organs to Aircraft Fabrication. J Mater Sci Technol 32:987–995. https://doi.org/10.1016/j.jmst.2016.08.011
49. (2016) The "FineSol" project: Hyper-fine solder powders leading the miniaturisation of printed circuit boards. Weld. Cut.
50. Tsongas K, Tzetzis D, Karantzalis A, et al (2019) Microstructural, surface topology and nanomechanical characterization of electrodeposited Ni-P/SiC nanocomposite coatings. Appl Sci 9. https://doi.org/10.3390/app9142901
51. Friedewald M, Raabe O (2011) Ubiquitous computing: An overview of technology impacts. Telemat Informatics. https://doi.org/10.1016/j.tele.2010.09.001
52. Ingrams A (2015) Mobile phones, smartphones, and the transformation of civic behavior through mobile information and connectivity. Gov Inf Q. https://doi.org/10.1016/j.giq.2015.07.007
53. Narayanan A (1999) Quantum computing for beginners. In: Proceedings of the 1999 Congress on Evolutionary Computation, CEC 1999
54. D-Wave (2020) Quantum Computing. https://www.dwavesys.com/quantum-computing
55. Dogutan DK, Nocera DG (2019) Artificial Photosynthesis at Efficiencies Greatly Exceeding That of Natural Photosynthesis. Acc Chem Res. https://doi.org/10.1021/acs.accounts.9b00380
56. Zhu Z, Joshi S, Pelz B, Moddel G (2013) Overview of optical rectennas for solar energy harvesting. In: Next Generation (Nano) Photonic and Cell Technologies for Solar Energy Conversion IV
57. Murabula HM, Kanyua K (2017) Factors influencing performance of private electricity mini grid projects in Kenya: A case of Kirinyaga county. Int J Adv Res Manag Soc Sci 62–89
58. Mwaura FM (2012) Adopting electricity prepayment billing system to reduce non-technical energy losses in Uganda: Lesson from Rwanda. Util Policy 23:72–9. https://doi.org/10.1016/j.jup.2012.05.004
59. Pasti F (2019) 2018 State of the Industry Report on Mobile Money. GSMA. https://doi.org/10.1002/9781118290743.wbiedcs023
60. Hughes N, Lonie S (2007) M-PESA: Mobile Money for the "Unbanked" Turning Cellphones into 24-Hour Tellers in Kenya. Innov Technol Governance, Glob 2:63–81. https://doi.org/10.1162/itgg.2007.2.1-2.63
61. Barrie J, Cruickshank HJ (2017) Shedding light on the last mile: A study on the diffusion of Pay As You Go Solar Home Systems in Central East Africa. Energy Policy 107:425–36. https://doi.org/10.1016/j.enpol.2017.05.016
62. Barron M, Torero M (2017) Household electrification and indoor air pollution. J Environ Econ Manage 86:81–92. https://doi.org/10.1016/j.jeem.2017.07.007
63. Buyinza F, Kapeller J (2018) Household Electrification and Education Outcomes: Panel evidence from Uganda. ICAE Work Pap Ser
64. Bos K, Chaplin D, Mamun A (2018) Benefits and challenges of expanding grid electricity in Africa: A review of rigorous evidence on household impacts in developing countries. Energy Sustain Dev 44:64–77. https://doi.org/10.1016/j.esd.2018.02.007
65. Lewis J, Severnini E (2020) Short- and long-run impacts of rural electrification: Evidence from the historical rollout of the U.S. power grid. J Dev Econ. https://doi.org/10.1016/j.jdeveco.2019.102412
66. van Gevelt T (2014) Rural electrification and development in South Korea. Energy Sustain Dev 23:179–87. https://doi.org/10.1016/j.esd.2014.09.004

67. Ding H, Qin C, Shi K (2018) Development through electrification: Evidence from rural China. China Econ Rev 50:313–28. https://doi.org/10.1016/j.chieco.2018.04.007
68. Cabraal RA, Barnes DF, Agarwal SG (2005) PRODUCTIVE USES OF ENERGY FOR RURAL DEVELOPMENT. Annu Rev Environ Resour 30:117–44. https://doi.org/10.1146/annurev.energy.30.050504.144228
69. GIZ (2016) Photovoltaics for Productive Use Applications
70. Chakravorty U, Emerick K, Ravago M-L (2016) Lighting Up the Last Mile: The Benefits and Costs of Extending Electricity to the Rural Poor. SSRN Electron J 16–22. https://doi.org/10.2139/ssrn.2851907
71. Pode R, Pode G, Diouf B (2016) Solution to sustainable rural electrification in Myanmar. Renew Sustain Energy Rev 59:107–18. https://doi.org/10.1016/j.rser.2015.12.320
72. Bhatia M, Angelou N (2015) Beyond Connections - Energy access redefined
73. IEA World Energy Outlook 20182018
74. Palit D, Chaurey A (2011) Off-grid rural electrification experiences from South Asia: Status and best practices. Energy Sustain Dev 15:266–76. https://doi.org/10.1016/j.esd.2011.07.004
75. GOGLA (2018) Global Off-Grid Solar Market Report—Semi-Annual Sales and Impact Data. Utrecht, The Netherlands
76. Kyriakarakos G, Papadakis G (2019) Multispecies swarm electrification for rural areas of the developing world. Appl Sci 9. https://doi.org/10.3390/app9193992
77. Dagnachew AG, Lucas PL, Hof AF, et al (2017) The role of decentralized systems in providing universal electricity access in Sub-Saharan Africa – A model-based approach. Energy 139:184–95. https://doi.org/10.1016/j.energy.2017.07.144
78. Chaurey A, Kandpal TC (2010) A techno-economic comparison of rural electrification based on solar home systems and PV microgrids. Energy Policy 38:3118–29. https://doi.org/10.1016/j.enpol.2010.01.052
79. Bhattacharyya SC, Palit D (2016) Mini-grid based off-grid electrification to enhance electricity access in developing countries: What policies may be required? Energy Policy. https://doi.org/10.1016/j.enpol.2016.04.010

A Circular Precision Farming System Towards the Optimization of Dairy Value-Chains

Maria Lampridi, Theodora Angelopoulou, Aristotelis C. Tagarakis, and Dionysis D. Bochtis

1 Introduction

Conventional farming practices regarding livestock and associated crops pose significant issues that could lead to inappropriate use and application of resources such as water for livestock (water use per kg of beef produced may range from 27 to 200,000 L) [1], feed, irrigation water, fertilizers and agrochemicals. This can result in non-optimized value chains in the production systems, and insufficient data handling and processing. Agriculture contributes the 10% of total greenhouse gas emissions in the EU-28 with main sources the enteric fermentation from ruminants, soil nitrification and denitrification and manure decomposition [2]. Specifically, 94% of ammonia emissions is due to agriculture and since 2012 these emissions have increased. It has been estimated that the total annual cost to the EU of nitrogen pollution's environmental impacts is between 70 billion and 320 billion euros [3]. The Paris Agreement confirmed that "all sectors need to contribute" to a reduction in overall Greenhouse Gas (GHG) emissions [4]. Under the policy framework for climate and energy, a GHG emission reduction of 40% by 2030 and 80% by 2050 compared to the 1990 emissions levels, has been proposed [5]. In addition to the above the mechanization of farm and field production has resulted in an increased direct and indirect energy consumption that is mostly dependent on fossils fuels. Although the direct energy accounted for 3.2% of final energy consumption in the EU in 2018, agriculture, plays a significant role to the depletion of non-renewable energy resources and to global warming through energy-related emissions [6, 7].

M. Lampridi (✉) · T. Angelopoulou · A. C. Tagarakis · D. D. Bochtis
Institute for Bio-economy and Agri-technology (iBO), Centre for Research and Technology Hellas (CERTH), Thessaloniki, Greece
e-mail: m.lampridi@certh.gr

D. D. Bochtis et al. (eds.), *Information and Communication Technologies for Agriculture—Theme IV: Actions*, Springer Optimization and Its Applications 185,
https://doi.org/10.1007/978-3-030-84156-0_4

Agricultural practices are related to various emissions and run-offs of pollutants to land, water and air [8]. Specifically, the intensity of fertilizer use in combination to the low use efficiency of fertilizers, mainly nitrogen, by the crops lead to severe environmental impacts [9]. The inefficient application of nutrients leads to nutrient run-off from fields that could result to increased nitrate and phosphorus levels in ground and surface water reservoirs causing eutrophication, acidification [10], and soil heavy metal pollution from phosphate rock which is used for the production of phosphorous based fertilizers [11]. Nitrogen can also be lost from farms to the atmosphere in the form of gaseous, nitrogen-based compounds, i.e., ammonia and nitrogen oxides that are harmful for both humans, the local ecosystem, and the environment in general [12]. Regarding animal production, there are also critical issues that need to be addressed in order to increase the overall system's sustainability. Livestock contributes to GHG emissions through methane (CH_4), nitrous oxide (N_2O) and carbon dioxide (CO_2) production.

Overall, the global livestock sector contributes up to 72% of the agricultural N_2O emissions and about 80% of agricultural CH_4 [13]. These emissions result directly from enteric fermentation and manure management and indirectly mainly from land use changes, such as livestock-induced desertification that causes carbon oxidation and could result in 100 million tons of CO_2 emissions, from feed production and on farm fossil fuel use [14]. Agriculture also requires large amounts of water for irrigation, livestock consumption, cleaning and other uses and while the amount used varies with region, it can reach up to nearly 90% of all water consumption as occurred in regions in Southern Europe [15]. However, improvements related to water conservation in crop and livestock management (e.g., crop/livestock integration with the use of crop by-products by livestock) are feasible. In addition, precision irrigation, a modern irrigation management concept, aims at supporting the optimum water needs of plants through spending minimum resources and energy. Finally, another environmental friendly agricultural practice is the controlled traffic farming (CTF) [16–19]. CTF employs permanent parallel tracks on the field area to where the machinery traffic is restricted. CTF minimizes soil compaction, an effect that is directly connected with machinery power requirements, cultivation energy requirements, and the corresponding CO_2 emissions, and in terms of plant health soil compaction generates difficulties in plant emergence and growth leading to reduced yields (reduced yield is translated to increased environmental footprint per product mass unit). Beyond increased yield potential and energy savings, other recorded benefits derived from CTF implementation include the educed loss of CO_2 and the minimization of water runoff [20–23]. One challenge for the implementation of CTF, is the requirement for a new view on the operations planning and execution [24–26]. However, tools developed in operations management of field machinery make it possible for CTF systems to efficiently apply to the majority of cropping systems [24, 27–31].

Nowadays, consumers are more interested in high quality products that are safe and have been produced in an environmentally and socially responsible way [32, 33]. In order to achieve sustainable crop production, the implementation of good farming practices that follow ecosystem-based approaches need to be adopted

[34, 35]. These practices integrate knowledge on biodiversity and ecosystem management utilizing information and communication technology (ICT) [36–40]. Their implementation is primarily at a local scale, and include agricultural management practices, approaches and technologies towards high yield production and resource optimization while maintaining and/or enhancing environmental sustainability [41, 42]. From livestock perspective to achieve the highest production efficiency with limited environmental impact, livestock should be managed in a way to meet the animals' needs according to their genetic potential (optimum feed for minimum manure and highest productivity) [43].

The implementation of Precision Farming approaches has enabled the management of variability at field and farm level in order to improve the economic benefit while reducing the environmental impact [44]. Combing Precision Farming approaches with the principles of circular economy [45], by means of creating close-loop systems that minimize the use of inputs and reduce pollution, waste and carbon emissions, the concept of Circular Precision Farming (CPF) could arise. These approaches aim to enable farmers to plan, measure and monitor all aspects of the farming system in order to calculate the appropriate nutritional requirements of crops and animals and apply in the most cost-effective and environmentally friendly way.

2 The Importance of Circular Precision Farming

Conventional agricultural practices along with inadequate operation and data processing have proved to lead to non-optimal value chains, with adverse consequences to the producers and consumers [46–48]. Especially in the dairy industry, a large number of different inputs is involved, taking into account that milk production includes both crop (used for feed) and livestock farming [49]. Considering the above, dairy farming shows great potential for improvement in various production processes aiming at input reduction, taking advantage of the rapid development of ICT technologies. Consequently, it is important to investigate the mitigation potential of the environmental impacts of the entire production process of agricultural systems, both in the crop and dairy section, their by-products, and the management of agricultural waste in a circular system. More specifically, all the important issues related to ineffective and unsustainable agricultural practices that lead to inappropriate or less efficient use of resources (such as water and fertilizers) should be addressed.

Water is one of the most important resources in agricultural production [50], while agriculture is one of the most water demanding sectors requiring more than 70% of the total water consumption globally (FAO, 2015). Thus, the need for limiting its use has triggered several potential actions and developments. More specifically innovative irrigation techniques can be investigated aiming at reducing inputs and subsequently the environmental cost of cultivating field crops [51]. Additionally, modern irrigation methods with high irrigation efficiency, such as deficient irrigation and Fixed Partial Root zone Drying (FPRD) using modern irrigation systems such as drip irrigation [52, 53], subsurface irrigation and advanced management practices such as smart irrigation, should be investigated further. The effectiveness of irrigation is an issue examined in almost all commercial crops; however, it has still not been thoroughly investigated under an integrated assessment system. Lastly, the relationship between the physiological and morphological characteristics of crops grown for feed, as well as optimal fertilization and irrigation levels should be examined in order to determine the ways in which these characteristics can be used for effective use of water. For example, the selection of varieties that are more resistant to water deficiency based on specific morphological and agronomical characteristics, would lead to increased water use efficiency and higher yields in dry climates [54].

Basic aim of the above is to decrease the environmental impact of the activities of the entire production chain of dairy units. This can be achieved through the utilization of modern technologies for reduction of inputs and improved waste management, in both crop and animal production, as well as in the processing and standardization of dairy products. Several economically sustainable advanced "smart farming" and "smart processing" technologies and tools are being developed towards that direction [55]. These tools incorporate multiple levels of decision making in agricultural holdings, examining and combining business intelligence, engineering and computer science systems [56]. All these systems can be incorporated in an integrated milk production management system which sets the basis for the development of a Circular Precision Farming system that supports circular farming decision making.

2.1 Chapter's Focus

In the context of this work, the development of a system based on algorithms for optimizing crop yield and livestock productivity is discussed. Several data from various sources can be used as inputs to this system; positioning data (GPS), geographic information system's data (GIS), remote and proximal sensing data, data from crop, soil [57], and weather sensors (form Internet of Things - IoT systems) as well as data from agricultural machinery (as applied maps of irrigation, seeds, fertilizers, pesticides etc.) [31, 58–62], and a wide range of other sources. The aforementioned data, as inputs to a circular precision farming system, are appropriately translated and used in order to analyze production efficiency. Such systems, along with the development of the appropriate algorithms, can also support the use of waste and the sub-products in the production process, offering the producer an integrated decision support tool for circular farming [63, 64]. The integration of all the different levels of information may be an issue; however, data fusion algorithms offer the potential to combine information from many different sources (GPS, GIS, IoT, crop and soil sensors etc.). Emphasis is also given to the ease of use, considering that such systems should be aimed at people that do not often have the required familiarity with complex technologically advanced systems [65, 66].

Concluding, the present chapter focuses on the development of a circular precision farming system which aims at addressing the following basic issues:

- Develop eco-efficient and sustainable farming processes to minimize inputs in crop and dairy production while offering support in decision making.
- Optimize milk and crop production with respect to quantity, quality, corporate social responsibility standards, and satisfaction of the upcoming consumer requirements.
- Minimize the environmental impact of milk and crop production process and prepare mitigation strategies in order to tackle climate change.

3 Conceptualization of a Circular Precision Farming System

Recent research has identified that there is lack of systematic research on circular agriculture. Key innovation gaps appear in issues related to soil fertility, crop management, animal behavior and feed distribution, parameters that limit the sustainable optimization of integrated dairy production systems. Hence, in order to develop an integrated circular precision farming system, several methodological steps should be taken. As mentioned above, this work discusses the concept of developing a system for optimizing the yield of crops grown for feed, and the productivity of livestock, in vertical milk producing dairy farms. A general representation of the system's flow is presented in Fig. 1.

Focusing on the milk production sector, the examination of the entire value chain requires the consideration of a large number of different research subjects. These research subjects include aspects related to feed production, feed supply chain, feed nutritional value, animal management, precision nutrition, optimization of animal housing conditions, production process optimization, production of "green" products, and management of waste and by-products.

The description of the data and the criteria for selecting the inputs and outputs at the life cycle stages of milk production, can be divided in three categories with respect to the production processes, namely initial, main, and final (Fig. 2).

- **Initial processes** include the inputs of raw materials and energy to produce 1 L of raw milk from dairy cows. These comprise mainly inputs regarding crop production. More specifically, they include the crop management applications and harvesting techniques used in feed production; irrigation water, energy, and fuel consumption for field cultivation, the amount of seeds, fertilizers and pesticides used, transportation of the produced feed, and the emissions of gaseous pollutants from machinery.

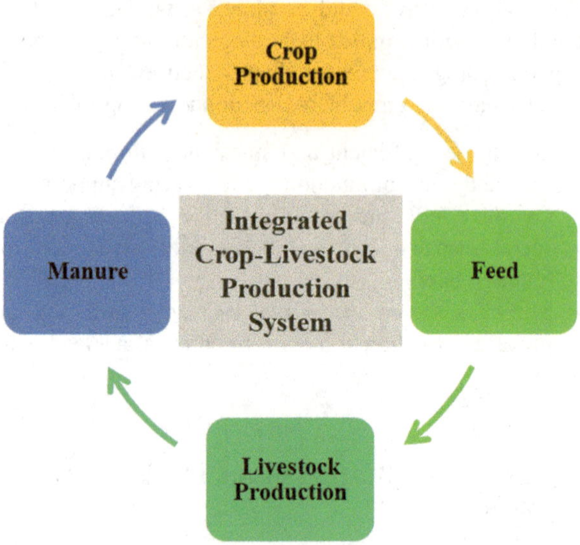

Fig. 1 Representation of the circular management system for vertical milk production in dairy farms

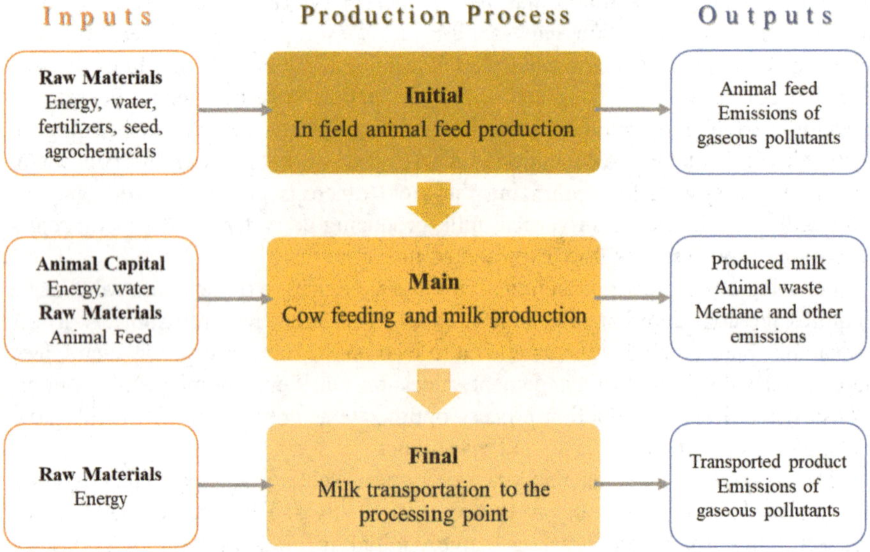

Fig. 2 Milk production system boundaries

- **Main processes** include the ones involved in the feeding of dairy cows for delivering the final product (raw cow's milk), and can be summarized as follows; first, is the on-farm feed production, the energy consumption, the amount of water used, and the emissions from intestinal fermentation. In addition, the external transport of raw materials used on the holding (main and complementary feed purchased) is assessed. Also, the materials used on the farm for cleaning, straw

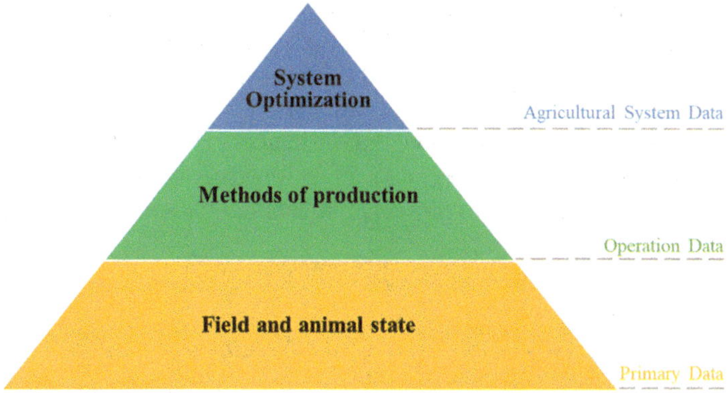

Fig. 3 Levels of data Collection of an integrated circular precision farming system

used as a bedding, maintenance procedures for the machinery and equipment used are considered. Lastly, the treatment of the waste generated on the farm and the impact on energy production and fuel used by the holding is included.

- **Final processes** include the transport of fresh cow's milk to the collection station of the processing unit. The energy and fuel consumption as well as the effects of their use are assessed. It should be noted that the processes of the production system do not include the production of equipment and construction of buildings. However, the production of raw materials used as well as the manufacturing processes can be included.

Considering the above, the development of an integrated circular precision farming system strongly depends on the appropriate collection of data. The data collected should map the entire production chain starting with animal feed production and following all the processes, the inputs and outputs of the system until the final production of milk products. Additionally, data modeling of the production procedures can be distinguished in modeling of primary data related to the field, crop, and animals status, and operational level data modeling which concerns the methods of production (Fig. 3). The above can be integrated into a holistic agricultural information system that leads to optimization and holistic assessment of farm information.

In that light the primary data collection begins at field level and concerns data related to soil and crop management in-field applications, and crop yield and quality. With the rise of IoT and sensing technologies in agricultural systems, a variety of sensors are made available providing detailed spatial information in the framework of precision farming. Thus, mapping of soil and crop properties, using remote and proximal sensors can provide detailed information producing large amount of data that can feed management information systems.

In the integrated circular precision farming system for dairy production, the quality of the produced yield at field level, is directly related to the quality of the produced animal feed. The next step is the modeling of the procedures at animal level, with the monitoring of the feed intake along with the animal living conditions and all the parameters that affect their behavior. Data modeling at operational level, includes the assessment of fertilizing and the harvesting programming in the field level, while from the livestock side it includes the management of the animal building environment. Lastly the modeling of the data at agricultural system level, includes the optimization and integration in an advanced farm management information system. All the above are presented in detail in the sections that follow.

3.1 Model Data at Animal and Crop Production Level

The required areas of data modeling in the animal and crop level are summarized in Fig. 4. The assessment of milk production chain begins with the production of animal feed. Cultivating crops is an important source of feed required for milk producing animals. Thus, yield should be monitored, and yield estimation should be attempted in early stages as possible, in order to safeguard the adequacy of feed

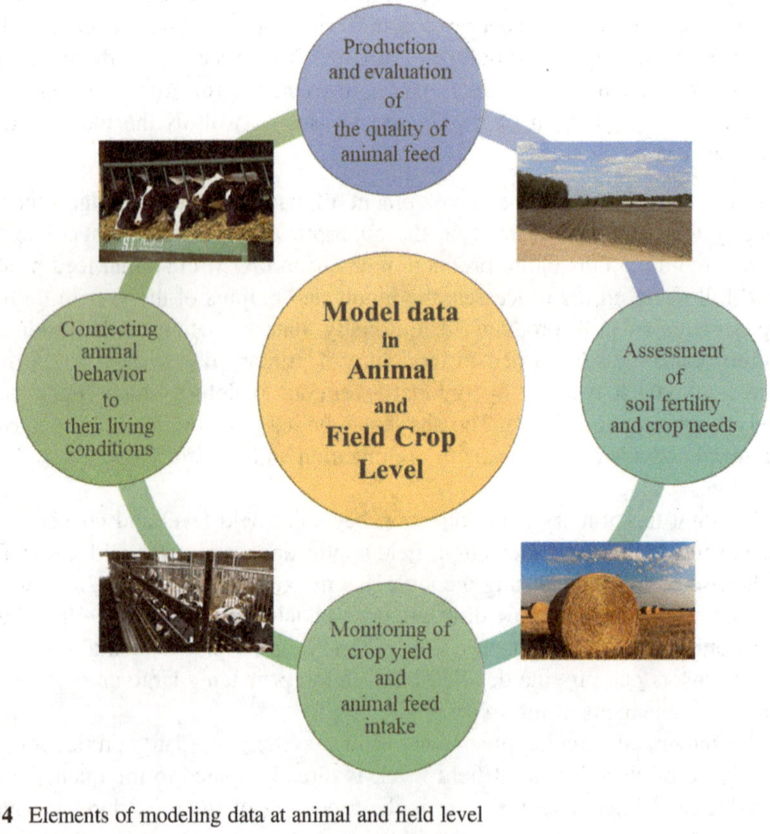

Fig. 4 Elements of modeling data at animal and field level

for the animals. Nevertheless, the quality assessment of the harvested crop is particularly important since it is related to the protein content of the produced feed, and consequently the animals' nutrition. The quantity and the quality of the produced feed is affected by soil fertility and crop management practices.

Application of alternative sources of fertilizer, such as green manure and organic residues, can decrease greenhouse gas emissions reducing the environmental cost. Consequently, the effect of chemical fertilization should be studied along with the use of green manure and other organic residues in fields cultivated for feed production. The assessment of soil nutrient status is widely performed through the collection and analysis of soil samples. The consequence of this practice is the increased cost. Therefore, a need arises for the development of automated and cost-effective methods for determining the spatial variation of nutrients and other properties related to soil fertility aiding in the assessment of the effect of fertilization. Further, adding to the reduction of the negative environmental impacts related to agriculture, the utilization of irrigation water should be assessed with the basic aim of improving water use efficiency.

The use of deficient irrigation and partial wetting of the rhizosphere to reduce water inflows should be investigated. In addition, advanced monitoring systems using wireless sensors have been developed and used for more efficient irrigation management. These systems, as part of the Internet of Things (IoT), can remotely and constantly monitor basic properties such as soil water content, solar radiation, precipitation, evapotranspiration etc., and provide information for localized weather forecast. Thus, the optimal timing and amount of irrigation can be assessed.

The combination of soil quality data along with the mapping of crop growth (using proximal crop sensors mounted on tractors or unmanned aerial vehicles—UAVs) and yield can lead to better crop management and provide all the valuable information with respect to the available feed produced in-farm. Furthermore, the feed intake of cattle can be monitored using microphones and accelerometers. The above data are related to the nutritional care of animals, however, the variations in animal behavior in relation to their living environment should also be monitored and assessed. This can be achieved with the help of precision livestock farming technologies that emphasize in the management of the environment of animal housing, since their behavior is affected to a great extent by thermal comfort and their overall conditions of living, along with the availability of animal feed.

The animal housing conditions and feeding program should be monitored and recorded. Specific wireless sensors are available on the market making IoT part of the advanced systems utilized in modern farms to remotely monitor the key parameters for securing optimal housing conditions for the animals. Additionally, the individual daily milk production should be recorded in order to determine the total milk production. With respect to the evaluation of the animal feeding process the following recording and monitoring procedures can be followed in regular intervals:

(a) Examination of the feed during feeding, to determine the optimal size of silage pieces and assess the nutritional behavior of cows.

(b) Assessment of rumination by observation or with the use of dedicated sensing devices such as rumination collars.
(c) Stool assessment (color, consistency, content) by observation and use of a special filter for chemical analysis (nitrogen, phosphorus, starch).
(d) Assessment of the mobility of animals by observing their exit from the milking parlor.
(e) Milk sampling to examine its chemical composition and determination of the number of somatic cells.

3.2 *Model Data at Operational Level*

The operational level concerns the assessment of the operating parameters related to milk production (Fig. 5). The operating parameters affect the inputs as well as the generated emissions of the production process. With respect to crop production, the optimization of fertilizer application, can result in nutrient losses, emission and packaging waste minimization through optimal routing and machine control. In addition, optimization of routing planning during harvest based on the biomass capacity of the plants, along with other operational restrictions, can result to increase in the effectiveness along with input reduction (e.g. fuel). Regarding the environmental impacts related to animal living, buildings for housing cattle and housing

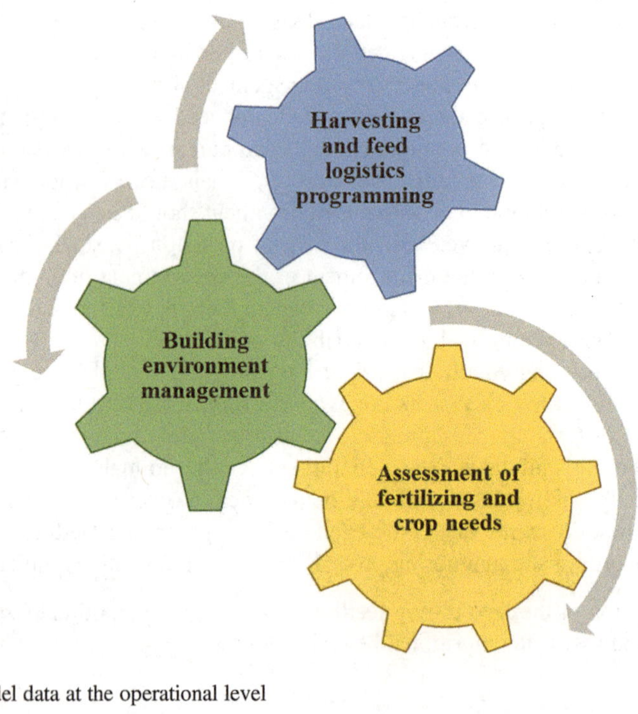

Fig. 5 Model data at the operational level

conditions, can contribute to NH_3 emissions [67]. Thus, there is a need to study the behavior of livestock in relation to their thermal comfort and NH_3 emission levels in relation to the automation of animal housing buildings (floor cleaners, building openings).

Adding to the above, the produced waste should be assessed by calculating the total waste volume of the dairy unit. The relationship between the generated waste and the methods of their collection, transportation and storage should be examined depending on the feeding method applied. The waste management system should be considered since it affects the health of the cattle and the production of high-quality milk. It is also related to surface and underground water pollution in case the waste is improperly discarded, collected or stored. Ideally, waste from livestock farms, should be evaluated with respect to its energy content, with the procedure of anaerobic digestion, in order to examine the potential for biogas production. Additionally, there is potential of producing manure to be applied as organic fertilizer, returning to the system as feed for the plants.

At operational level, the assessment of the aforementioned serious environmental impacts, related to milk production, in the entire supply chain of milk production is imperative. These impacts must be examined for all the stages of the production process including the feed cultivation and production, the energy demand for the housing and the feeding of cattle and the final production and transportation of milk. The environmental impact assessment can be carried out throughout the product's life cycle and more specifically through the modeling of the production process and the use of Life Cycle Analysis [68].

3.3 Modeling Data at Agricultural System Level

The combination of all the in-field and operational parameters provides the required information for optimization of the agricultural system on the basis of sustainability. To that end, the basic sustainability parameters and thresholds as well as the optimal operational methods that must be followed should be determined. Mapping the milk supply chain and recording the stages that are mostly responsible of the environmental burden caused (hotspots), as presented in the previous sections, are essential for the optimization of the entire agricultural system examined. Supply chain optimization can be achieved using simulation based on environmental and economic performance criteria, utilizing the data collected in real time along with the use of a relevant information system.

Agricultural System modelling includes algorithms that integrate the operation and primary data through the development of a digital tool for data integration at various levels of the production process and the operation of precision farming algorithms. This Circular Farm Management Information System incorporates all the information related to the supply chain providing guidelines for the optimization of the production system. The development of such a system includes several stages as presented in Fig. 6. First stage is the determination of operational and technical

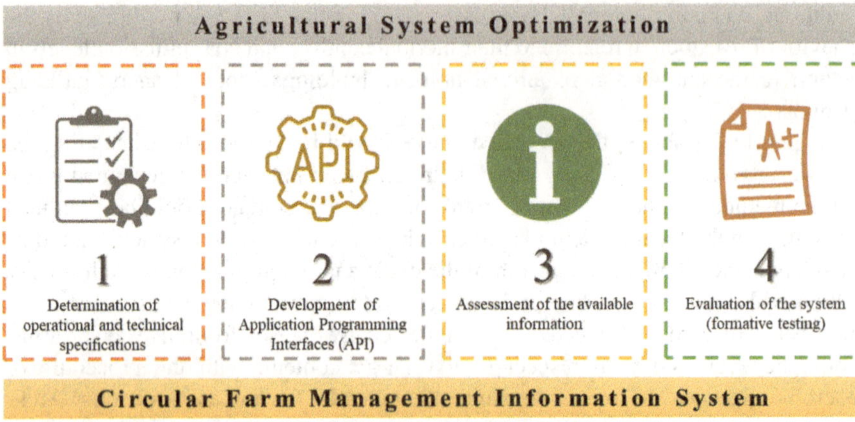

Fig. 6 Stages of development of a circular precision farming system

specifications of a circular precision farming system. The operational characteristics and the architecture of the system is based on extensive research on the available Information and Communication Technologies (ICT).

Next, the application programming interface (API) is developed, for the communication with other software and/or the exchange of data. To access the data, the API interface should comply with the latest safety technologies regarding the personal data protection of the final users. For the management of the collected information, a Graphic User Interface (GUI) should be developed; a graphic environment which is used for the assessment of the Internet of Things (IoT) platforms and the extraction of the different data sources that can be managed by the system. Through the system's GUI, the user makes a request and receives the results in easy to view and interpret form. Its development can be based on modern web technologies (HTML5, CSS3, JavaScript, Websockets) in order to provide the end user with an enhanced web 3.0 experience. Finally, for the verification of the system an internal quality assurance procedure should be followed to ensure that the software meets the operational and technical specifications and minimize the number of operational errors.

4 Expected Outputs

Circular Precision Farming is an innovative concept aiming at utilizing the modern technological revolution and the possibilities of ICT and IoT to interconnect all respective processes in the dairy production and other related agricultural practices. The benefits from the implementation of CPF, lay on the development of state-of-the-art software and tools based on modern ICT that will help farmers and dairy companies in making decisions at strategic, tactic, and operational level. More specifically it helps setting the ground for vertical integration of production to produce competitive and sustainable dairy products with low environmental impact (green products). This will be achieved by improving the efficiency of agricultural production in terms of quantity, quality, environmental performance, and corporate social responsibility. It is also expected to promote synergies between the stakeholders and the different stages in the agricultural economy while it creates highly educated human resources, trained in precision agriculture technologies. Finally, analysis of consumers' needs will lead to products that better meet their requirements by mapping the milk market, emphasizing on environmentally certified products while evaluating market viability.

More specifically, through modelling data at animal and field crop level, the development of automated technologies that could be utilized from the first steps of crop cultivation can be enabled. Starting from soil monitoring, by utilizing governmental databases and sensor crop monitoring using remote and proximal sensing, high resolution digital soil and crop maps can be created. In addition, IoT weather devices can monitor weather and soil water content. The analysis of these datasets will assist in crop planning and subsequently animal feed production. The applications based on this planning, may include precision seeding at different spatio-temporal conditions and optimization of the fertilization procedure with site specific application and optimal irrigation management, minimizing inputs. In the next stage of the dairy production process, linking data on feed quality, feed intake and animal performance can help optimize feed delivery strategies. Controlling the building environment of animal housing through ventilation system specifications with real time management of the microclimate, promotes animal welfare and reduces energy consumption. The development of prototype algorithms for monitoring animal health and welfare, can result in controlling the overuse of antibiotics with the development of a system for recognition of animal motion parameters for diseases warning in early stages.

Integrating all these aspects throughout the steps of the milk production chain, will lead to the development of an advanced Farm Management Information System that facilitates sustainable oriented benefits to operations within a dairy farming system, addressing possible threats associated with the production goals. The integration of information about logistics, waste management, biomass exploitation and other parameters, results in the optimization of reverse supply chain networks, and better exploitation of animal waste. In addition, life cycle assessment analysis also assists in the development of the decision support tool. This in turn can provide with sustainability indicators identification on the primary sector. Integrating a diverse set of data in the farm management system, could provide valuable information, specific for their respective crops and dairy production, in timely manner.

The applicability and reliability of the previously mentioned systems was demonstrated in the case of a real vertical milk production. It is worth mentioning that the selection of a vertical Small Medium Enterprise (SME) specialized in milk production, allows for the replicability of the model and its adoption from a large number of enterprises in the sector. Considering the above, the competitiveness of the production units is expected to increase through the increase of their efficiency (due to lower inputs) and the attractiveness of the products (environmental certification potential). At the same time, a competitive strategy based on the growing sensitivity of the public in matters of environmental protection and sustainability, as well as animal welfare (ethical buying), is an effective response to the growing pressure applied to the industry by cheap imported products in the market.

5 Conclusions

Conventional farm management systems can lead to environmental degradation and are not considered cost effective. To date, there are still issues that need to be addressed as farmers are not always sufficiently aware of optimal livestock and associated crops production management practices. There is also lack of low cost and effective technologies and methods to manage the complexities of the entirety of farm resources. Farmers require multidisciplinary expertise to transform data into knowledge suitable for decision-making, to monitor and interpret essential, strongly interrelated indicators, such as soil readiness, nutrient efficiency, product quality and animal welfare. Tools currently available on the market, mainly focus on partial solutions, without providing meaningful and actionable recommendations on a systematic approach that can use the most efficient means to generate consistent, optimum global results. This is also dominant for smallholders who work with very few inputs, and limited resources and machinery.

The implementation of Circular Precision Farming approaches can be considered as the path for achieving sustainable development in agriculture. Following the basic principles of this concept by reducing the inputs and waste during the life cycle of agricultural products the benefits could be multidimensional regarding the environment and socio-economic impact. There are many efforts towards these goals; however, though the implementation of integrated approaches that take into consideration the whole dairy production system, is at a very early stage with no readily developed solutions.

Acknowledgments This research has been co-financed by the European Union and Greek national funds through the Operational Programme Competitiveness, Entrepreneurship and Innovation, under the call RESEARCH—CREATE—INNOVATE (project code: T1EDK-03987) "BioCircular: Bioproduction System for Circular Precision Farming".

References

1. Peters, G.M.; Wiedemann, S.G.; Rowley, H. V.; Tucker, R.W. Accounting for water use in australian red meat production. *Int. J. Life Cycle Assess.* **2010**, *15*, 311–320.
2. Eurostat *Agriculture, forestry and fishery statistics*; 2013; ISBN 9789279330056.
3. European Commission Nitrogen Pollution and the European Environment: Implications for Air Quality Policy. *Sci. Environ. Policy IN-DEPTH Rep.* **2013**, 28.
4. Paris Agreement.

5. European Commission *A policy framework for climate and energy in the period from 2020 up to 2030*; 2014
6. European Union *Agriculture, forestry and fishery statistics*; 2019; ISBN 9789279330056.
7. Achillas, C.; Bochtis, D. Toward a Green, Closed-Loop, Circular Bioeconomy: Boosting the Performance Efficiency of Circular Business Models. *Sustainability* **2020**, *12*, 10142.
8. Adegbeye, M.J.; Ravi Kanth Reddy, P.; Obaisi, A.I.; Elghandour, M.M.M.Y.; Oyebamiji, K.J.; Salem, A.Z.M.; Morakinyo-Fasipe, O.T.; Cipriano-Salazar, M.; Camacho-Díaz, L.M. Sustainable agriculture options for production, greenhouse gasses and pollution alleviation, and nutrient recycling in emerging and transitional nations - An overview. *J. Clean. Prod.* 2020, *242*, 118319.
9. Sopegno, A.; Rodias, E.; Bochtis, D.; Busato, P.; Berruto, R.; Boero, V.; Sørensen, C. Model for Energy Analysis of Miscanthus Production and Transportation. *Energies* **2016**, *9*, 392.
10. de Wit, M.; Behrendt, H.; Bendoricchio, G.; Bleuten, W.; van Gaans, P. The contribution of agriculture to nutrient pollution in three European rivers, with reference to the European Nitrates Directive. *Eur. Water Manag. Online* **2002**, *19*.
11. Mehmood, T.; Chaudhry, M.M.; Tufail, M.; Irfan, N. Heavy metal pollution from phosphate rock used for the production of fertilizer in Pakistan. *Microchem. J.* **2009**, *91*, 94–99.
12. Aneja, V.P.; Roelle, P.A.; Murray, G.C.; Southerland, J.; Erisman, J.W.; Fowler, D.; Asman, W. A.H.; Patni, N. Atmospheric nitrogen compounds II: emissions, transport, transformation, deposition and assessment. *Atmos. Environ.* **2001**, *35*, 1903–1911.
13. Gerber, P.J., Steinfeld, H., Henderson, B., Mottet, A., Opio, C., Dijkman, J., Falcucci, A. & Tempio, G. *Tackling climate change through livestock – A global assessment of emissions and mitigation opportunities*; 2013; ISBN 9789251079201.
14. Sejian, V.; Bhatta, R.; Malik, P.K.; Madiajagan, B.; Al-Hosni, Y.A.S.; Sullivan, M.; Gaughan, J.B. Livestock as Sources of Greenhouse Gases and Its Significance to Climate Change. In *Greenhouse Gases*; InTech, 2016.
15. EEA *Water resources across Europe – confronting water scarcity and drought. EEA Report 2/2009*; 2009; ISBN 9789291679898.
16. Tullberg, J. Tillage, traffic and sustainability—A challenge for ISTRO. *Soil Tillage Res.* **2010**, *111*, 26–32.
17. Tullberg, J.N.; Yule, D.F.; McGarry, D. Controlled traffic farming—From research to adoption in Australia. *Soil Tillage Res.* **2007**, *97*, 272–281.
18. Mouazen, A.; Palmqvist, M. Development of a Framework for the Evaluation of the Environmental Benefits of Controlled Traffic Farming. *Sustainability* **2015**, *7*, 8684–8708.
19. Bochtis, D.D.; Sørensen, C.G.; Jørgensen, R.N.; Green, O. Modelling of material handling operations using controlled traffic. *Biosyst. Eng.* **2009**, *103*.
20. Hamza, M.A.; Anderson, W.K. Soil compaction in cropping systems: A review of the nature, causes and possible solutions. *Soil Tillage Res.* **2005**, *82*, 121–145.
21. Tullberg, J.N.; Ziebarth, P.J.; Li, Y. Tillage and traffic effects on runoff. *Aust. J. Soil Res.* **2001**, *39*, 249.
22. McPhee, J.E.; Braunack, M.V.; Garside, A.L.; Reid, D.J.; Hilton, D.J. Controlled Traffic for Irrigated Double Cropping in a Semi-arid Tropical Environment: Part 2, Tillage Operations and Energy Use. *J. Agric. Eng. Res.* **1995**, *60*, 183–189.
23. Bennett, J.M.; Roberton, S.D.; Jensen, T.A.; Antille, D.L.; Hall, J. A comparative study of conventional and controlled traffic in irrigated cotton: I. Heavy machinery impact on the soil resource. *Soil Tillage Res.* **2017**, *168*, 143–154.
24. Samuel Gan-Mor, S.; Rex L.; Clark, R.L. DGPS-Based Automatic Guidance - Implementation and Economical Analysis. In Proceedings of the 2001 Sacramento, CA July 29-August 1,2001; American Society of Agricultural and Biological Engineers: St. Joseph, MI, 2001; p. 1.
25. Bochtis, D.D.; Sørensen, C.G.; Green, O.; Moshou, D.; Olesen, J. Effect of controlled traffic on field efficiency. *Biosyst. Eng.* **2010**, *106*.

26. Bochtis, D.D.; Sørensen, C.G.; Busato, P.; Hameed, I.A.; Rodias, E.; Green, O.; Papadakis, G. Tramline establishment in controlled traffic farming based on operational machinery cost. *Biosyst. Eng.* **2010**, *107*, 221–231.
27. Jensen, M.A.F.; Bochtis, D.; Sørensen, C.G.; Blas, M.R.; Lykkegaard, K.L. In-field and inter-field path planning for agricultural transport units. *Comput. Ind. Eng.* **2012**, *63*, 1054–1061.
28. Hameed, I.A.; Bochtis, D.D.; Sørensen, C.G.; Nøremark, M. Automated generation of guidance lines for operational field planning. *Biosyst. Eng.* **2010**, *107*.
29. Bochtis, D.D.; Sørensen, C.G.; Vougioukas, S.G. Path planning for in-field navigation-aiding of service units. *Comput. Electron. Agric.* **2010**, *74*, 80–90.
30. Hameed, I.A.; Bochtis, D.D.; Sørensen, C.G.; Jensen, A.L.; Larsen, R. Optimized driving direction based on a three-dimensional field representation. *Comput. Electron. Agric.* **2013**, *91*, 145–153.
31. Hameed, I.A.; Bochtis, D.D.; Sørensen, C.G.; Vougioukas, S. An object-oriented model for simulating agricultural in-field machinery activities. *Comput. Electron. Agric.* **2012**, *81*.
32. Cheah, I.; Phau, I. Attitudes towards environmentally friendly products: The influence of ecoliteracy, interpersonal influence and value orientation. *Mark. Intell. Plan.* **2011**, *29*, 452–472.
33. Achillas, C.; Bochtis, D.D.; Aidonis, D.; Folinas, D. *Green Supply Chain Management*; Routledge, 2018;
34. Imadi, S.R.; Shazadi, K.; Gul, A.; Hakeem, K.R. Sustainable crop production system. In *Plant, Soil and Microbes: Volume 1: Implications in Crop Science*; Springer International Publishing, 2016; pp. 103–116 ISBN 9783319274553.
35. Bochtis, D.; Sørensen, C.A.G.; Kateris, D. *Operations management in agriculture*; Elsevier, 2018; ISBN 9780128097861.
36. Saidu, A.; Clarkson, A.M.; Adamu, S.H.; Mohammed, M.; Jibo, I. Application of ICT in Agriculture: Opportunities and Challenges in Developing Countries. *Int. J. Comput. Sci. Math. Theory* **2017**, *3*, 8–18.
37. Sørensen, C.G.; Pesonen, L.; Bochtis, D.D.; Vougioukas, S.G.; Suomi, P. Functional requirements for a future farm management information system. *Comput. Electron. Agric.* **2011**, *76*.
38. Sørensen, C.G.; Bochtis, D.D. Conceptual model of fleet management in agriculture. *Biosyst. Eng.* **2010**, *105*, 41–50.
39. Sørensen, C.G.; Fountas, S.; Nash, E.; Pesonen, L.; Bochtis, D.; Pedersen, S.M.; Basso, B.; Blackmore, S.B. Conceptual model of a future farm management information system. *Comput. Electron. Agric.* **2010**, *72*, 37–47.
40. Angelopoulou, T.; Tziolas, N.; Balafoutis, A.; Zalidis, G.; Bochtis, D. Remote Sensing Techniques for Soil Organic Carbon Estimation: A Review. *Remote Sens.* **2019**, *11*, 676.
41. Bochtis, D.D.; Sørensen, C.G.C.; Busato, P. Advances in agricultural machinery management: A review. *Biosyst. Eng.* 2014, *126*, 69–81.
42. Rodias, E.; Berruto, R.; Busato, P.; Bochtis, D.; Sørensen, C.G.; Zhou, K. Energy savings from optimised in-field route planning for agricultural machinery. *Sustain.* **2017**, *9*.
43. Berckmans, D. General introduction to precision livestock farming. *Anim. Front.* **2017**, *7*, 6–11.
44. McBride, W.D.; Daberkow, S.G. Information and the Adoption of Precision Farming Technologies. *J. Agribus.* **2003**, *21*, 21.
45. Zink, T.; Geyer, R. Circular Economy Rebound. *J. Ind. Ecol.* **2017**, *21*, 593–602.
46. Peano, C.; Tecco, N.; Dansero, E.; Girgenti, V.; Sottile, F. Evaluating the sustainability in complex agri-food systems: The SAEMETH framework. *Sustain.* **2015**, *7*, 6721–6741.
47. Iakovou, E.; Bochtis, D.; Vlachos, D.; Aidonis, D. Sustainable Agrifood Supply Chain Management. In *Supply Chain Management for Sustainable Food Networks*; John Wiley & Sons, Ltd: Chichester, UK, 2016; pp. 1–39 ISBN 9781118937495.
48. Iakovou, E.; Bochtis, D.; Vlachos, D.; Aidonis, D. *Supply Chain Management for Sustainable Food Networks*; 2015; ISBN 9781118937495.

49. Cecchini, L.; Venanzi, S.; Pierri, A.; Chiorri, M. Environmental efficiency analysis and estimation of CO2 abatement costs in dairy cattle farms in Umbria (Italy): A SBM-DEA model with undesirable output. *J. Clean. Prod.* **2018**, *197*, 895–907.
50. Noya, I.; González-García, S.; Berzosa, J.; Baucells, F.; Feijoo, G.; Moreira, M.T. Environmental and water sustainability of milk production in Northeast Spain. *Sci. Total Environ.* **2018**.
51. de Vito, R.; Portoghese, I.; Pagano, A.; Fratino, U.; Vurro, M. An index-based approach for the sustainability assessment of irrigation practice based on the water-energy-food nexus framework. *Adv. Water Resour.* **2017**.
52. Samperio, A.; Moñino, M.J.; Vivas, A.; Blanco-Cipollone, F.; Martín, A.G.; Prieto, M.H. Effect of deficit irrigation during stage II and post-harvest on tree water status, vegetative growth, yield and economic assessment in "Angeleno" Japanese plum. *Agric. Water Manag.* **2015**.
53. Romero, P.; Muñoz, R.G.; Fernández-Fernández, J.I.; del Amor, F.M.; Martínez-Cutillas, A.; García-García, J. Improvement of yield and grape and wine composition in field-grown Monastrell grapevines by partial root zone irrigation, in comparison with regulated deficit irrigation. *Agric. Water Manag.* **2015**.
54. Ali, Q.; Ashraf, M. Exogenously applied glycinebetaine enhances seed and seed oil quality of maize (Zea mays L.) under water deficit conditions. *Environ. Exp. Bot.* **2011**.
55. Sørensen, C.A.G.; Kateris, D.; Bochtis, D. ICT Innovations and Smart Farming. In Proceedings of the Communications in Computer and Information Science; 2019.
56. Tsolakis, N.; Bechtsis, D.; Bochtis, D. Agros: A robot operating system based emulation tool for agricultural robotics. *Agronomy* **2019**.
57. Angelopoulou, T.; Balafoutis, A.; Zalidis, G.; Bochtis, D. From laboratory to proximal sensing spectroscopy for soil organic carbon estimation-A review. *Sustain.* 2020, *12*.
58. Bochtis, D.D.; Sørensen, C.G. The vehicle routing problem in field logistics part I. *Biosyst. Eng.* **2009**, *104*, 447–457.
59. Bochtis, D.D.; Sørensen, C.G. The vehicle routing problem in field logistics: Part II. *Biosyst. Eng.* **2010**, *105*, 180–188.
60. Jensen, M.F.; Bochtis, D.; Sørensen, C.G. Coverage planning for capacitated field operations, part II: Optimisation. *Biosyst. Eng.* **2015**, *139*.
61. Jensen, M.F.; Nørremark, M.; Busato, P.; Sørensen, C.G.; Bochtis, D. Coverage planning for capacitated field operations, Part I: Task decomposition, *Biosyst. Eng.* **2015**, *139*, 136–148.
62. Busato, P.; Sørensen, C.G.; Pavlou, D.; Bochtis, D.D.; Berruto, R.; Orfanou, A. DSS tool for the implementation and operation of an umbilical system applying organic fertiliser. *Biosyst. Eng.* **2013**, *114*.
63. Menexes, I.; Kolorizos, V.; Arvanitis, C.; Banias, G.; Kateris, D.; Bochtis, D. Robotics applications in agriculture with the use of an integrated information system. In Proceedings of the 11th National Conference of the Hellenic Society of Agricultural Engineers; Volos, Greece, 2019.
64. Bochtis, D.D.; Sørensen, C.G.; Busato, P.; Berruto, R. Benefits from optimal route planning based on B-patterns. *Biosyst. Eng.* **2013**, *115*.
65. Moysiadis, V.; Tsolakis, N.; Katikaridis, D.; Sørensen, C.G.; Pearson, S.; Bochtis, D. Mobile Robotics in Agricultural Operations: A Narrative Review on Planning Aspects. *Appl. Sci.* **2020**, *10*, 3453.
66. Marinoudi, V.; Sørensen, C.G.; Pearson, S.; Bochtis, D. Robotics and labour in agriculture. A context consideration. *Biosyst. Eng.* **2019**, *184*, 111–121.
67. Groot Koerkamp, P.W.G.; Metz, J.H.M.; Uenk, G.H.; Phillips, V.R.; Holden, M.R.; Sneath, R. W.; Short, J.L.; White, R.P.; Hartung, J.; Seedorf, J.; et al. Concentrations and emissions of ammonia in livestock buildings in Northern Europe. *J. Agric. Eng. Res.* **1998**.
68. Lampridi, M.G.; Sørensen, C.G.; Bochtis, D. Agricultural Sustainability: A Review of Concepts and Methods. *Sustainability* **2019**, *11*, 5120.

An Analysis of Safety and Health Issues in Agriculture Towards Work Automation

Lefteris Benos and Dionysis D. Bochtis

1 Introduction

Agriculture is regarded as the sector which is going to shoulder the burden of some of the most crucial challenges that the humanity ever faced with. These challenges contain the increasing international call for food, on account of the remarkable explosion of the global population [1], the negative consequences originated from the climate change, including global warming [2, 3], loss of biodiversity [4], agricultural production stability [5, 6], and resources depletion, such as groundwater depletion and soil degradation [7, 8]. However, a very important factor that is usually ignored or underestimated, especially in developing countries, is the safety and health of the agricultural workers. With approximately one third of the labor force around the earth being employed in agriculture, this sector is ranked as second pertaining to occupational illnesses and injuries [9]. Epidemiological studies that have been conducted in both developed and developing countries have recognized several illnesses related to agricultural engagement. These illnesses extend (but not limited) to respiratory and dermatological diseases, hearing loss, cancer cases, circulatory, motor and sensory disorders, such as the vibration white finger syndrome [10–13].

Among the most common non-fatal illnesses, the musculoskeletal disorders (MSDs) stand out, which are hard to be early identified and, definitely, undermine the farmers' quality of life and their ability to labor. MSDs is an umbrella term consisting of a number of disorders taking place in the whole body, such as strains, sprains, tears, tendinitis, hip and knee osteoarthritis, back pain, tension neck

L. Benos (✉) · D. D. Bochtis
Institute for Bio-economy and Agri-technology (iBO), Centre for Research and Technology Hellas (CERTH), Thessaloniki, Greece
e-mail: e.benos@certh.gr

D. D. Bochtis et al. (eds.), *Information and Communication Technologies for Agriculture—Theme IV: Actions*, Springer Optimization and Its Applications 185,
https://doi.org/10.1007/978-3-030-84156-0_5

syndrome, hand-arm vibration syndrome, carpal tunnel syndrome [14, 15]. MSDs are accompanying with high costs to farm owners, including lost of productivity, absenteeism and augmented expenses concerning disability, health care and worker's compensation. The partial mechanization of the agricultural sector, particularly in the developed countries, has managed to increase productivity and reduce the intensive manual agricultural labor [16, 17]. The intensive agricultural activities include, for instance, harvesting, weeding, pruning, carrying and lifting of harvesting lugs. Nonetheless, machinery introduction caused additional problems to the farmers' health coming from the whole-body vibration, engine fumes and accidents during the operation of them, to mention but a few.

Agriculture takes place within a peculiar environment, which can be characterized as dynamic and complex being vulnerable to uncertainty, heterogeneity and unpredictable situations and environmental conditions [18–20]. The wide range of factors involved in agricultural activities, such as working both outdoor and indoor, the different kinds of tasks, machines and crops, affect substantially the risk awareness levels and attitudes for preventing diseases and accidents. However, prevention necessitates controlling exposures. Four ways for controlling hazards during the agricultural activities exist, according to [21], which are:

- Reengineering of the process of agricultural production: This corresponds to rethinking the chemicals, equipment, tools and machines that are utilized in agriculture. On the one hand, in developing countries, this can signify introducing of machinery, for example, to lessen the labor-intensive operations that are usually carried out manually. On the other hand, in developed countries, this can stand for redesigning of human-machine interaction [22]. Also, reducing the strong reliance of the agricultural production on chemicals agrochemicals, such as pesticides and fertilizers, constitutes a comprehensive approach, provided that the management programs are environmentally and economically sustainable [23].
- Administrative controls: These measures involve enforcement of public health protections, ranging from controls regarding pesticide distribution to child labor restrictions, for example.
- Working environment controls: This includes safety interventions for placing barriers between the worker and the hazard, such as shock absorbers and multi-axial tractor's seat suspension systems [24].
- Worker's behavior controls: This corresponds to giving, for instance, to the farmer access to appropriate training and adequate information so as to reduce the hazardous exposures.

In a nutshell, an integrated strategy has to be adopted through the development of national programs and policies for occupational safety and health focused particularly on agriculture to promote an effective management and safety culture. A safety culture in agriculture can yield several benefits, such as preventing workers from musculoskeletal injuries and accidents, saving money from medical allowances and new training costs [25], while it may attract young people in agriculture [26]. As far as young workers are concerned, a special handling is needed. In this case, both employers and young farmers must be aware of the increased risks which are related to the lack of experience and physical strength of performing certain tedious manual tasks or operating agricultural machineries. First of all, young workers should be supervised during their agricultural activities. Moreover, their ability for performing a task should be adequately evaluated by following the laws and regulations of the national competent authority. Specific training and instructions must be received giving them the competence to execute such kind of work in as safe manner. Finally, employers must guarantee that workers below the national legal employment age are not employed.

In this chapter, an overview of the potential sources of danger that may negatively affect the farmers' health is given in tandem with the main choices for controlling hazards. Since each source of danger has unique characteristics and, reasonably, needs different control measures, each of them is analyzed in a separate section for the sake of clarity. Emphasis is put on human factors and ergonomics, as this discipline can offer valuable solutions in terms of designing the working environment and protective equipment in addition to developing effective management policies.

2 Identification of Potential Hazards and Risk Assessment

One key element of occupational safety and health management policies is the hazard identification and risk assessment. The term "hazard" refers to the potential source of causing harmful consequences, whereas "risk" is the possibility of someone to be exposed to a hazard. For the purpose of identifying the hazards within a working environment and practice, possible illnesses or injuries with respect to the activity, crop or chemical substances should be identified and evaluated. Individual characteristics like age, experience and past injuries should also be taken into consideration. Particular emphasis should be put on hazards coming from rarely maintained tools and equipment, such as rare pruner maintenance [27, 28]. Also, the identification procedure should carefully consider the design of the working sites, equipment, plant, processes and any alterations that may occur as well as the management regarding the disposal of dangerous materials for workers' health in the farms.

Concerning the risk, it can be roughly seen as the product between the likelihood (of the particular hazard to provoke harm) and the severity (of the consequences), namely "$Risk = Likelihood \times Severity$". By defining the level of risk, which is related to each hazard, the areas of priority action can be recognized. In particular, the higher the risk is, the more imperative are the actions to minimize the worker's exposure to that hazard. For instance, a frequently occurred practice having hazardous consequences is going to determine the highest level of priority action. In contrast, a rarely occurred practice having inconsiderable consequences will correspond to the lowest priority. A simplified table for determining the risk and, subsequently, determining the areas of intervention is that of the International Labour Organisation (ILO), which is depicted in Fig. 1. ILO tries to bring together

Severity \ Likelihood	Rare: 1	Unlikely: 2	Moderate: 3	Likely: 4	Almost certain: 5
Insignificant: 1	1	2	3	4	5
Minor: 2	2	4	6	8	10
Moderate: 3	3	6	9	12	15
Major: 4	4	8	12	16	20
Catastrophic:5	5	10	15	20	25

Fig. 1 A simplified table for determining the degree of risk according to ILO [21]. The red areas correspond to high risk and, consequently, high level of priority action, whereas the orange and green areas correspond to moderate and small risk, requiring moderate and low level of priority action, respectively

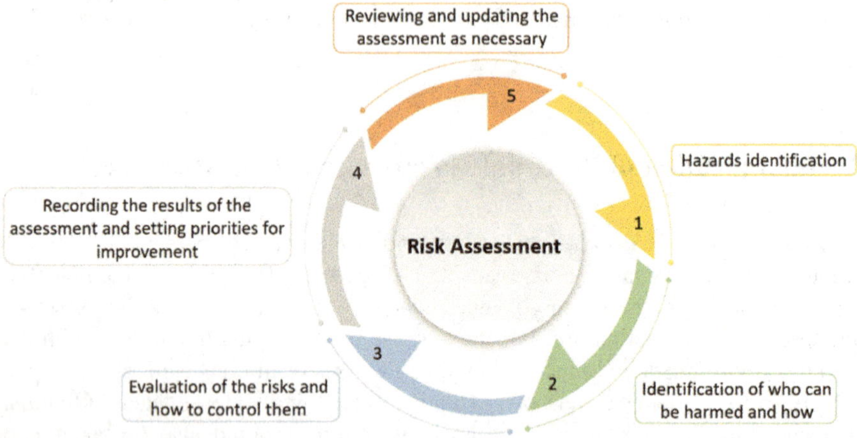

Fig. 2 The five steps needed for carrying out a risk assessment analysis

governments, workers and employers with the intention of developing policies, setting labor standards and programs establishing decent labor for all people around the world [29].

A risk assessment analysis includes a thorough investigation in an attempt to detect potential hazards, in terms of organizational, ergonomic, physical, biological and chemical aspects, and assess the harm that they can cause to the worker. In short, a risk assessment analysis can be executed in five steps, as can be seen in Fig. 2.

Relying on the outcomes of the risk assessment analysis, the employer should then determine the objectives regarding the lessening of the identified risks (so far as are reasonably practicable), apply the appropriate control interventions and approve a safe workplan before any operation at the farm starts. The implemented measures should be reviewed at a regular base and revised if required. Revision is essential, especially after an accident or when new evidence exists in relation to the risks or the implemented control measures.

3 Main Sources of Danger

As said above, one of the most key steps during a risk assessment analysis is, definitely, the identification of the hazards. Overall, this phase addresses two main questions:

- Have the agricultural activity, working environment and equipment the potential to provoke a hazard?
- If so, what kind of adverse health problems can be caused?

In this section, the main sources of danger in agricultural sector are described along with the possible health problems that can provoke and the control strategies to attenuate them.

3.1 Agrochemicals

Agrochemicals, namely the chemical products which are utilized in the agricultural sector for several management practices, comprise a wide range of substances, such as pesticides, commodity chemicals, growth regulators, soil conditioners, fertilizers, on-farm veterinary products and feed additives for the animal husbandry. Policy makers have recognized that the excessive usage of some agrichemicals constitutes a considerable barrier towards establishing a sustainable agriculture, while it can threat the environment and human health [11, 30]. Therefore, their use is appraised on the basis of both increasing production yield and production costs reduction as well as their possible environmental and health side effects [31].

Extensive research has been conducted on the health effects originated from pesticides. Pesticides, which are widely used agrochemicals, are utilized to decrease or prevent losses coming from pests and can contribute to the quality and nutritional value of the products. They are usually classified in relation to the target pest. Target pests can refer, for example, to weeds (herbicides), insects (insecticides), fungi (fungicides), bacteria (bactericides), algae (algicides), mites (miticides), nematodes (nematicides), rodents (rodenticides), snails (molluscicides) and birds (avicides) [32, 33]. In spite of their benefits, serious concerns have been reported regarding the possible health risks resulted from the extensive farmers' pesticide exposure. By their nature, the majority of pesticides are related to toxicity, as they have been fabricated for the purpose of eliminating organisms. In addition, some pesticides have proven to be counterproductive, since they may eradicate species that are natural opponents of pests, while they can lead to pest resistance to the applying pesticides. Moreover, despite the benefits of using fertilizers (natural or chemical materials for increasing the fertility of soils and, thus, improving the growing of plants), the overuse of chemical fertilizers may decrease the content of soil organic matter. This can result in reducing of agricultural soil quality and also in soil acidification, air and water pollution [34], and so on.

Focusing on human health side effects, there are three possible ways that exposure of a farmer to agrochemicals can take place, namely via direct contact with the skin, inhalation and ingestion. Exposure to the farmer's hands can occur, while also facial exposure is very common during spraying and mixing [33]. Moreover, exposure to the torso can take place when farmers carry agrochemicals on their backs, or to the legs with recently sprayed foliage, a fact that is quite usual in fields having very little space between crop rows, like greenhouses. In particular, working in greenhouses can be very harmful due to possible inhalation of volatile compounds of chemicals. Ingestion can also be an important contributor if food is eaten, for instance, after contact with agrochemicals and prior to hands' washing. Certain groups, such as young farmers, pregnant women and overweight persons, are mostly vulnerable to agrochemicals [35]. Within the human body, pesticides, for example, can be metabolized or accumulated within the body fat [36]. The individual health status, the kind of the pesticide and also the route and duration of the exposure are additional determining factors for possible health problems.

A plethora of negative health issues have been linked with pesticides, such as respiratory, dermatological, neurological, reproductive, endocrine and gastrointestinal diseases [37]. Organophosphorus and carbamate insecticides, as indicative examples, seem to be responsible for several acute pesticide poisonings. The reported effects range from symptoms like headaches and nausea to respiratory distress. Epidemiological studies have also revealed an alarming positive relationship between the exposure of farmers to pesticides and some cancers. Notably, prostate, brain and kidney cancers have been reported by Bassil et al. [38] as well as non-Hodgkin lymphoma and leukemia. Other forms of cancers have also been connected with pesticide exposure, like multiple myeloma, cancer of breast and ovary, although the epidemiologic evidence is not conclusive. Insecticides containing arsenic, for example, have been recognized as carcinogens, while many other pesticides have been under suspicion [21].

Other chronic effects on human health are related to neurotoxicity, thyroid and liver disease as well as allergic dermatitis [35]. Such kind of side effects are usually linked with specific pesticides. As a consequence, information must be provided in special labels for each chemical. Nevertheless, additional effort should be put on the detailed assessment of farmers' exposure to individual pesticides, considering also factors like the working practices and protective equipment usage (e.g., gloves and safety masks). Chemical fertilizers constitute also a toxic hazard for farmers and can cause irritation of the skin and possibly severe respiratory effects through inhalation. Some veterinary products, such as veterinary medicines, are also toxic and, similarly to fertilizers, special attention should be taken when handling them for the sake of minimizing skin exposures. Moreover, agrochemicals have the potential to be hazardous not only to farmers that are in close proximity with them, but to people being near the farms as well. Finally, many farmers have insufficient awareness of the associated risks regarding the proper application and the essential precautions [39, 40].

In order to address the above issues, legislation should be enforced according to international standards to ensure safety during the handling of hazardous

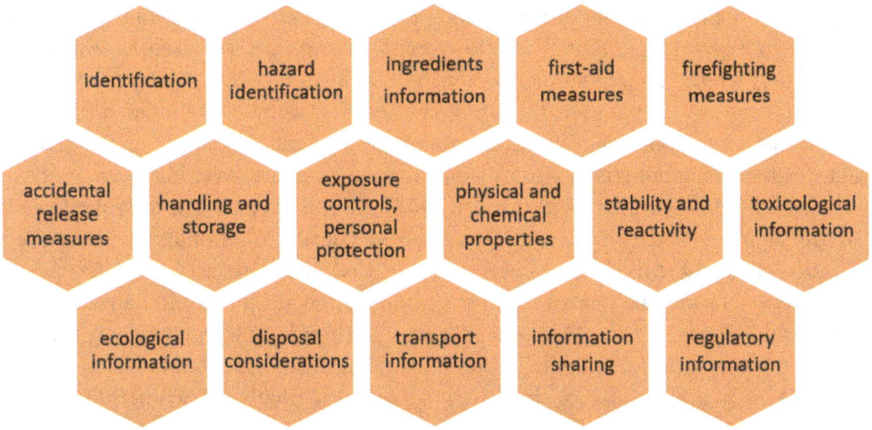

Fig. 3 Required information that should be provided on agrochemical safety data sheets

agrochemicals. To this end, chemical safety data sheets must be provided by health services on agrochemicals, which should be written in simple and understandable language so that even non-experts are able to recognize all the risks coming from the usage of the chemical substances. In addition, the labels have to be easily readable and contain pictograms to help farmers that cannot read the labels. The labels must also be durable so that they remain available to everyone throughout the supply chain and practice. According to the ILO code of practice [35], these data sheets must provide information under the headings illustrated in Fig. 3.

For agricultural activities involving the usage of agrochemicals, the hazards identification along with the risk assessment has to be done at the earliest phase. In particular, the risks should be assessed regarding the full agrochemicals' life-cycle, ranging from their transport, storing, mixing and using them to the cleaning of tools and the disposal of empty containers. Finally, engineering control measures to protect the workers include completely enclosed systems or partial enclosure allowing for locally available exhaust ventilation and isolation of operator from the hazardous practices. As far as the administrative control measures are concerned, they include decreasing of the exposure period and number of the exposed workers as long as possible, proper maintenance of the equipment and cleaning, especially in case of an accidental contamination, and efficient management throughout the entire agrochemicals' life-cycle [35, 41].

3.2 Adverse Weather Conditions

Studies concerning the effect of weather and climate conditions commonly focus on how crop yield production is affected, thus, overlooking the potential negative consequences on health of agricultural workers. Actually, agriculture may expose workers to high levels of humidity, temperature, wind and solar radiation. The increase in global temperatures, due to climate change, is anticipated to contribute to the "heat stress" phenomenon. In particular, heat stress refers to the excessively received heat as compared with normal values that human can tolerate without impairment. In short, ambient air temperature beyond 24–26 °C is related to decreased labor productivity. At 33–34 °C, the workers working with modest intensity lose 50% of their work capacity [42]. Excessive heat can provoke health risks resulting in headaches, heatstroke and even to death. Additionally, farmers over 50 years old are more likely to suffer from cardiovascular diseases [42, 43]. Extended exposure to sun may be responsible for skin malignancies. It should be mentioned that combination of high ambient temperatures and humidity with intensive physical labor can result in life-threatening conditions. Hence, farmers should use specially designed body and head coverings. Also, keeping agricultural workers hydrated (with proper electrolytes addition) and guaranteeing a minimum of 10 min rest break per hour have been reported to decrease the risk, preferably in a cool and shade area [21]. Besides, alcohol consumption should be prohibited, as it dehydrates the body rendering it more vulnerable to the heat stress. Finally, as the majority of agricultural activities routinely take place under full sunlight, farmers are exposed to ultraviolet radiation, which can provoke damage to eyes and the skin. Both intense and cumulative exposures have been linked with skin wrinkling and lesions and skin cancer. Control measures include similar actions as mentioned above in conjunction with sunglasses and sunscreen lotions that screen out ultraviolet radiation.

On the other hand, also exposure to a cold environment, especially when it is combined with moisture, cold winds and rain, may lead to hypothermia [44]. In mild hypothermia, there exists some form of shivering. In moderate hypothermia, confusion increases, while in severe hypothermia even heart stopping may occur [45]. As in hot environment exposure, employers must ensure proper protecting clothing, gloves and boots as well as rest periods in specially designed heated areas.

3.3 Noise

In general farmlands are conjectured to be quiet places. However, if someone works in farms, they would assert that this is not always the case. Surveys, such as [46, 47], have investigated farm's noise levels during ordinary agricultural operations. They have demonstrated that noise originated from tractors without cabs, old cabbed tractors, shearing sheds, silage blowers, small motors (e.g., pumps, chainsaws and augers), cotton module presses, grain dryers, skid-steer loaders, forage harvesters and other heavy machineries constitute sources of harmful noise for farmers. Remarkably, also the pigs' squeal is considered to be one of the typical sources, as this sound can become loud enough for farmers making them wear earmuffs or remove themselves from pigs when they are feeding. Prolonged and everyday exposure to high levels of noise can induce temporary or permanent hearing loss [48].

In general, sound has two important properties, namely intensity and frequency which are measured in decibels (dB) and Hertz (Hz), respectively. Intensity corresponds to the loudness of the sound, while frequency stands for the number of waves passing a fixed point per second. The perceived noise that a farmer experiences relies on both magnitudes in tandem with how close the farmer is to the source as well as the individual's ear health. Hearing loss can be stemmed from either cumulative experience of harmful noise or single intense exposure. Standards regarding health and safety have established the safe noise levels within working sites, although some farmers may be vulnerable to noise causing hearing loss at values below the standards. High levels of noise may also interfere with communication and prevent farmers from listening to the instructions and warnings. Moreover, they can also cause irritability and fatigue to farmers, leading to reduced performance.

As in any risk assessment, employers should be able to identify the operations, equipment and agricultural machinery that produce considerable levels of noise, evaluate the risk of hearing impairment and that of mental workload. Towards that direction, measurements should be conducted to detect the sources of noise, the intensity and duration of the exposure and compare the values with the established exposure limits. Based on the above measurements, a noise map has to be created for rendering the employers and workers aware of the risky areas and agricultural activities. Finally, assessment of the necessity of controls measures should be made along with evaluation of their effective implementation.

The main strategies include the reduction or the isolation of the noise, personal protective equipment and administrative controls [49]. Noise reduction can also be succeeded by properly maintaining the agricultural equipment, as worn machinery parts have the potential to increase the levels of decibels throughout operation. Towards reducing the noise, regular lubrication of the identified parts can diminish friction and, consequently, reduce the noise levels. Also, since larger machineries tend to produce more noise, operation of them under lower speeds is indicated. As regards noise isolation, it may be accomplished by using sound reducing cabs at skid-steers and tractors, while acoustical materials can be utilised on walls to enclose sound. As far as personal protective equipment is concerned, it can be in the form of ear plugs or muffs and result in effective noise reduction. It should be ensured by employers that the personal protective equipment is properly maintained and replaced if necessary. In addition, in working sites that have been recognised to be dangerous in terms of noise exposure, zones of hearing protection should be combined with appropriate signs. Finally, other administrative controls include monitoring the exposure of individuals to noise sources, by also adopting practices like job rotation. For instance, one worker can operate a noisy machinery for a specific time interval and afterwards work in a quieter working site, while another worker can, in turn, operate that machinery [49].

3.4 Vibration

According to the recent review study of Benos et al. [15] on disorders originated from agricultural mechanized operations, the machineries with driving seats (e.g., tractor, all-terrain vehicles) are mainly related to low-back pain, whereas handheld machines (e.g., grass trimmer, handheld olive beater, power tiller), to upper extremities disorders. The main origins of the above disorders are the whole-body vibration (WBV) and the hand-arm transmitted vibration (HATV). The former is transmitted by standing or sitting on vibrating parts of the machine, such as the seat of the tractor, whereas the latter is transmitted when using hand-held vibrating equipment, like hedge trimmers, handles of power tillers and chainsaws. Similar to all objects in nature, the parts of the body have a specific resonant frequency. Once this characteristic vibration frequency is reached, the body part oscillates with a considerable large amplitude owing to the phenomenon of resonance. Considering that several body parts exist, each of them is stimulated at different resonant frequency [50]. Furthermore, the values of the resonant frequencies differentiate among individuals. As the magnitude of vibrations increases, the muscles have a tendency to extraordinarily be tensed in order to dampen the vibration. Overall, the effects of vibration exposure are based on several factors including its direction, its magnitude, its duration and the body posture.

The vibrations can be produced as a result of the ground roughness and engine operation. Then, they can be transferred towards the operator through the contact between the body with the seat or the cabin floor, for instance [51]. The effects of WBV tend to be worse when driving on rough ground [52]. Moreover, WBV related to on-road vehicles is mostly along the vertical axis. In contrast, for off-road vehicles WBV is generated and transmitted along all axes [24]. Short-term vibration exposure in not considered to be very detrimental and can lead to headaches, loss of balance and heart rate increase. On the contrary, long-term exposure to WBV may give rise to degenerative joints' changes, particularly of lumbar spine [53]. Given that the intervertebral disks serve as dampers for energy dissipation from WBV, the spine (especially the region of low back) seems to be affected mostly. The two reported mechanisms for energy dissipation are stiffening (following shocks) and softening (following the further increase of the magnitude of vibration) [15]. Surveys have indicated that the low-back's resonant frequency is at about 4.5 Hz. Around this value, the fatigue of the muscles at low back occurs. This fatigue changes the muscles' response to sudden loads, rendering them more susceptible to injury [54]. In terms of microstructure of spine, extended movement of the intervertebral discs, as a result of vibration exposure, may stress the annular fibers. Afterwards, pressure increases within the disks that may lead the material to fail. This fail of the disk's material can result in protruding of the disk from the vertebral system. Low-back discomfort can be developed because of this movement, as it can press the spinal nerves [55]. The vibration exposure for a long period of time has the potential to result in the damage of the soft tissues at the low back, as mentioned above, and neck. These impairments of the soft tissues can constitute a "precursor of musculoskeletal injuries" [56].

Another usually observed evidence in agriculture is the HATV due to working with vibrating tools, which is responsible for the well-known HAVS. Initially, this syndrome was called "vibration white finger", but the name was altered to HAVS, since also other symptoms were recognised as a consequence of HATV. These symptoms refer to neurological, vascular and other disorders. The diagnosis of vibration-induced white finger currently relies mainly on the identifying of symptoms, like finger blanching due to cold [57]. Characteristic symptoms of HAVS are the vasospasm of the fingers, tingling and paresthesia, loss of sensitivity, reduced hand function as well as inability to carry out complex everyday activities. Furthermore, diseases, including hand and elbows cysts, wrist osteoarthritis and carpal tunnel syndrome, have been connected with HATV [58]. As reported also in the case of WBV, duration and magnitude of vibration exposure are the main risk factors. Provided that the vibration exposure is not continuous, a partial recovery may take place [58].

For the purpose of identifying the sources of vibration along with adequately quantifying the levels of vibration and its duration, direct measurement methods are needed. To this end, tri-axial accelerometers are used which are small-sized portable measurement instruments that can measure simultaneously the acceleration in all directions so as to quantify the levels of vibration experienced by the operator. Inertial Measurements Units (IMUs) are also applied, which are also small-sized devices and can integrate data measured through accelerometers, magnetometers and gyroscopes. Therefore, IMUs take precedence over individual sensors, as they can exploit the advantages of one device to compensate the limitations of the others [59]. Accelerometers and IMUs have been implemented by several researchers for quantifying the levels of vibration is recent studies such as those concerning tractors [60–63], quad bikes [64–67], grass trimers [68] and handheld olive beaters [10, 69].

Subsequently, the values of vibrations that are measured are compared with nationally or internationally established exposure limits. If the measured values exceed the limits, appropriate interventions should be implemented so as to decrease the exposure to the minimum standards. However, if the levels of exposure remain still high enough, employers should be informed about how to reduce such exposures. A first approach is to minimize the time of operating the harmful equipment. Also, personal protective equipment like anti-vibration gloves is proposed, which, in turn, should be checked for their suitability on a regular basis. As regards engineering controls, they involve vibration damping as far as it is practicable, for example by using suspended tractor cabs instead of unsuspended ones, anti-vibration mounts, shock absorbers and multi-axial suspension systems for the seats of agricultural vehicles [24, 66, 70, 71]. Overall, by reducing the levels of vibration on the seat is of central importance. The design of the seats should assure the lowermost vibration transmission. Concerning the HATV originated from vibrating handles, rubber handles have been proposed for the vibration reduction [72]. Overall, when it comes to buying new agricultural equipment or machinery, employers should have in mind as a key criterion the low vibration emission, so that they conform to the relevant standards. As in the case of noise protection, employers should assure the regular maintenance of the tools and machines, as worn parts can intensify the levels of vibration. Consequently, particularly worn components must be replaced with new ones. Finally, as described in the section referring to risk assessment, evaluation of the effectiveness of the applied prevention measures is required.

3.5 Physically Demanding Agricultural Activities Carried Out Incorrectly

Agricultural tasks include a plethora of activities that sometimes are performed manually, because of either the inability of the farmer to purchase the corresponding machinery that performs this work or the specificity of the crop and the process that require gentle handling. In fact, manual operations, such as pruning, digging, sorting, harvesting, weeding, lifting and carrying, include a number of risk factors which can provoke MSDs. For instance, during pruning, the non-dominant hand commonly holds the branches with awkward wrist flexion, while the dominant hand makes repetitive gripping and experiences large stresses from the shears, especially the non-ergonomically ones. Pruning has been associated with pains and injuries at the regions of wrists, elbow and shoulder [73]. Additionally, weeding is an ordinary hazardous agricultural activity, as it involves prolonged or repetitive trunk bending (usually called stooping), which has been linked with discomfort at the back, particularly at the low-back region. As a matter of fact, low-back pain have reached alarming epidemic proportions among agricultural workers [74–76], as there is a strong evidence for low-back pain originated also from prolonged WBV exposures, as analyzed above. Also, lifting and carrying of heavy loads, which are common tasks in agriculture, are responsible for complaints with reference to pains of low-back, shoulder and wrist by farmers. During lifting, the weight and shape of the object, the distance, the way of lifting (e.g., stooping, squat, semi-squat) and the potential trunk twisting are predisposing factors for causing MSDs. Moreover, carrying baskets on uneven farms appears to accelerate the onset of MSDs [77]. Another prevalent MSD among farmers is osteoarthritis of hip and knee [78]. Osteoarthritis is the most ordinary kind of arthritis and its painful consequences increase with age. Repetitive kneeling and stooping, which are very usual during manual agricultural operations, accelerate osteoarthritis [79]. As an indicative example, the harvesting of low-growing crops, like vegetables, force farmers to kneel and bend in a repetitive manner, hence, contributing to the aggravation of osteoarthritis and low-back pain. The above tasks can be hazardous even when they are not combined with heavy loads, since sustained awkward postures or activities executed in an iterative manner can increase the risk of injury.

In conclusion, heavy carrying and lifting, awkward body postures of the trunk, wrists, shoulders and neck and repetitive actions have been identified in the relative literature such as [14, 80]. Concerning the awkward body postures, farmers use them because either they have not been learned how to correctly carry out the task or they are forced to use the wrong posture as a result of the improper design of the working environment [26]. In fact, fatigue, namely the way the part of the body indicates that the posture should be changed, can be caused both by static postures and repetitive postures. Also, other risk factors have been associated with MSDs including age, gender, anthropometric characteristics, obesity, working hours and rarely maintained tools [14]. In general, older farmers are more vulnerable to musculoskeletal injuries, because of some degenerative changes of the soft tissues or due to the decline of some physical and cognitive skills with aging. In contrast, the experience gained over years can contribute to handle a manual operation by using a more ergonomic body posture, as for example kneeling during lifting an object instead of purely stooping. Another remarkable fact in agricultural sector is the engagement of young workers with labor-intensive manual operations, as they seem to be prone to MSDs on account of the developing musculature. Also, the gender factor appears in several epidemiological studies [81]. In effect, women have a tendency to osteoarthritis, while they usually tend to excessively activate their muscles in order to compensate their smaller strength or performing tasks using equipment that has been designed mainly for men anthropometric characteristics [82]. Anthropometric characteristics constitutes a usually reported risk factor, as according to many studies, such as [83, 84], there is an inconsistency among different countries, which can result in risky over-extension.

As a means to cope with the above issues, guidelines have been developed by the competent authorities for assuring the safety standards implementation pertaining to the design and manual handling of the agricultural equipment and machines and the transport of the products. Such safety standards are based on recognized international practices and scientific criteria. As it has been stressed by investigations, such as [85], even "*small changes make big differences*". For example, by implementing simple interventions, such as using ladders with smaller space between the rungs, smaller harvest tubs and frequent rest breaks, the reported fatigue of the farmers was observed to be reduced, without affecting the agricultural productivity. Ergonomic interventions are also extended to the use of ergonomically designed bucket carriers, grip handles [86] and bags worn around the waist for coffee harvesting [87] as well as load transfer devices [88], which are personal devices aiming at reducing the load at the lumbar spine.

Interestingly, recent advances in collaborative robots have allowed to overcome some issues appearing when implementing purely human or mechanized work [22, 89]. In effect, the cognitive characteristics of human associated with perception, reasoning and decision making can be integrated with the repeatable accuracy and strength of robots with the aim of coping with the unpredictable and dynamic agricultural working environments. As in any working ecosystem, we need to be conscious of the safety issues that human-robot interaction might cause both in terms of preventing accidents originated from mechanical risks (e.g., undesired contact with the robot's mechanical parts and obstacles existing in the working site as well as mistakes due to wrong operation) and minimizing musculoskeletal injuries occurrence [26].

4 Human Factors and Ergonomics

Throughout the chapter, phrases, such as “ergonomics”, “ergonomic principles”, “ergonomic risks”, “ergonomic body postures” and “ergonomic interventions”, were mentioned without paying attention on what the word “ergonomics” really involves. Ergonomics is a scientific discipline that draws on many research fields to improve the interaction between the worker and the work environment. Another commonly used phrase for describing this scientific field is “human factors”. In fact, human factors and ergonomics are synonymous with the former being adopted in order to highlight, apart from biomechanics and physiology, also the cognitive attribute or social behavior of a person that can influence their work. According to the Health and Safety Executive (HSE), “*Human factors refer to environmental, organizational and job factors, and human and individual characteristics, which influence behavior at work in a way which can affect health and safety*” [90]. The definition given by HSE consists of three interconnected parts that should be taken into account, namely: (a) the job, (b) the individual and (c) the organization, which are graphically illustrated in Fig. 4. The first aspect considers the working ecosystem, the required workload, the nature of the agricultural task and processes and the design engineering controls. The design of the tasks that are going to be executed should rely on ergonomic principles for the purpose of considering the human strengths and limitations; both the physical and mental ones. In particular, mental features contain decision making, attentional and perceptual skills of each individual worker. Also, other individual skills should be taken into account including attitude, personality,

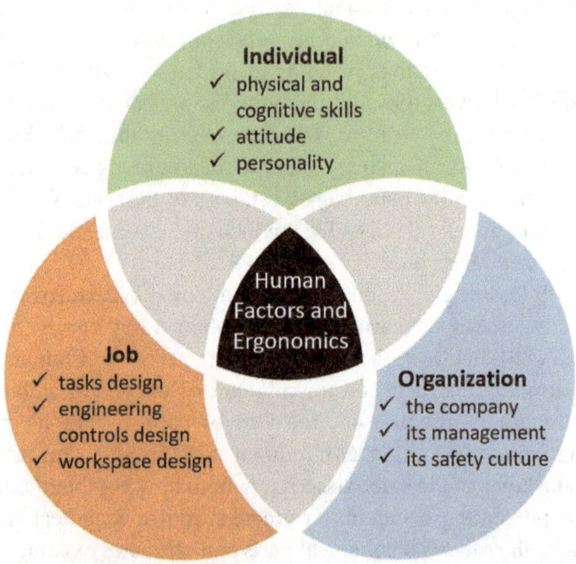

Fig. 4 The three key elements of human factors and ergonomics

reasoning and risk perception, since these characteristics can affect the working procedure. Finally, some significant factors that are frequently ignored throughout the design of the tasks are those related to the organization and may involve leadership, resources, etc.

The above aspects are closely correlated with the physical, cognitive and organizational ergonomics. More specifically, physical ergonomics pays attention on biomechanics along with human physiology, kinesiology and anthropometry. It identifies wrong postures and movements that can cause MSDs and focuses on providing effective solutions. Physical ergonomics usually utilizes self-reports, such as personal interviews and specific questionnaires like the Nordic one [91]. Moreover, observation-based methods are implemented with the use of cameras with the intention of recording farmers' activities and, afterwards, analyzing them. Finally, direct measurement methods are applied. To this end, experimental measurements can exploit various kinds of sensors and equipment such as accelerometers, gyroscopes, inertial measurement units, electromyography, handgrip dynamometers and optical markers [27, 64, 92]. Cognitive ergonomics should also be carefully taken into account. It considers the mental interaction among workers and also between workers and any element of the working system. This system may include agricultural robots and, thus, cognitive ergonomics is of major importance for the realization of a safe human-robot interaction [26, 93]. Finally, organizational ergonomics focuses on optimizing the socio-technical systems through implementing proper risk management and policies. It can include telework, resource management, cooperative work, participatory design, virtual organizations and design of working hours and communication [94].

The competent authorities should provide useful technical information and ergonomic reference manuals in simple language. Regarding employers, they should organize the agricultural operations on the basis of the available risk assessments and national guidelines and implement control measures, accordingly, so as to lessen the potential ergonomic risks for farmers. Based on the evidence presented in the above section, engineered processes should be implemented that minimize, wherever possible, non-neutral working body postures, lifting and carrying of heavy loads (at least larger than 23 kg), highly repetitive work usually accompanied by concurrent wrong posture of the back, knees or wrist. Furthermore, ergonomically designed tools have to be selected like pneumatically-powered shears and tools with shorter or longer shafts in order to be optimally adapted to the worker's anthropometric data. Also, regular maintenance of the agricultural tools and equipment should be assured as well as documentation of all tasks and processes that should be performed. Towards controlling the ergonomic hazards, employers should implement regular rest breaks for workers (to lessen the negative results of the viscoelastic creep) [95], while also a precursor activity, such as walking (to prepare the muscles for the imminent task), can prevent the onset of musculoskeletal injuries [96]. Additionally, routine task rotation among farmers should take place along with specific training to provide information on how to prevent themselves from MSDs.

In turn, agricultural workers, should be informed about the safe working techniques linked with each task, the standard limits, such as the maximum load that should be lifted, the right use of hand tools and machinery, as well as be aware of the risks related to awkward postures and repetitive actions. In addition, they should know how to properly use the personal protective equipment (e.g., load transfer devices, gloves, waist belts, wrist wraps) and identify when it needs to be replaced, having in mind that this measure is only a supplementary choice, as it is going to be elaborated next. Finally, they must be encouraged to report any pain, numbness, discomfort or injury to their employer without fearing for potential discrimination.

5 Discussion and Conclusions

In summary, as can be seen in Fig. 5, the most reported health hazards are originated from the use of agrochemicals, adverse weather conditions, occupational noise, vibrations and physically demanding repetitive agricultural tasks which are commonly performed improperly by using awkward postures. In addition, the potential health effects for each hazard are summarized. As can be gleaned from Fig. 5, considering that an agricultural worker usually takes part in several tasks, it turns out that they can be exposed to multiple hazards that cumulatively can cause a host of health disorders. For example, low-back pain can come from both the repetitive trunk bending, which is required during the execution of a lot of agricultural activities, and WBV, mainly owing to the use of machinery with driving seats. Furthermore, when the arduous tasks are carried out under adverse weather conditions and noisy environments, the possible health effects are intensified.

To make matters worse, agriculture is responsible for accidents originated from several sources. The number of severe accidents is still alarming in spite of the improvement on prevention measures and training programs. These accidents pertain to collision with parts of the machines or caught by rotating parts, falls form agricultural machineries, like quad bikes [67], and rollover of the machines. These

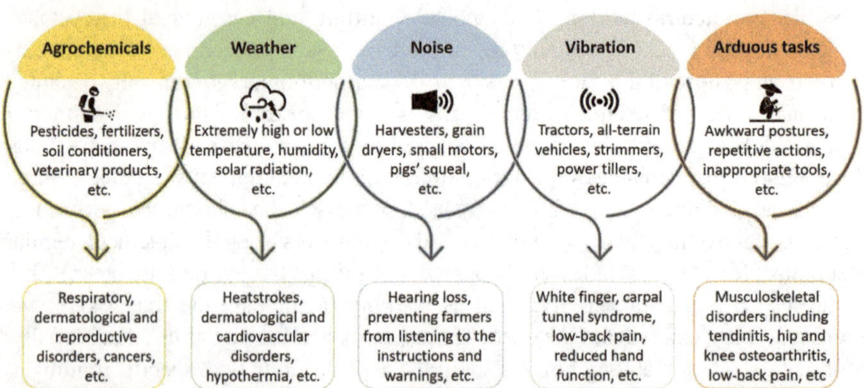

Fig. 5 Main sources of danger during agricultural activities along with the major health problems that may cause

accidents are mainly associated with human factors (e.g., improper equipment handling and misjudgment that can result in falling, slipping and tripping), machinery factors (e.g., incorrectly designed parts and possible technical defects) and environmental factors (e.g., steep and slippery ground conditions) [97]. Finally, agriculture is regarded to be among the most stressful areas [15]. Both farmers and also their families experience economic pressures regarding their livelihood, due the precarious agricultural market and weather conditions. Psychological distress has also a great potential to lead to misjudgement of the severity of the situation at hand and, consequently, to accidents occurrence during the operation of machineries. Also non-compliance with control measures can be observed concerning all the hazards mentioned in Fig. 5.

In this chapter, also multiple control measures were analyzing for the purpose of mitigating the negative consequences from the chemicals (e.g., chemical safety data sheets, ventilation, separation of the worker from the hazardous environment), adverse weather conditions (e.g., specially designed coverings, ultraviolet radiation filters), noise (e.g., regular lubrication of machinery parts, ear plugs, muffs), vibration (multi-axial suspension systems, anti-vibration gloves) and tiring agricultural activities (e.g., informing and training farmers about the wrong and right postures, the practical limits for each task, the risk factors associated with the commonly reported MSDs).

In this fashion, it should be stressed that control measures should follow a hierarchy with the intention of being both feasible and effective. By following the hierarchy, which is depicted in Fig. 6, the establishment of safe ecosystems can be assured, under normal circumstances, where the risk of getting ill or injured has been substantially reduced. This hierarchical structure can be considered as a step-by-step method to reduce risks, while it ranks the control measures from the highest protection and reliable level (elimination) to the lowest level (personal protective equipment). According to the National Institute for Occupational Safety and Health (NIOSH) [98], this hierarchy, in order of implementation, consists of measures with reference to:

1. Elimination
2. Substitution
3. Engineering controls
4. Administrative controls
5. Personal protective equipment

By far, the most effective control approach is to eliminate the hazard and the causing risk or, secondarily, substitute it with a less hazardous risk. Although these approaches are effective enough, their implementation is too challenging when it comes to an existing process, since major changes in both procedures and equipment are usually essential. In contrast, when a process is at the design phase, the above two control measures might be relatively inexpensive and simpler to be applied. Engineering controls, despite the fact that they are less effective than elimination and substitution, they are, in turn, superior to administrative controls and personal protective equipment as they are intended to eliminate the identified hazard at the

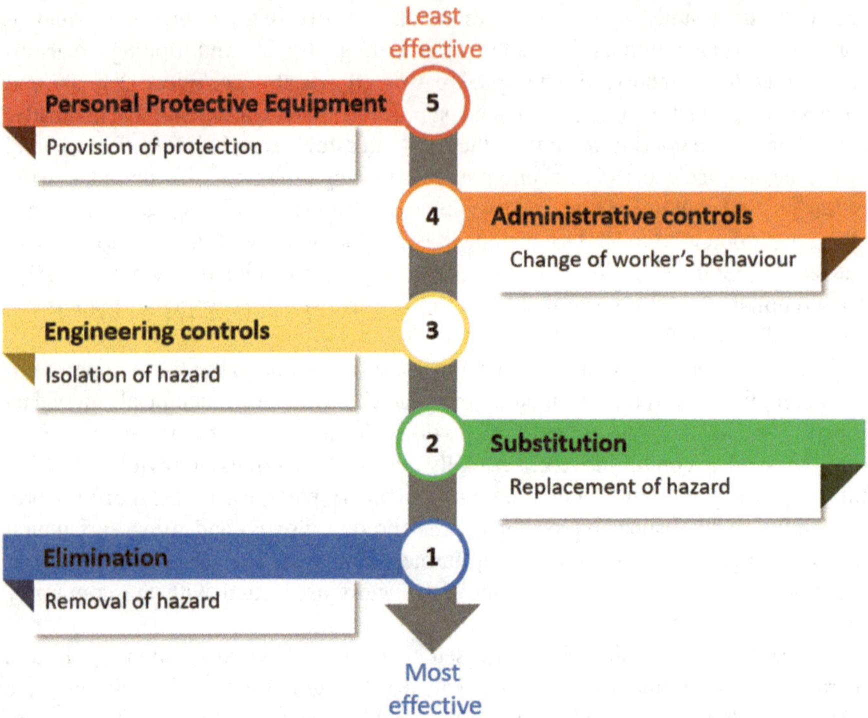

Fig. 6 The hierarchy of control measures along with their effectiveness for reducing the risk of injury

source of origin before contacting the agricultural worker. Furthermore, although engineering controls are more expensive at the initial stage than them, over a longer perspective, operating costs can be lower, while cost savings can be provided also in other stages of the procedure. Finally, administrative controls and personal protective equipment are often implemented in existing processes, when the hazards cannot be well-controlled. These approaches might not be expensive at a first stage, but they can be costly enough to be sustained over the longer term. These control measures have proven to be the least effective [98]. In particular, personal protective equipment can only be seen as a last resort as soon as no other control measures are available.

To conclude, managing risks in a sustainable way has plenty of benefits including:

1. A reduction of injuries and illnesses resulting in lower sickness payments as well as training costs for the new workers who come to replace the non-healthy workers
2. Less output loss because of the experienced workers being off work
3. The ability to perform weather-critical tasks at the right time
4. Reduced legal costs and insurance premiums

5. Healthier and, thus, happier workforce with job satisfaction leading to improved productivity and attraction of young workers in agricultural sector
6. Reduced risk of damaging the business reputation

A preventive occupational safe and health culture should be promoted in agriculture, a sector that has been long recognized as one of the most hazardous ones. To this end, outreach on occupational safety and health is a key element of any national or international effort. Outreach concerns a large variety of stakeholders ranging from inspectors, employers, workers, manufacturers, trade associations, community and training organizations to advice and information organizations (such as FAO and WHO). The competent authorities should cooperate with the organizations of employers and workers in order to raise awareness of safety and health risks in agricultural enterprises and promote a safety culture at local and national levels, especially in small and medium-sized enterprises. This can be accomplished through a variety of activities, such as promotional and training activities, seminars, conferences, meetings, media coverage and sharing information via websites and social media. Besides, the advent of safety and health to the agricultural population needs to be incorporated into the primary health care structure. In a nutshell, for the purpose of accomplishing a sustainable agricultural development, a balance should be guaranteed between not only the agricultural growth with the protection of environment, but also with the protection of safety and health of workers through integrated it into well-defined rural development policies.

References

1. Saiz-Rubio, V.; Rovira-Más, F. From Smart Farming towards Agriculture 5.0: A Review on Crop Data Management. *Agronomy* **2020**, *10*, 207.
2. Anderson, R.; Bayer, P.E.; Edwards, D. Climate change and the need for agricultural adaptation. *Curr. Opin. Plant Biol.* 2020, *56*, 197–202.
3. Achillas, C.; Bochtis, D.D.; Aidonis, D.; Folinas, D. *Green Supply Chain Management*; Routledge, 2018;
4. Dawson, T.P.; Jackson, S.T.; House, J.I.; Prentice, I.C.; Mace, G.M. Beyond predictions: Biodiversity conservation in a changing climate. *Science (80-.).* 2011, *332*, 53–58.
5. Agovino, M.; Casaccia, M.; Ciommi, M.; Ferrara, M.; Marchesano, K. Agriculture, climate change and sustainability: The case of EU-28. *Ecol. Indic.* **2019**, *105*, 525–543.
6. Lampridi, M.G.; Sørensen, C.G.; Bochtis, D. Agricultural sustainability: A review of concepts and methods. *Sustain.* 2019, *11*.
7. Keating, B.A.; Carberry, P.S.; Bindraban, P.S.; Asseng, S.; Meinke, H.; Dixon, J. Eco-efficient agriculture: Concepts, Challenges, And opportunities. *Crop Sci.* **2010**, *50*, S-109–S-119.
8. Angelopoulou, T.; Tziolas, N.; Balafoutis, A.; Zalidis, G.; Bochtis, D. Remote sensing techniques for soil organic carbon estimation: A review. *Remote Sens.* **2019**, *11*.
9. *Agricultural Medicine*; Lessenger, J.E., Ed.; Springer-Verlag: New York, 2006; ISBN 0-387-25425-0.
10. Calvo, A.; Romano, E.; Preti, C.; Schillaci, G.; Deboli, R. Upper limb disorders and hand-arm vibration risks with hand-held olive beaters. *Int. J. Ind. Ergon.* **2018**, *65*, 36–45.

11. Nicolopoulou-Stamati, P.; Maipas, S.; Kotampasi, C.; Stamatis, P.; Hens, L. Chemical Pesticides and Human Health: The Urgent Need for a New Concept in Agriculture. *Front. Public Heal.* **2016**, *4*, 1.
12. Piperigkou, Z.; Karamanou, K.; Engin, A.B.; Gialeli, C.; Docea, A.O.; Vynios, D.H.; Pavão, M. S.G.; Golokhvast, K.S.; Shtilman, M.I.; Argiris, A.; et al. Emerging aspects of nanotoxicology in health and disease: From agriculture and food sector to cancer therapeutics. *Food Chem. Toxicol.* **2016**, *91*, 42–57.
13. Bochtis, D.; Benos, L.; Lampridi, M.; Marinoudi, V.; Pearson, S.; Sørensen, C.G. Agricultural workforce crisis in light of the COVID-19 pandemic. *Sustain.* **2020**, *12*.
14. Benos, L.; Tsaopoulos, D.; Bochtis, D. A review on ergonomics in agriculture. part I: Manual operations. *Appl. Sci.* 2020, *10*, 1–21.
15. Benos, L.; Tsaopoulos, D.; Bochtis, D. A Review on Ergonomics in Agriculture. Part II: Mechanized Operations. *Appl. Sci.* **2020**, *10*, 3484.
16. Bochtis, D.D.; Sørensen, C.G.C.; Busato, P. Advances in agricultural machinery management: A review. *Biosyst. Eng.* **2014**, *126*, 69–81.
17. Kyriakarakos, G.; Balafoutis, A.T.; Bochtis, D. Proposing a paradigm shift in rural electrification investments in Sub-Saharan Africa through Agriculture. *Sustain.* **2020**, *12*.
18. Bechar, A.; Vigneault, C. Agricultural robots for field operations: Concepts and components. *Biosyst. Eng.* **2016**, *149*, 94–111.
19. Bechar, A.; Vigneault, C. Agricultural robots for field operations. Part 2: Operations and systems. *Biosyst. Eng.* **2017**, *153*, 110–128.
20. Moysiadis, V.; Tsolakis, N.; Katikaridis, D.; Sørensen, C.G.; Pearson, S.; Bochtis, D. Mobile robotics in agricultural operations: A narrative review on planning aspects. *Appl. Sci.* **2020**, *10*.
21. Perry, M.J. Agricultural Health and Safety. In *International Encyclopedia of Public Health*; Elsevier Inc., 2016; pp. 38–44 ISBN 9780128037089.
22. Marinoudi, V.; Sørensen, C.G.; Pearson, S.; Bochtis, D. Robotics and labour in agriculture. A context consideration. *Biosyst. Eng.* **2019**, *184*, 111–121.
23. Kogan, M. Integrated pest management: Historical perspectives and contemporary developments. *Annu. Rev. Entomol.* 1998, *43*, 243–270.
24. Kim, J.H.; Dennerlein, J.T.; Johnson, P.W. The effect of a multi-axis suspension on whole body vibration exposures and physical stress in the neck and low back in agricultural tractor applications. *Appl. Ergon.* **2018**, *68*, 80–89.
25. Bevan, S. Economic impact of musculoskeletal disorders (MSDs) on work in Europe. *Best Pract. Res. Clin. Rheumatol.* 2015, *29*, 356–373.
26. Benos, L.; Bechar, A.; Bochtis, D. Safety and ergonomics in human-robot interactive agricultural operations. *Biosyst. Eng.* **2020**, *200*, 55–72.
27. Berrio, S.; Barrero, L.H. Effect of Time Elapsed since Last Pruner Maintenance on Upper-Extremity Biomechanics during Manual Flower Cutting. *J. Agromedicine* **2018**, *23*, 166–175.
28. Bochtis, D.; Sørensen, C.A.G.; Kateris, D. *Operations management in agriculture*; Elsevier, 2018; ISBN 9780128097861.
29. About the ILO.
30. Prüss-Ustün, A.; Vickers, C.; Haefliger, P.; Bertollini, R. Knowns and unknowns on burden of disease due to chemicals: A systematic review. *Environ. Heal. A Glob. Access Sci. Source* 2011, *10*, 9.
31. Udeigwe, T.K.; Teboh, J.M.; Eze, P.N.; Hashem Stietiya, M.; Kumar, V.; Hendrix, J.; Mascagni, H.J.; Ying, T.; Kandakji, T.; Kandakji, T.; et al. Implications of Leading Crop Production Practices on Environmental Quality and Human Health. *J. Environ. Manage.* **2015**, *151*, 267–279.
32. Aktar, W.; Sengupta, D.; Chowdhury, A. Impact of pesticides use in agriculture: Their benefits and hazards. *Interdiscip. Toxicol.* **2009**, *2*, 1–12.
33. Damalas, C.A.; Eleftherohorinos, I.G. Pesticide Exposure, Safety Issues, and Risk Assessment Indicators. *Int. J. Environ. Res. Public Health* **2011**, *8*, 1402–1419.

34. Angelopoulou, T.; Balafoutis, A.; Zalidis, G.; Bochtis, D. From Laboratory to Proximal Sensing Spectroscopy for Soil Organic Carbon Estimation—A Review. *Sustainability* **2020**, *12*, 443.
35. ILO Safety and health in the use of agrochemicals: A guide.
36. Pirsaheb, M.; Limoee, M.; Namdari, F.; Khamutian, R. Organochlorine pesticides residue in breast milk: a systematic review. *Med. J. Islam. Repub. Iran* **2015**, *29*, 228.
37. Mostafalou, S.; Abdollahi, M. Pesticides and human chronic diseases: Evidences, mechanisms, and perspectives. *Toxicol. Appl. Pharmacol.* 2013, *268*, 157–177.
38. Bassil, K.L.; Vakil, C.; Sanborn, M.; Cole, D.C.; Kaur, J.S.; Kerr, K.J. Cancer health effects of pesticides: systematic review. *Can. Fam. Physician* **2007**, *53*, 1704–1711.
39. Yassin, M.M.; Abu Mourad, T.A.; Safi, J.M. Knowledge, attitude, practice, and toxicity symptoms associated with pesticide use among farm workers in the Gaza Strip. *Occup. Environ. Med.* **2002**, *59*, 387–393.
40. Damalas, C.A.; Theodorou, M.G.; Georgiou, E.B. Attitudes towards pesticide labelling among Greek tobacco farmers. *Int. J. Pest Manag.* **2006**, *52*, 269–274.
41. Iakovou, E.; Bochtis, D.; Vlachos, D.; Aidonis, D. Sustainable Agrifood Supply Chain Management. In *Supply Chain Management for Sustainable Food Networks*; John Wiley & Sons, Ltd: Chichester, UK, 2016; pp. 1–39 ISBN 9781118937495.
42. Kjellstrom, T.; Maître, N.; Saget, C.; Otto, M.; Karimova, T. Working on a warmer planet: The effect of heat stress on productivity and decent work. **2019**. https://www.ilo.org/global/publications/books/WCMS_711919/lang--en/index.htm
43. Lundgren, K.; Kuklane, K.; Gao, C.; Holm Eacute, R.I. Effects of Heat Stress on Working Populations when Facing Climate Change. *Ind. Health* **2013**, *51*, 3–15.
44. Khounsy, S.; Nampanya, S.; Inthavong, P.; Yang, M.; Khamboungheung, B.; Avery, M.; Bush, R.; Rast, L.; Windsor, P.A. Significant mortality of large ruminants due to hypothermia in northern and central Lao PDR. *Trop. Anim. Health Prod.* **2012**, *44*, 835–842.
45. Brown, D.J.A.; Brugger, H.; Boyd, J.; Paal, P. Accidental Hypothermia. *N. Engl. J. Med.* **2012**, *367*, 1930–1938.
46. McBride, D.I.; Firth, H.M.; Herbison, G.P. Noise Exposure and Hearing Loss in Agriculture: A Survey of Farmers and Farm Workers in the Southland Region of New Zealand. *J. Occup. Environ. Med.* **2003**, *45*.
47. Depczynski, J.; Franklin, R.C.; Challinor, K.; Williams, W.; Fragar, L.J. Farm noise emissions during common agricultural activities. *J. Agric. Saf. Health* **2005**, *11*, 325–334.
48. Von Essen, S.G.; McCurdy, S.A. Health and safety risks in production agriculture. *West. J. Med.* **1998**, *169*, 214–220.
49. Murphy, D.; William, H. Noise Induced Hearing Loss in Agriculture.
50. Seidel, H.; Blüthner, R.; Hinz, B. Application of finite-element models to predict forces acting on the lumbar spine during whole-body vibration. *Clin. Biomech.* **2001**, *16*.
51. Langer, T.H.; Ebbesen, M.K.; Kordestani, A. Experimental analysis of occupational whole-body vibration exposure of agricultural tractor with large square baler. *Int. J. Ind. Ergon.* **2015**, *47*, 79–83.
52. Caffaro, F.; Lundqvist, P.; Micheletti Cremasco, M.; Nilsson, K.; Pinzke, S.; Cavallo, E. Machinery-related perceived risks and safety attitudes in senior Swedish farmers. *J. Agromedicine* **2018**, *23*, 78–91.
53. Servadio, P.; Marsili, A.; Belfiore, N.P. Analysis of driving seat vibrations in high forward speed tractors. *Biosyst. Eng.* **2007**, *97*, 171–180.
54. Pope, M.; Wilder, D.; Magnusson, M. Possible mechanisms of low back pain due to whole-body vibration. *Journal Sound Vib.* **1998**, *215*, 687–697.
55. Jensen, M.C.; Brant-Zawadzki, M.N.; Obuchowski, N.; Modic, M.T.; Malkasian, D.; Ross, J.S. Magnetic resonance imaging of the lumbar spine in people without back pain. *N. Engl. J. Med.* **1994**, *331*, 69–73.
56. Thomsen, G.F.; Johnson, P.W.; Svendsen, S.W.; Kryger, A.I.; Bonde, J.P.E. Muscle fatigue in relation to forearm pain and tenderness among professional computer users. *J. Occup. Med. Toxicol.* **2007**, *2*.

57. Ye, Y.; Griffin, M.J. Assessment of two alternative standardised tests for the vascular component of the hand{\textendash}arm vibration syndrome (HAVS). *Occup. Environ. Med.* **2016**, *73*, 701–708.
58. Pelmear, P.L.; Leong, D. Review of occupational standards and guidelines for hand-arm (Segmental) vibration syndrome (havs). *Appl. Occup. Environ. Hyg.* **2000**, *15*, 291–302.
59. Schall, M.C.; Fethke, N.B.; Chen, H.; Oyama, S.; Douphrate, D.I. Accuracy and repeatability of an inertial measurement unit system for field-based occupational studies. *Ergonomics* **2016**, *59*, 591–602.
60. Vallone, M.; Bono, F.; Quendler, E.; Febo, P.; Catania, P. Risk exposure to vibration and noise in the use of agricultural track-laying tractors. *Ann. Agric. Environ. Med.* **2016**, *23*, 591–597.
61. Taghizadeh-Alisaraei, A. Analysis of annoying shocks transferred from tractor seat using vibration signals and statistical methods. *Comput. Electron. Agric.* **2017**, *141*, 160–170.
62. Deboli, R.; Calvo, A.; Preti, C. Whole-body vibration: Measurement of horizontal and vertical transmissibility of an agricultural tractor seat. *Int. J. Ind. Ergon.* **2017**, *58*, 69–78.
63. Gialamas, T.; Gravalos, I.; Kateris, D.; Xyradakis, P.; Dimitriadis, C. Vibration analysis on driver's seat of agricultural tractors during tillage tests. *Spanish J. Agric. Res.* **2016**, *14*.
64. Kociolek, A.M.; Lang, A.E.; Trask, C.M.; Vasiljev, R.M.; Milosavljevic, S. Exploring head and neck vibration exposure from quad bike use in agriculture. *Int. J. Ind. Ergon.* **2018**, *66*, 63–69.
65. Mani, R.; Milosavljevic, S.; Ribeiro, D.C.; Sullivan, S.J. Effects of agricultural quad bike driving on postural control during static, dynamic and functional tasks - A field study. *Int. J. Ind. Ergon.* **2014**, *50*, 158–169.
66. Milosavljevic, S.; Bergman, F.; Rehn, B.; Carman, A.B. All-terrain vehicle use in agriculture: Exposure to whole body vibration and mechanical shock. *Appl. Ergon.* **2010**, *41*, 530–535.
67. Milosavljevic, S.; McBride, D.I.; Bagheri, N.; Vasiljev, R.M.; Carman, A.B.; Rehn, B.; Moore, D. Factors associated with quad bike loss of control events in agriculture. *Int. J. Ind. Ergon.* **2011**, *41*, 317–321.
68. Hao, K.Y.; Ripin, Z.M. Nodal control of grass trimmer handle vibration. *Int. J. Ind. Ergon.* **2013**, *43*, 18–30.
69. Lenzuni, P.; Deboli, R.; Preti, C.; Calvo, A. A round robin test for the hand-transmitted vibration from an olive harvester. *Int. J. Ind. Ergon.* **2016**, *53*, 86–92.
70. Exposure to Whole-Body Vibration and Mechanical Shock: A Field Study of Quad Bike Use in Agriculture. *Ann. Occup. Hyg.* **2011**.
71. Romano, E.; Pirozzi, M.; Ferri, M.; Calcante, A.; Oberti, R.; Vitale, E.; Rapisarda, V. The use of pressure mapping to assess the comfort of agricultural machinery seats. *Int. J. Ind. Ergon.* **2019**.
72. Chaturvedi, V.; Kumar, A.; Singh, J.K. Power tiller: Vibration magnitudes and intervention development for vibration reduction. *Appl. Ergon.* **2012**, *43*, 891–901.
73. Meyers, J.M.; Miles, J.A.; Faucett, J.; Janowitz, I.; Tejeda, D.G.; Kabashima, J.N. Ergonomics in agriculture: Workplace priority setting in the nursery industry. *Am. Ind. Hyg. Assoc. J.* **1997**, *58*, 121–126.
74. F. A. Fathallah; B. J. Miller; J. A. Miles Low Back Disorders in Agriculture and the Role of Stooped Work: Scope, Potential Interventions, and Research Needs. *J. Agric. Saf. Health* **2008**, *14*, 221–245.
75. Osborne, A.; Blake, C.; Fullen, B.M.; Meredith, D.; Phelan, J.; McNamara, J.; Cunningham, C. Prevalence of musculoskeletal disorders among farmers: A systematic review. *Am. J. Ind. Med.* 2012, *55*, 143–158.
76. Jain, R.; Meena, M.L.; Dangayach, G.S.; Bhardwaj, A.K. Association of risk factors with musculoskeletal disorders in manual-working farmers. *Arch. Environ. Occup. Heal.* **2018**, *73*, 19–28.
77. Stuempfle, K.J.; Drury, D.G.; Wilson, A.L. Effect of load position on physiological and perceptual responses during load carriage with an internal frame backpack. *Ergonomics* **2004**, *47*, 784–789.
78. Kirkhorn, S.; Greenlee, R.T.; Reeser, J.C. The Epidemiology of Agriculture-related Osteoarthritis and its Impact on Occupational Disability. *Wis. Med. J.* **2003**, *102*, 38–44.

79. Maetzel, A.; Mäkelä, M.; Hawker, G.; Bombardier, C. Osteoarthritis of the hip and knee and mechanical occupational exposure--a systematic overview of the evidence. *J. Rheumatol.* **1997**, *24*, 1599–607.
80. Schuman, S.H. Ergonomics in agriculture: Commentary and literature review. *J. Agromedicine* **2002**, *8*, 9–18.
81. Gjesdal, S.; Bratberg, E.; Mæland, J.G. Gender differences in disability after sickness absence with musculoskeletal disorders: five-year prospective study of 37,942 women and 26,307 men. *BMC Musculoskelet. Disord.* **2011**, *12*, 37.
82. Mugisa, D.J.; Katimbo, A.; Sempiira, J.E.; Kisaalita, W.S. Anthropometric characteristics of female smallholder farmers of Uganda - Toward design of labor-saving tools. *Appl. Ergon.* **2016**, *54*, 177–185.
83. Dewangan, K.N.; Owary, C.; Datta, R.K. Anthropometry of male agricultural workers of north-eastern India and its use in design of agricultural tools and equipment. *Int. J. Ind. Ergon.* **2010**, *40*, 560–573.
84. García-Cáceres, R.G.; Felknor, S.; Córdoba, J.E.; Caballero, J.P.; Barrero, L.H. Hand anthropometry of the Colombian floriculture workers of the Bogota plateau. *Int. J. Ind. Ergon.* **2012**, *42*, 183–198.
85. Fathallah, F.; Duraj, V. Small Changes Make Big Differences: The role of Ergonomics in Agriculture. *Resour. Mag.* **2017**, *24*, 12–13.
86. Fathallah, F.A.; Tang, S.C.H.; Waters, T. Development and Evaluation of Ergonomic Interventions for Bucket Handling on Farms. *Hum. Factors* **2016**, *58*, 758–776.
87. Silverstein, B.A.; Bao, S.S.; Russell, S.; Stewart, K. Water and coffee: A systems approach to improving coffee harvesting work in Nicaragua. In Proceedings of the Human Factors; 2012; Vol. 54, pp. 925–939.
88. Ulrey, B.L.; Fathallah, F.A. *Effect of a personal weight transfer device on muscle activities and joint flexions in the stooped posture*; 2013; Vol. 23.
89. Lampridi, M.G.; Kateris, D.; Vasileiadis, G.; Marinoudi, V.; Pearson, S.; Sørensen, C.G.; Balafoutis, A.; Bochtis, D. A Case-Based Economic Assessment of Robotics Employment in Precision Arable Farming. *Agronomy* **2019**, *9*, 175.
90. HSE Human factors/ergonomics – Introduction to human factors.
91. Deakin, J.M.; Stevenson, J.M.; Vail, G.R.; Nelson, J.M. The use of the Nordic questionnaire in an industrial setting: a case study. *Appl. Ergon.* **1994**, *25*, 182–185.
92. Çakmak, B.; Ergül, E. Interactions of personal and occupational risk factors on hand grip strength of winter pruners. *Int. J. Ind. Ergon.* **2018**, *67*, 192–200.
93. Matt, D.T.; Rauch, E. SME 4.0: The role of small-and medium-sized enterprises in the digital transformation. In *Industry 4.0 for SMEs: Challenges, Opportunities and Requirements*; Palgrave Macmillan, 2020; pp. 3–36 ISBN 9783030254254.
94. Glander-Dolo, S.M. Organizational Ergonomics. In *Global Encyclopedia of Public Administration, Public Policy, and Governance*; Springer International Publishing, 2016; pp. 1–10.
95. Shin, G.; Mirka, G.A. An in vivo assessment of the low back response to prolonged flexion: Interplay between active and passive tissues. *Clin. Biomech.* **2007**, *22*, 965–971.
96. Hudson, D.S.; Copeland, J.L.; Hepburn, C.G.; Doan, J.B. Stooped Postures Are Modified by Pretask Walking in a Simulated Weed-Pulling Task. *J. Agromedicine* **2014**, *19*, 27–34.
97. Robert, K.; Elisabeth, Q.; Josef, B. Analysis of occupational accidents with agricultural machinery in the period 2008-2010 in Austria. *Saf. Sci.* **2015**, *72*, 319–328.
98. NIOSH Hierarchy of Controls | NIOSH | CDC.

Part II
Agriculture Digital Transformation Around the World

Smart Farming as a Game-Changer for Regional-Spatial Planning

Stella Agostini

1 Introduction

Rapid population growth, as reported at NASA Socioeconomic Data and Applications Center—SEDAC—on 2018, addressed new challenges for sustainable development, referring to the emerging food policy agenda. In 2019, the world's population was projected to grow from 7.7 billion to 8.5 billion in 2030 (10% increase), and to increase further to 9.7 billion in 2050 (by 26%) and to 10.9 billion in 2100 (by 42%) [1].

The Food and Agriculture Organization (FAO) indicated that 821 million people were undernourished in 2017, involving approximately one out of every nine people in the world. The Global Report on Food Crises 2020 [2], analysing 55 countries and territories, epitomised the increasing number of people facing crisis-level food insecurity, up from almost 80 million in 2015 to 135 million in 2019. Around 183 million people in 47 countries were classified in stressed conditions, at risk of slipping into crisis.

Due to the devastating socio-economic impacts of COVID-19 pandemic, more than 71 million people have been pushed into extreme poverty in 2020 [3, 4]. A condition primarily concerning fragile contexts which include the particularly vulnerable workers involved in the informal agriculture as well one-third of the economically active population depending on agriculture for survival. Given these data, increasing total food availability as well as providing higher quality standards in particular for food safety and for a healthier environment are the major challenges in front of the agri-food sector.

S. Agostini (✉)
Department of Environmental Science and Policy, University of Milan, Milan, Italy
e-mail: stella.agostini@unimi.it

D. D. Bochtis et al. (eds.), *Information and Communication Technologies for Agriculture—Theme IV: Actions*, Springer Optimization and Its Applications 185,
https://doi.org/10.1007/978-3-030-84156-0_6

Considering that agriculture provides the principal basis for livelihood of one-third of the world's population, the International Assessment of Agricultural Knowledge, Science and Technology for Development [5] highlighted that agriculture and food is by far the world's largest business. Globally, agricultural expansion has been stabilized over the past 20 years at around 4.9 billion hectares (ha) [6, 7].

Intensive agriculture processes have led to the degradation of natural resources and contributed to climate change causing loss of forests, depletion of groundwater and sources, and erosion of biodiversity [8]. To address the problem, Agenda 21 [9], required that farmers made a sustainable use of land resources. These include soil, water, animals and plants for the production of goods to meet the changing human needs while ensuring the long-term productive potential of these resources and the maintenance of their environmental functions.

The key challenges that sustainable agriculture should address, have been identified by FAO [10] as:

1. Sustainably improve agricultural productivity to meet increasing demand
2. Ensure a sustainable natural resource base
3. Address climate change and intensification of natural hazards
4. Prevent transboundary pests and diseases
5. Eradicate extreme poverty and reduce inequality
6. End hunger and all forms of malnutrition
7. Improve income-earning opportunities in rural areas and address the root causes of human migration
8. Build resilience to protracted crises, disasters and conflicts
9. Make food systems more efficient, inclusive and resilient
10. Meet the need for coherent and effective national and international governance.

The Agenda 2030 for Sustainable Development stressed these needs [11–13] directly involving agriculture in implementing actions necessary to achieve sustainable development world-wide. Within the set of 17 Sustainable Development Goals (SDGs) in the Post-2015 Development Agenda [14], the agri-food system can meet the following SDGs:

- SDG 1, focuses on eradicating extreme poverty for all people everywhere. Improving access to productive resources is a priority action.
- SDG 2, is related to ending hunger, achieving food security and promoting sustainable agricultural practices. It requires that all food systems, including both production and consumption, must be pursued from a holistic and integrated perspective, while smallholder productivity and income has to be increased.
- SDG 3, targets on guaranteeing good health and well-being also through National Sustainable Development Strategies (NSDS).
- SDG 6, is centered on ensuring availability and sustainable management of water and sanitation for all. It is related to the UN Water Action Decade launched as a platform for meeting water issues at all levels in 2016 [15].
- SDG 7, is addressed to deal affordable and clean energy, as to limit global greenhouse gas emissions which are dominant contributor to climate change.

Table 1 The economic size of farms farmland in the EU farm structure survey (source: Eurostat)

Size of farms	Standard output
Very small/semi-subsistence	<2,000
Small	2,000–8,000
Medium	8,000–25,000
Large agricultural enterprises	25,000–100,000
Very large agricultural enterprises	>100,000

- SDG 12, is fastened on ensuring sustainable consumption and production patterns. This goal is identified by the World Summit on Sustainable Development [16] as one of the essential requirements for sustainable development, together with poverty eradication and the management of natural resources in order to foster economic and social development.
- SDG13, is directed to face climate change and its impacts.
- SDG 15, is focused on protecting life on land through a sustainable forests management, as well as stopping and reversing desertification, land degradation and biodiversity loss.

Enhancing the efficiency of food production while reducing adverse impacts of agriculture on the environment are also the goals for a sustainable European agriculture [17–19]. The EU's agricultural industry is broadly distinguished according to the economic size of farms. A farm structure survey within the European Union recognised five distinct groups of farms, as set out in Regulation (EC) No 1166/2008 [20, 21] (Table 1).

In 2016, there were 10.5 million farms in the EU-28, with 175 million hectares of utilised agricultural area (UAA)[1] (some 40.0% of the total land area), giving an average size of 16.1 hectares (ha) per agricultural holding [22]. The overwhelming majority (96.0% in 2016) of the EU's farms were classed as being family farms and four million as very small. About three-quarters of such farms in the EU consumed more than one-half of their own production. They were responsible for only 1% of the EU's total agricultural economic output [23]. In Italy, the persistence of many very small-sized agricultural holdings over the last 70 years has made the productive structure of agriculture particularly fragmented. From 2000 to 2018, the number of agricultural holdings underwent an overall decrease of 19,685 units (1.8% per year), particularly involving farms in the 10–20 hectare-size while the number of farms in the over 30 hectare-size class increased.[2]

In addition to the possible incomplete transfer of the UAA from very small farms to those larger than 20 ha, the decrease of farms is also linked to land consumption due to residential, productive and infrastructural uses, emphasizing the vulnerability of agricultural holdings towards urbanisation pressures. The COVID-19's impacts

[1] The UAA includes arable land, permanent grassland, permanent crops and other agricultural land such as kitchen gardens.

[2] Source: elaboration on ISPRA and ISTAT data—Structure and production of farms 2013, 2016 and 2020.

on food supply and demand has highlighted the need to re-orient sustainable development strategies to allow the agri-food system to meet the SDGs.

The aim of this work is to understand how smart farming and its future development can be affected by strategy options of policy-makers and planners. We refer in the text in a general manner to regional/spatial planning, as integrative and comprehensive planning process on making decisions relating to draw the best development of land, as expression to the economic, social, cultural and ecological policies of a society.

After giving an overview of the key challenges linked to the Agenda 2030 for smart farming, the effects of spatial planning on agri-food systems are considered together with current and future demanding tasks to support farm holdings in their digital revolution. Eventually, the gap, between regional spatial planning and global development strategies, that can shape the framework conditions compromising the evolution of smart farming, is addressed. The case study is centered on approach to strategic spatial planning to exemplify a possible approach for decision makers and planners to meet these challenges.

2 Smart Farming and Regional/Spatial Planning

Food security will continue to be a key driver of socio-political priorities at global, regional and national levels. Tackling strong competition with the associated countries able to produce at lower costs, as well as dealing with climate change, environmental degradation, protecting biodiversity and an excessive volatility of food prices that adversely affect food security, represent other challenges for Europe-27. Special attention is required to maintain a steady capacity of the soil to function as a vital living system, which includes the control of substance and energy cycles within the ecosystems [24].

Given that a global sustainable development is closely dependant on the necessary solutions to such deals, the European Parliament pointed out how Smart Farming could represent an effective answer to these demanding tasks.

Aiming at improving the environmental and social sustainability of their production, farmers have always invested on their business, from the Precision Agriculture which allowed to target agronomic interventions, to Agriculture 4.0 of the latest generation using interconnected technologies, which improve the yield and sustainability of cultivations, production and processing quality as well as food traceability, as focused by many authors [25–28].

Agriculture 4.0 represents the Fourth Industrial Revolution, developing the new digital technologies in four directions. The first direction concerns the use of data, computing power and connectivity, comprising Big Data, Open Data, Internet of Things (IoT), Machine-to-Machine and Cloud Computing for the centralization of information and their storage. IoT processes data from different sensors to transform them into decisions and hence into actions based on: data collection, connection between the various sensors, and data processing for decision making through a

Decision Support System (DSS). By crossing environmental, climatic and cultural factors, the quality of products is increased with savings on pesticides and nutrients [29]. The second direction is the process of "Machine Learning", where machines improve their performance by learning from the data gradually collected and analysed.

The third is the interaction between man and machine, which involves touch interfaces. The fourth direction deals with the transition from digital to real, which includes new technologies to rationalise costs and energy by optimising performance, such as Robotics, Machine-to-Machine interactions etc.

As Sørensen et al. [30] focused, automation technologies for field operations management activities can be implemented in an open environment (arable farming and forestry domain); semi-structured environments (controlled traffic farming systems, open air horticulture, vineyards, orchards etc.); and controlled environments (greenhouses, urban farming, animal production units, processing plants for agro-food, wineries) that are also in need of a task and operation management.

Where the global agriculture 4.0 market is worth 7.8 billion dollars worldwide [31], the Italian agriculture represents 5% of the world's market. A big side of agricultural world is made up of small scale food agri-producers, with an average size of farms of 12 ha, and only 1% of the utilised agricultural area is managed with ICT techniques.

According to a survey of Lombardy Region, in the provinces of Brescia, Cremona and Mantua, out of a sample of 135 farms, 81% have invested in innovation in the last 3 years. Fifty per cent of farms adopt precision agriculture and 4.0 technologies, both on their own or through third parties. The technologies they adopt on their own cover business management and activity monitoring through software, whereas the use of variable dosage operating machines and tractors with assisted driving and/or automatic Global Positioning System (GPS) are entrusted to contractors. Agritech tools include: monitoring systems (39%), management software (20%), natively connected machinery (14%), remote sensing for crop monitoring (10%), land mapping systems (9%) and decision support tools (8%).

Moving from input-intensive models of ever-increasing agricultural production to a more sustainable pathway that protects natural resources, Smart Farming based on Information and Communication Technology (ICT) represents a paradigmatic shift in agriculture [32–34]. In order to support innovation in agriculture, the Italian Government [35, 36] focused rural sustainable development on four strategic areas:

1. Economic efficiency, profitability, sustainability and stability of agricultural systems, with reference to crops, livestock and forestry activities in the various contexts;
2. Conservation and reproduction of natural resources and biodiversity and provision of environmental services, including the mitigation of climate change;
3. Production (controlled and constant) of safe and high-quality foods;
4. Relationship between agriculture and local communities able to ensure the quality of life in rural areas.

Table 2 Land Take in EU-28 (Source: CORINE Land Cover provided by European Environment Agency)

Period	Land take (hectares per year)
2000–2006	102,171
2006–2012	86,014
2012–2018	53,933

The national strategies for Industry 4.0, launched within a shared Action Plan in cooperation with France and Germany (Platform Industrie 4.0, 2016), offered to farmers an opportunity to invest in high quality machinery, providing a tax discount, in the form of a tax credit.

Beyond incentives, the production structure and size of the business of agricultural holdings are the key factors for investing in new technologies which are strictly linked to the availability of land, healthy soils, water and plant genetic resources. Smaller dimensions of agricultural holdings determine a lower economic availability of farmers and when soil and water are lost for urbanisation, the effects result in decreased investment in sustainable agricultural practices, difficult management and farm crisis. Dealing with these concerns and limiting the consumption of non-renewable resources is one of the key issues of sustainable spatial planning.

3 How Does Smart Farming Grow with Land Take?

The UN Resolution adopted on 20 December 2017 [37] invited Planners to support sustainable and efficient food production systems facing the scarcity and reduced quality of land and water resources. Due to sprawl of urbanisation, exploitation of natural resources around the metropolitan regions has significantly changed in the last decades and a wide range of approaches to assess productive farm and forest land in land-use planning has been studied [38–40].

The consumption of natural resources due to urbanisation is summarised in the "land take" indicator, which assesses the increase of artificial surfaces due to the pressures of social and economic activities, including areas sealed by construction and urban infrastructure, as well as urban green areas, and sport and leisure facilities [41].

Land take intensity is calculated as a percentage of land converted to urban areas (artificial surfaces) in the given period or as a percentage of the total urban area in the year. In a sustainable development, the reduction of land take represents a pivotal policy target for EU-27 and it concerns the economic use of soil resources and the avoidance of unnecessary urban sprawl. Table 2 shows as land take has been improving in Europe, though the 78% of land take affected agricultural areas, i.e. arable lands and pastures, and mosaic farmlands. The main drivers of land take during 2000–2018 were industrial and commercial land use as well as extension of residential areas and construction sites. When these surfaces are also ‘sealed’, by covering of the ground by an impermeable material, the soil degradation affects irreversibly fertile agricultural land, puts biodiversity at risk, increases the risk of flooding and water scarcity [42].

According to the National System for Environmental Protection, in Italy the agricultural damage due to land take caused a potential loss of three million quintals of produce between 2012 and 2018. In 2018, the consumption of soil had summarised 2139.786 ha and in 2019, the soil sealing increased by another 5700 ha at the rate of 2 m^2 s^{-1}, with a per capita consumption of 354.5 m^2 [43].

In order to minimise agricultural land consumption regional territorial planning, basic principles are:

- the rejection of a dissipative concept of land use;
- the incentive to reuse the existing buildings;
- the completion and compacting of settlements;
- the polarization on urban contexts characterized by better accessibility conditions;
- the sustainable qualification of transformations;
- the protection of large green open spaces;
- the enhancement of farmlands.

In EU-27, policies to avoid further *land take* and sealing on their best agricultural soils have been established, i.e. protecting special landscapes from infrastructure developments in France and the Netherlands; strictly controlling building activities within the first 500 m from the sea in Spain, or requiring the payment of a fee for farmland consumption in the Czech Republic and Slovakia. In Italy, building activities are under a strict law control in areas of historic, natural, environmental and cultural values, as well as limited in agricultural areas. Problems arise when land use pressures are considered as the only solution to achieve better progress. Getting out of an economic crisis requires quick action and lower expenditure. Costs are often referred to as the outlay to be incurred for the work. The fragility of existing resources and the values erased from territories are not considered. In this context, protecting farmland becomes harder.

Since 2001, to relaunch its economy, the Italian Government has ruled to simplify the construction of major infrastructure works similarly to the United States and China. As works of national strategic interest, they can be built by overcoming any environmental-landscape constraint. In the last years, the simplification of design implementation plan has led to the rapid development of large infrastructures, which means a growing amount of taken land. Such road systems have been built underestimating the value of the land crossed and the damage caused to agricultural holdings and agri-food systems. The only construction of the motorway connecting the cities of Brescia-Bergamo-Milan (Bre.Be.Mi.) involved 400 farms. Two million square metres of farmland had been lost. This amount does not include the land dedicated to the secondary road network and logistics. Two hundred and sixty-five ha of agricultural land disappeared in the territory of the province of Lodi for building a High-speed Rail ("TAV"). This amount is equal to 20% of the consumption of land registered in Italy between 1999 and 2007.

It is also the case of the Cispadana motorway linking the cities of Parma and Ferrara. Despite the opposition of the Ministry for Cultural Heritage, it will erase 65 km of historically consolidated agricultural landscape, causing a large consumption of land to the detriment of farm holdings. Another project, the new Vigevano Magenta ring road, will cross restricted areas of great rural, environmental and landscape value notwithstanding opposition and protests by farmers and local organisations and communities.

It is still the case of the construction of the Pedemontana motorway in the Veneto region, insisting on the territories of the provinces of Vicenza and Treviso. It has a total extension of about 94.5 km, in addition to ordinary connecting roads for a further extension of over 50 km. In 2018, for its construction, Veneto was the Italian region with the most consumption of soil. The land take amounted to 227,368 ha, equal to 12.40% of the total land area, against a national average of 7.64% [44].

Spatial planning plays a fundamental role in finding effective solutions by interconnecting development and sustainability, combining land values with economic, environmental and climatic emergencies. Unaware choices can irreversibly compromise the survival of small farms. Thus, we forget that proximity to a big infrastructure:

- fragments the land of an agricultural holding
- compromises its functionality by moving it away from irrigation sources
- poses sources of immediate pollution closer to the farm
- induces future environmental and landscape pollution, and ecological imbalance through the proliferation of logistics.

The frequent alarms raised by farmers and environmental organisations highlight that the farmland values are not fully understood yet in 2020.

4 Smart Farming Up, Farmland Down: A Gap to Be Solved

The lack of willingness in investing in innovation in agriculture can be connected to the vulnerability and uncertainty of future for agricultural holdings. The implementation plan of a new infrastructure involves an immediate expropriation of farm holding land and imposes constraints on every development project of farmers, even when the infrastructure project remains blocked for many years [45, 46].

The aim of territorial planning is to design a decision-making process focused on environmental, economic, cultural and social goals through the development of spatial visions, strategies and plans implemented in a set of policy principles, regulatory procedures and tools. Therefore, it becomes essential for spatial planning to know how to give the right value to farm holding resources to avoid erasing any prospect of possible development for farmers. It seems to be paradigmatic that when

Table 3 The farmland consumption in the provinces of Lombardy region between 1999 and 2019 (source: elaboration on data CRCS—Centro di Ricerca sui Consumi di Suolo and ISPRA/SNPA)

Provinces of Lombardy Region	Land area (km^2)	Farmland (ha)	Farmland consumption (ha)		
		1999	1999–2015	2018	2019
Bergamo	2,755	82,428.78	6,620.92	35,329	35,525
Brescia	4,786	180,205.87	10,593.14	55,285	49,527
Como	1,279	20,276.80	1,789.82	10,374	15,615
Cremona	1,770	153,767.77	4,024.46	20,043	18,450
Lecco	805.6	14,359.67	1,444.19	16,627	9,648
Lodi	782	64,478.87	2,201.44	10,218	9,492
Mantua	2,341	197,717.81	6,153.39	26,346	24,639
Milan	1,575	87,393.97	8,537.00	50,443	49,742
Monza and Brianza	405.5	16,118.28	2,365.82	16,627	16,456
Pavia	2,969	225,253.66	4,129.16	32,631	28,104
Sondrio	3,196	25,387.16	1,557.52	10,324	8,444
Varese	1,198	19,176.97	2,274.11	26,518	25,099
TOTAL	23,844	1,086.545.60	51,391.01	310,912	290,741

the whole world is engaged in land grabbing, as in food security at a national and regional level the emergency of land consumption is still growing [47, 48].

This happened in Lombardy, region of northern Italy, where though minimizing land take had been one of the territorial planning issues since 2005, 51,391 ha of farmlands have been eroded by settlements and infrastructures between 1999 and 2019 as summarised in Table 3.

This discrepancy between global environmental issues and local development seems to be based on a badly placed question. Farmers are required by environmental policies to develop sustainable agricultural practices, but they don't have any support to achieve this goal by new planning strategies reducing land and water resources with loss and damage to ecosystem services. As Gargan and Sokolow emphasised, land use planning should proactively address sustainable economic development, not only care the development control and this issue is particularly relevant in rural contexts [49].

Assessing sustainability of territorial development towards the agri-food system requires understanding of what farms need to be a valuable environmental resource, considering the inhomogeneity of rural spaces [50–52]. This inhomogeneity is emphasized in "urban influenced" agriculture located in the midst and/or on the fringes of urbanised areas where farms have to face growing scarcity and reduced quality of land and water resources day by day.

Among the main threats to a sustainable development, the Commission of the European Communities [53] pointed out the soil loss and declining fertility which have been eroding the viability of agricultural land in recent decades.

Today, policy makers are asked to highlight the role of smart farming in the future scenarios of territory transformation. Two opposing perspectives represent critical issues in regional planning with respect to the evolution of Smart farming. The first is the approach to read development only as new urbanisation, the second concerns limiting the protection of the territory to the conservation of agricultural landscapes without considering farming evolution.

In this perception Smart farming epitomises the industrial side of farming, which erases the identity and cultural values of territories, something to prevent rather than to support. To solve this gap, two aspects are crucial: considering the rural system in its integrity and reviewing the environmental function of innovation in agriculture tackling the challenges posed by the Agenda 2030 The imperative evocated by the United Nations Conference on Trade and development [54] still resonates: "Wake up before it is too late!".

5 Challenges in the Metropolitan Area

As pointed out by the Strategic Research Agenda of ICT-AGRI ERA-NET, [55] Europe has a huge potential to produce high quality and safe food for a continuously growing market thanks to its leading role in the agricultural engineering sector. A potential that can be enhanced or dispersed relating how the value of farmland and agricultural holdings is considered in development strategies planning. In Italy spatial planning operates through different levels, from national and regional policy strategies to local municipalities. In 2005, the revision of the Title V in the Italian Constitution led to a general redrafting of the Regional planning powers on agricultural areas, weakening the policies of safeguarding agricultural soils against urbanization pressures.

To exemplify how the rural system can be threatened or enhanced in sustainable planning, a small town in the metropolitan area of Milan City, Melzo, was chosen as a case study. Since the 1950s, the outskirt of the Milan City had been affected by an intensive urbanisation compromising a large part of farmland. To protect both the environmental and economic value of the lands from the urban sprawl, the South Milan Agricultural Park was established in 1990, bordering the southern part of the City. It includes 61 municipalities with a green belt of 47,000 ha; the UUA is 39,900 ha with 1400 agricultural holdings scattered on its territory (Fig. 1).

The 51% of the town of Melzo is inserted in the South Milan Agricultural Park. In 2014, the UAA still covered 403.56 ha, with 16 agricultural holdings devoted to zoo-technical cereal production and a multifunctional agriculture. The landscape is characterised by corn crops, reaching as far as the urbanised fringes, as well as autumn-winter crops and beetroot. The architectural model is historically represented by industrial big farms designed for intensive agriculture, living on rice growing and cattle breeding, meat and dairy production. The lands can take advantage from a rich surface irrigation system with resurgences, waterways, streams and canals. In recent years, new infrastructures have been added, i.e. the

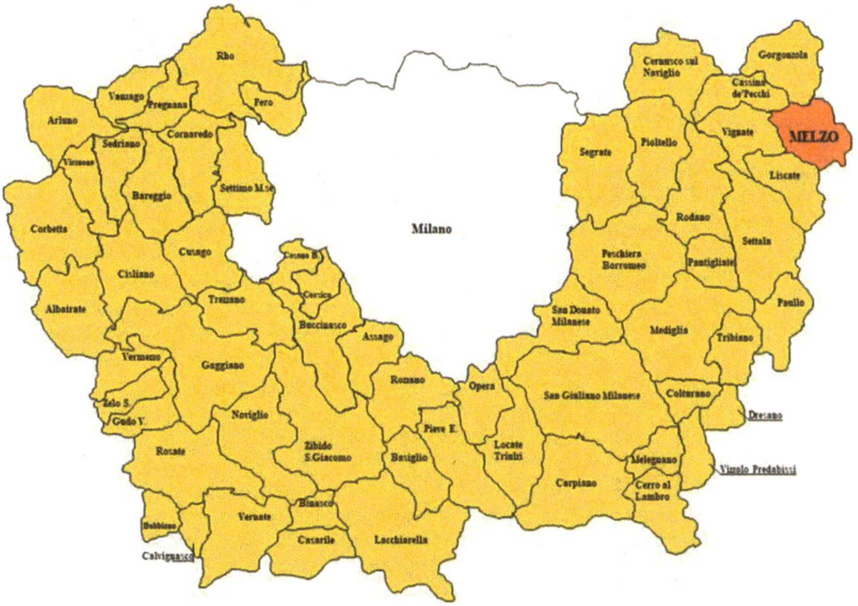

Fig. 1 The municipalities involved in South Milan Agricultural Park. In evidence the case study location

32-km long East Milan Outer Ring Road (TEEM) connecting Melegnano (A1 Milan-Bologna-Rome-Naples Motorway) to Agrate Brianza (A4 Turin-Milan-Venice-Trieste Motorway). The TEEM has redefined the extra-urban layout of the countryside.

To prevent this unfavourable situation for farmers, the cited case study compared the regional and local spatial planning scenarios with the territorial characteristics needed for the development of agricultural holdings. Thus, information about land use, livestock, labour force and other selected structural variables was gathered from the agricultural holdings.

The survey of land use included: use of areas according to main uses and purposes, to urbanisation; open areas; agricultural areas (crops, plant groups, species, and cultivation types etc.). The variables included are: type, size and location of the holding, ownership and type of tenure, irrigated and irrigable UAA, organic farming; installations for the use of renewable energies, and other gainful activities linked to diversification. The analysis also included necessary aspects to consider the agri-environmental and cultural territorial resources: e.g., the evolution of urban morphology, the value of agricultural areas (Fig. 2a) and related settlements; the landscape.

The rural development sustainability was assessed comparing the impact of urban spatial planning (Fig. 2b) with the valuable and critical elements connected to the development of farms. The survey evaluated the agricultural holdings on parallel

Fig. 2 Graphical illustration of sustainability assessment of planning process as analysis of farmlands values (**a**), infrastructures development in City Plan (**b**), assessment of farms vulnerability (**c**), sample alternative scenario with mitigation measures for farms (**d**)

layers, in terms of utilisation and production, considering in pairs, their relationships with territory and landscape, buildings and environment (Fig. 2c).

The results tried to define alternative scenarios for the improvement and conservation of local agricultural values that had strategic value for the territorial policy and were consistent with the forecasts with prevailing effectiveness at the supra-municipal level (Fig. 2d).

Priorities as identified in this planning process were:

- Identify local/regional needs and specializations of farms,
- Map and build on existing initiatives,
- Interact with farmers,
- Consider agricultural and environmental concerns,
- Identify barriers that prevent farmers to adopt the best management practices,
- Designate agricultural soils with development constraints,
- Consider "strategic farms" to design alternative scenarios during the planning process.

In fact, the agricultural territory has been affected by the road system thus facilitated on a regional scale. The new infrastructure has been built according to

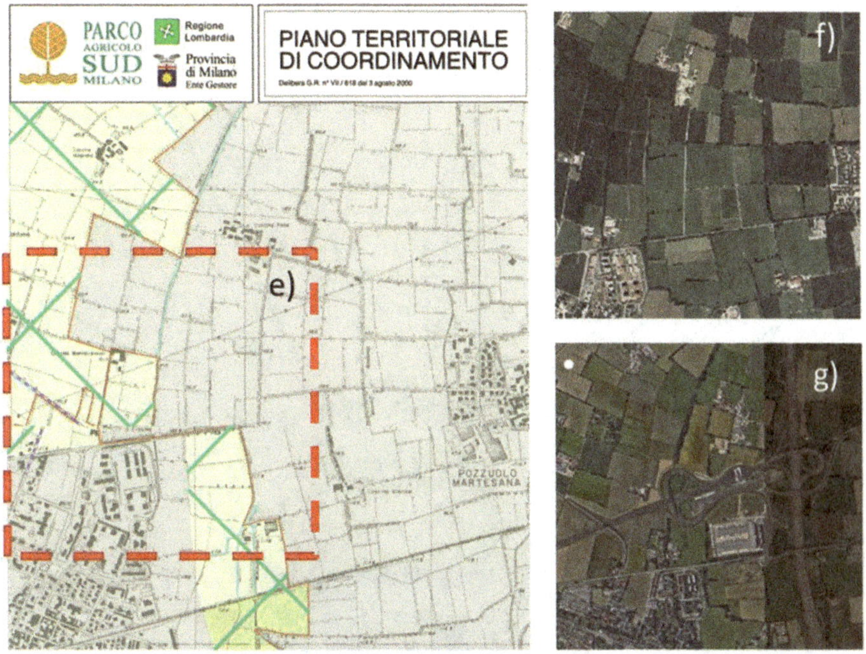

Fig. 3 The transformation triggered by TEEM through the farmlands: the district area involved in the protected area of the South Milan Agricultural Park (**d**), the image satellite of the area before starting works in 2000 (**e**), the same area in 2020 (**f**)

the Program Agreement defined by the Lombardy Region, the Provinces of Milan, Lodi and Monza and Brianza together with the 34 Municipalities directly lapped by the motorway route. Figure 3 reveals how the new infrastructures impact on farmlands also due to the activities connected to them. Not only has the land configuration been changed, but also the farmers' perspectives and autonomous planning. In 2020, seven farms remain from 16 surveyed in 2014.

Providing effective coordination across development approaches to improve growth of the environmental benefits of connected agri-food sector facing the consumption of farmland remains an important issue and challenge for metropolitan governance.

Considering agriculture as a generic use of land leads to the construction of houses, industries, services and infrastructures, as if the rural territory were an indifferent basis to urbanization effects. The risk is to consider rural space as an indefinite urban space.

Sustainable planning asks the new generation of plans to carefully read agri-food systems [56]. In reverse, the growing emergency of farmland consumption signals a lack awareness in assessing their territorial values. If smart farming is key to developing sustainable agriculture [57], sustainable spatial planning is key to developing smart farming (Fig. 4).

Fig. 4 Implementing smart farming to achieve SDGs faces a range of spatial planning challenges

The smart farming needs a change of perceiving agriculture. The availability of new technologies offers new perspectives to spatial planning. The interconnected open access data can guide development choices by considering the ecological, ergonomic, economic and energy factors related to the agricultural holdings. This can be achieved thanks to sharing and collaborating with farmers and local stakeholders to find elements of convergence and a common vision, to better inform smart farming development. Each agricultural holding ensuring food security, provides also territorial protection through the conservation of no-renewable resources—such as habitats, biodiversity and water—the control of floods, drought, soil erosion, and it is, also, a source of enjoyment of the landscape.

Agricultural holdings adopting ICT should be identified as "Strategic Farms" for achieving food security and promoting sustainable agricultural practices, as fundamental resources to safeguard in the sustainable development, fostering collaboration between policy makers and farmers [58]. The key words is cooperation between territories, farmers and planners by creating new concepts in territorial development strategies.

6 Conclusions

The chapter showed that regional/spatial planning has a key role to play in allowing Smart farming putting the 2030 Agenda for Sustainable Development and the 2050 EU's vision into practice. As a result of the hypotheses of this study, it was analysed how the emerging trend of smart farm in agriculture can be compromised by unawareness strategies development only focused on urbanisation locking up any future for farmers.

If innovative farmers are investing on improving both the environmental and the social sustainability of their products, their efforts deserve careful attention by spatial planning as well as land take needs to be limited. In the last decades, the main drivers of land take were industrial and commercial land use as well as extension of residential areas and construction sites.

To support the development of digital agriculture, preventing soil sealing, conserving land, water, plant and animal genetic resources, biodiversity and ecosystems, limiting the change in the area of agricultural, forest and other semi-natural land taken for urban and other artificial land development are priority issues. These aims are crucial for smallholder farmers and they are also fundamental cornerstones of spatial planning, at different levels. On reverse, the study pointed out how the gap between global development strategies and regional spatial planning in Italy has a direct negative impact on the chances for agriculture to achieve the SGDs.

The case study introduced, approaching a complex system for urban and regional planning, sought to provide a framework for decision makers and planners to meet these challenges. The method described in this chapter suggests including Smart Farming in regional/spatial planning through the identification of "Strategic Farms" as rural drivers of sustainable development. Recognizing the agri-food system as a territorial system allows understanding how spatial planning strategies can influence the economic availability and willingness of farmers necessary to invest in innovation. Each area destined for agricultural production, as well as ensuring food security, is at the same time an instrument of territorial protection, control of water resources, floods, drought, soil erosion, and a source of enjoyment of the landscape. Further steps forward are to be taken in establishing coherent measures at global, regional and local level so that spatial planning can support the 4.0 revolution in agriculture. The new revolutionary technologies for farming can make the boundaries between agriculture and food systems dissolve. Prospecting vertical farm as the iconic future of agricultural industry, without considering the incidence of land take on the evolution of smart farming can trace an irreversible pathway. In this way agriculture will meet its food goals but it will miss its environmental goals.

References

1. United Nations *World Population Prospects, Comprehensive table*; 2019; Vol. I.
2. FAO 2020 *Global Report On Food Crises Joint Analysis For Better Decision*; 2020.
3. Bochtis, D.; Benos, L.; Lampridi, M.; Marinoudi, V.; Pearson, S.; Sorensen, C.G. Agricultural workfoce crisis in light of the COVID-19 pandemic. *Sustain.* **2021**, *12*.
4. United Nations, *Policy Brief: The Impact of Covid-19 on Food Security and Nutrition*; 2020.
5. Herren, H.R.; Haerlin B.; IAASTD, Ed. *Transformation of our Food System, The Making of a Paradigm* Shift, Zukunftsstiftung Landwirtschaft Publishing: Gilching, 2019; ISBN 978-3-00-066209-6.
6. FAO *World Agriculture: towards* 2015/2030; Multimedia Service, Information Division Publ.: Rome, 2002; ISBN 92-5-104761-8.
7. OECD/FAO. *Agricultural Outlook* 2021–2030; OECD Publ.: Paris, 2021.
8. OECD *Agriculture and Biodiversity: Developing Indicators for Policy Analysis*. OECD Publ.: Paris, 2003.
9. United Nations. *Conference on Environment & Development Rio de Janerio, Brazil, 3 to 14 June 1992 AGENDA 21*; United Nations DESA; 1992.
10. FAO *Environmental and social management guidelines*; FAO: Rome, 2015.
11. WSSD UN *Plan of Implementation of the World Summit on Sustainable Development Contents*. Johannesburg, 2002.
12. United Nations *Back to our Common Future: Sustainable Development in the 21st Century* (SD21); 2012.
13. United Nations *Global Sustainable Development Report*; 2015.
14. United Nations *Transforming our world: the 2030 Agenda for Sustainable Development*; 2015.
15. United Nations International Decade for Action, 'Water for Sustainable Development', 2018–2028. Resolution adopted by the General Assembly on 21 December 2016; 2017; Vol.22804.
16. WSSD UN *Plan of Implementation of the World Summit on Sustainable Development Contents*; Johannesburg, 2002.
17. European Parliamentary Research Service *Precision agriculture and the future of farming in Europe Scientific Foresight Study*; Brussels, 2016.
18. Allen, B.; Bas-Defossez, F.; Weigelt, J. *Feeding Europe: Agriculture and sustainable food systems, Think 2030;* Institute European Environmental Policy: Brussels-London 2018.
19. EU Agricultural Markets Briefs *Agriculture and Rural Development (2019) 'Global food supply and demand, consumer trends, trade challenges;* European Commission, 2019.
20. European Commission *Regulation (EC) No 1166/2008 of the European Parliament and of the Council of 19 November 2008 on farm structure surveys and the survey on agricultural production methods and repealing Council Regulation (EEC) No 571/88;* 2008.
21. Eurostat *Key figures on Europe, Statistical Books;* European Union; Belgium; 2017.
22. Eurostat *Agriculture, forestry and fishery statistics;* Publications Office of the European Union: Luxembourg, 2018; ISBN 9789279947582.
23. Eurostat *Agriculture, forestry and fishery statistics;* Publications Office of the European Union: Luxembourg, 2019; ISBN 9789279330056. GPS 2015.

24. FAO *Keep soil alive, protect soil biodiversity. Global symposium on soil biodiversity,* 19–22 April 2021 – Outcome document: Rome, Italy; 2021.
25. Papajorgji, P.; Pinet, F. *Innovations and Trends in Environmental and Agricultural Informatics;* IGI Global (Advances in Environmental Engineering and Green Technologies): Hershey PA, USA, 2018. ISBN . 9781522559795.
26. Sponchioni, G.; Vezzoni, M.; Bacchetti, A.; Pavesi, M.; Renga F. *The 4 . 0 revolution in agriculture: a multi-perspective definition.* XXIV Summer School "Francesco Turco" – Industrial Systems Engineering Web Proc.: Brescia, 2018, 143–149. ISSN 2283–8996.
27. Lioutas, E. D.; Charatsari, C. Land Use Policy Smart farming and short food supply chains: Are they compatible?. Land Use Policy 2020, 94, 104541.
28. Yoon, C.; Lim, D.; Park, C. Computers in Human Behavior Factors affecting adoption of smart farms: The case of Korea, *Computers in Human Behavior;* Elsevier Ltd, **2020**, *106309.*
29. AIOTI - *Alliance for Internet Of Things Innovation Innovation Hubs: democratising digital technologies in agriculture;* 2017, 26.
30. Sørensen, C. A. G., Kateris, D. and Bochtis, D. ICT *Innovations and Smart Farming. Communications in Computer and Information Science*; Springer: Verlag, 2019, 1–19.
31. FAO *The future of food and agriculture: trends and challenges, The future of food and agriculture: trends and challenges*: Rome, 2017.
32. Adinarayana, J. Azmi, S.; Tewari, G.; Sudharsan, D. GramyaVikas: A distributed collaboration model for rural development planning, *Computers and Electronics in Agriculture* **2008**, 62(2), 128–140.
33. Brewster, I. Roussaki, N. Kalatzis, K. Doolin, K. Ellis, IoT in Agriculture: Designing a Europe-Wide Large-Scale Pilot, IEEE Communications Magazine **2017**, 55, 26–33.
34. Kamariotou, M.; Kitsios, F.; Madas, M.; Manthou, V.; Vlachopoulou M. (2019) *Strategic Decision Making and Information Management in the Agrifood Sector.* Salampasis M., Bournaris T. (eds) Information and Communication Technologies in Modern Agricultural Development. HAICTA 2017. Communications in Computer and Information Science;. Springer: Cham, 2019, vol 953, 97–109. ISBN 978-3-030-12997-2.
35. MIPAAF –Ministero Politiche agricole e Forestali, *Piano strategico per l'innovazione e la ricerca nel settore agricolo alimentare e forestale (2014–2020);* 2015.
36. MIPAAF –Ministero Politiche agricole e Forestali, *Linee guida per lo sviluppo dell'Agricoltura di Precisione in Italia;* 2017.
37. United Nations, *Agriculture development, food security and nutrition, Report of the Secretary-General Summary;* 2018.
38. Gosnell, H; Kline, J.D.; Chrostek, G.; Duncan, J. Is Oregon's land use planning program conserving forest and farm land? A review of the evidence. *Land Use Policy* **2004**, 185–192.
39. Prokop, G; Jobstmann, H.; Schönbauer, A.; Environment Agency Austria *Report on best practices for limiting soil sealing and mitigating its effects. Technical Report 2011–050,* European Commission, European Communities,; 2011.
40. Krauze, K.; Wagner, I. From classical water-ecosystem theories to nature-based solutions - Contextualizing nature-based solutions for sustainable city. *Sci. Total Environ.* **2019**, 655, 697–706.
41. European Environment Agency *Land accounts for Europe 1990–2000: Towards integrated land and ecosystem accounting;* EEA: Copenhagen, 2006;Vol. 11; ISBN 9291678880.
42. European Commission *Guidelines on best practice to limit, mitigate or compensate Soil sealing,* European Union: Luxembourg, 2012; ISBN 9789279262104.
43. Munafó, M., Ed. *Consumo di suolo dinamiche territoriali e servizi ecosistemici.* Edizione 2019. Report SNPA 8/19; 2019.
44. Munafó, M. Ed., *Consumo di suolo, dinamiche territoriali e servizi ecosistemici.* Edizione 2020. Report SNPA 15/20.; 2020.
45. Agostini, S. Infrastructures versus agriculture: critical challenges in Lombardy planning. *Sci. del Territ.* **2014**, 2,27–34.

46. Agostini, S. Identitá territoriali della filiera agroalimentare. La produzione del luogo e del patrimonio', in *Cibo e identitá locale. Sistemi agroalimentari e rigenerazione di comunitá;* Corti, M.; De La Pierre, S.; Agostini, S., Eds.; Centro Studi Valle Imagna: Sant'Omobono Terme (Bg), 2015; 431–500; ISBN 978-88-6417-066-4.
47. Borras, S.M.Jr .; Franco J.C. Global Land Grabbing and Trajectories of Agrarian Change: A Preliminary Analysis. *Journal of agrarian change* **2012**, *12,* 34–59.
48. Rudel, T. K.; Meyfroidt, P. Organizing anarchy: The food security-biodiversity-climate crisis and the genesis of rural land use planning in the developing world. *Land Use Policy,* **2014**, 36, 239–247.
49. Gargan, J.J.; Sokolow, A.D. Town and Township Government: Serving Rural and Suburban Communities. In *Handbook of Local Government Administration;:* Gargan, J.J. Ed.; Marcel Dekker incor.: New York 2019; 293–316; ISBN 0-8247-9782-5.
50. Rivera Escribano, M.J.; Mormont, M. Neo-rurality and the different meanings of the countryside. Les mondes ruraux á l'épreuve des Sci. Soc. **2006**, 17–19.
51. Woods, M. Rural Conflicts. In *Rural Geography: Processes, Responses and Experiences in Rural Restructuring;* Woods, M. , Ed.; SAGE Publications Ltd: 1 Oliver's Yard, 55 City Road, London EC1Y 1SP United Kingdom, 2011; 210–220.
52. Rusta, A. Deconstruction of Rural-Urban Divide in New Tools Prospective. *Mediterranean J. Soc. Sci.* 2019, 9, 157–163.
53. Commission of the European Communities *Communication from the Commission: A Sustainable Europe for a Better World: A European Union Strategy for 'Sustainable Development';* Bruxelles, 2001.
54. UNCTAD *Wake up before it is too late. Make agriculture truly sustainable now for food security in a changing climate.* Trade and environment review 2013; United Nations Pub, 2013.
55. Brewster, C,; Erpenbach, J.; Kelly, R.; Marchand, F.; Rakers, P.,; Thysen, I.; Vangeyte, J.; Haselberg, C. V.; & Willener, A. *Digital Technologies for a Sustainable Agrifood System: A Strategic Research and Innovation Agenda;* 2019.
56. Agostini, S. *Urbanistica periagricola. Pratiche di rigenerazione territoriale.* Maggioli: Rimini, 2018; ISBN 8891622396/97888891622396.
57. Walter, A.; Finger, R.; Huber, R.; Buchmann, N. Smart farming is key to developing sustainable agriculture, *Proc. Nat. Acad. Sci.* U.S.A. **2017**, 114, 6148–6150.
58. Agostini, S. Vincoli che trasformano la tutela in consumo. In *Per una citt´ socievole. Le alterne fortune di piani e progetti;* Treu M. C, Ed.; Palazzo Bonaretti Editore Srl: Milan, 2015; 29–43; ISBN 9788897288060.

Agriculture in Latin America: Recent Advances and Food Demands by 2050

Jesús Soria-Ruiz, Yolanda M. Fernández-Ordoñez, Guillermo Medina-García, Juan A. Quijano-Carranza, Martha E. Ramírez-Guzmán, Liliana Aguilar-Marcelino, and Leila M. Vazquez-Siller

1 Introduction

By 2020, the world population registered at 7.79 billion inhabitants, and it is expected to reach 9.77 billion people by 2050. For its part, by 2020 the estimated total population of Latin America and the Caribbean was 653 million inhabitants; with a high concentration in South America (438 million), Central America and the Caribbean (88 million) and Mexico—a country that is located in North America—with 127 million people [1].

The agricultural sector is diverse and full of contrasts. It represents a small proportion of the world's economy, but it remains central to the lives of millions of people. In 2010, approximately 2.6 billion people in the world were economically dependent on this sector [2]. Agriculture and livestock occupy around 40% of the planet's land surface; approximately 1.5 billion hectares of land are used for planting crops, while 3.5 billion are used for grazing [2, 3].

Agriculture is an essential activity in Latin America and the Caribbean, a region rich in resources that has the potential to become a great provider of food security for the planet. The region has a quarter of arable land and a third of the world's freshwater resources. Its exports of agricultural products account for 15% of the global total, while it is the world's largest net food exporting region.

J. Soria-Ruiz (✉) · G. Medina-García · J. A. Quijano-Carranza · L. Aguilar-Marcelino
National Institute of Research for Forestry Agricultural and Livestock (INIFAP), Zinacantepec, Estado de México, Mexico

Y. M. Fernández-Ordoñez · M. E. Ramírez-Guzmán
Postgraduate College in Agricultural Sciences (COLPOS), Texcoco, Estado de México, Mexico

L. M. Vazquez-Siller
Autonomous Agrarian University "Antonio Narro" (UAAAN), Saltillo, Coahuila, Mexico

D. D. Bochtis et al. (eds.), *Information and Communication Technologies for Agriculture—Theme IV: Actions*, Springer Optimization and Its Applications 185,
https://doi.org/10.1007/978-3-030-84156-0_7

This chapter briefly and summarily comments on the current situation of agriculture in Latin America, as well as on the recent advances to produce more and better, through new production schemes that in parallel mitigate the effects of climate change on agriculture. The challenge is to continue finding formulas to supply food to a constantly growing population and the corresponding demand for food by 2050. The effects of climate change in these latitudes of the world are discussed, as well as the strategies that have been followed to increase food production, the most suitable areas to practice agriculture, and food requirements for the year 2050 considering that in this region of the world, population growth ranks third after the Asian and African countries.

2 Mitigation and Adaptation of Climate Change in Agriculture

In economic terms, the agricultural sector will be the most affected by the negative effects of climate change [4, 5]. The Latin American region that could be most affected is Central America, especially regarding food security [6, 7].

The increase in the concentration of greenhouse gases has forced the agricultural sector to take mitigation and adaptation measures. Adaptive capacities are limited; therefore, climate change affects the availability and access to food together with the volatility of the prices of agricultural products in the Latin American region [6].

Changes in weather patterns reflected in extreme events such as prolonged droughts, floods, and frosts, both are recurrent with negative effects in the rural sector, particularly in agriculture. On the other hand, the incidence of pests and diseases manifested in latitudes where the attacks of pests and diseases on crops were not significant or simply did not occur. Currently, harmful insects and microorganisms have adapted to the current climate conditions, causing damage to crops, whose economic losses in agriculture are highly significant.

Agriculture is highly vulnerable to the effects of the weather, but it is also responsible for between 19 and 24% of greenhouse gas emissions globally, which constitutes it as a generator of global warming. The increase in temperature has direct consequences on agriculture, as well as the production and availability of food, putting at risk the food and nutritional security of the most vulnerable countries and societies. Extreme events have been recorded throughout the Latin American region, for example, east of the Andes. Rainfall in South America has shown an upward trend, while in several tropical regions of Central America and Mexico, rainfall is more intense in short periods. In the Caribbean, floods caused by hurricanes have had a strong impact on the region's economy in recent years, causing landslides, destroyed infrastructure, damage to agriculture and have exposed the population to recurrent risk [8].

In 2010, agriculture accounted for, on average, 29% of total Gross Domestic Product (GDP) in low-income countries (per capita income below $1005), while for middle- and high-income countries it represented 10.5 and 1.5% [2]. Estimates indicate that by 2050 there will be losses in the value of production in southern Mexico and Central America; while in northern Mexico, some regions of Argentina, Chile and Colombia could have increases in production, associated with a more stable climate [9]. In some agricultural areas of Central (El Salvador, Honduras and Nicaragua) and South America (Bolivian plateau, North of Chile, part of Ecuador and Peru) total losses of agricultural production caused by natural disasters are reported. The yields of crops such as corn, beans, wheat, potatoes, rice, and some fruit trees show increasing trends in regions with good production potential, and decreasing in regions with degraded soils, in addition to a pronounced instability, which will have negative impacts on farm income. It is expected that by 2050, the population will not only live in an increasingly hot region with less frequent rainfall in arid and semi-arid areas, but more incidences of hurricanes in the Caribbean and the Gulf of Mexico will be experienced as well. As a result, landscape modifications have occurred, due to the lesser availability of water and the advance of the desert and semi-desert that decreases the forest cover. Currently, the Intergovernmental Panel on Climate Change (IPCC) indicates that long-term policies are necessary for the growth and progress of agriculture and rurality, in areas such as smart industrialization that adds value to agriculture, social responsibility actions and smart agriculture in nutritional matters [9].

Agriculture in Latin America and the Caribbean is key to global food security, producing food for hundreds of millions. This region is home to 50% of the global biodiversity and provides 30% of the arable soils [9]. Agriculture, food systems, and rural areas are part of the solution to stimulate the development of the region and represent an opportunity that should be seized. It is urgent to promote rural development in this region of the world, because it offers agricultural, food and productive opportunities, as well as possibilities for a new energy development and to face poverty, hunger and the effects of climate change.

In Latin America, with the support of governments and agricultural research institutions, farmers have begun to implement some low-cost mitigation and adaptation measures to climate change, such as modifying planting and harvest dates, the use of varieties of plants tolerant to high temperatures, drought, pests and diseases, use of short-cycle varieties, irrigation infrastructure works, and the implementation of new production practices, among others. However, these measures will not be sufficient to cope with climate change and, therefore, the implementation of planned adaptation measures that include local, regional, national and even international components will be necessary [3]. Other efficient traditional adaptation measures are to increase the level of knowledge of farmers in the efficient management of their crops, create and introduce varieties of plants that are tolerant to high temperatures [10]. A new category of ecosystem-based adaptation practices known as EbA (Environmental based Adaptation), which consists of the establishment of protected areas and payments for environmental services [7], to improve the capacities that ecosystems have to isolate human communities from the adverse effects of climate change through the provision of environmental services; a typical example is the protection of mangroves from storms and hurricanes.

In recent years, advances have been made in research that contributes to the increase in agricultural production in Mexico and Latin America, among which are:

(a) **Advances in computer and communication technologies in agriculture.** The tenet of this section is to exemplify how informatics is already addressing problems and building computerized solutions related to the 3Fs, namely: food demand, food production and food security. The advancements of modern information and communication technologies (or ICTs), which delve in computer science are still seeing unequal application in different countries, especially so in Latin America. Recent analyses of the reach of modern-day informatics confirm the existence of an important gap in the use of web information resources [11]. For instance, 5G technology, which promises universal reach to wireless internet access, will see slow or null penetration for some time in many food production enclaves in the region, such as rural home-based agriculture and small farms or agricultural production units. This undoubtedly affects productive, innovation and knowledge management processes that could contribute to solve issues related to 3Fs.

Academic programs in agro-ecology, which target among other themes, food security through sustainable management of natural resources based on local indigenous agriculture knowledge, are seemingly not taking full advantage of informatics tools [12]. Somewhat paradoxically, ownership of or access to powerful computing equipment is becoming less of a necessity, due to cloud computing services accessible via devices such as intelligent mobile telephones. Even though in deprived Latin American environments mobile telephones are becoming more common, there is still an important access void to web resources. App developments to support agriculture or targeting any of the 3Fs are scarce. In several Latin American countries, the incidence of web access to data, information and knowledge to support food production and food demand will remain geographically and socially unequal for some time. In contrast, a rather recent literature review supports the view that "the increase in ICT investment explains an important part of the acceleration growth in the US since 1995" [13]. The complex interplay of issues and dilemmas of agriculture in comparison with the situation in developed countries has been discussed extensively from angles other than informatics advances, as they relate to a country's development, per capita income and population growth [14, 15]. The relevance of informatics in food traceability, strongly relates to food consumption security. Food security undoubtedly adds value to agricultural products. Japan illustrates this with the development of comprehensive food traceability systems, which rely on web access and highlight the need to involve the public and private sectors in these endeavors [16].

Powerful commercial enterprises are already developing and buying information systems and data suited to their particular interests in the USA [17]. In Latin America government agencies, civil organizations, academia, technicians, farmers and other interested parties should not trail behind in addressing the precarious 3F panorama. Training and education in informatics are needed so no stakeholders would stay unknowingly outside what is available to support modern decision-making [18].

Decision-making is an everyday activity in the agricultural environment [51]. It is performed solely by humans or aided by a variety of instruments. In modernized regions of the world, these instruments include computers, data from sensors or third-party information services. In order to obtain stand-alone computing components such as for mobile phone apps or entire integrated systems, informatics specialists guide themselves by workflows or process models. Workflows start with conceptualizing problems and possible solutions, and then proceed to gathering data, organizing, analyzing and summarizing it as information and finally producing knowledge (DIK). Parts of any workflow should be discussed among stakeholders, because there are many issues involved and compromises have to be evaluated. Software construction is an engineering activity, its outputs are relevant to decision making in many endeavors [19]. This is why education and training are relevant [20].

Precision agriculture (PA), clearly exemplifies the DIK workflow. Instruments gather terrain data, mobile phones may supply weather information, agricultural services and products are obtained after knowingly evaluating providers. Much of the process can be automated and embedded in smart components onboard programmable vehicles. Under what conditions can this sort of computerized solution be a solution to what 3F problems in Latin America? What is at stake? PA experiments were reported in Argentina since 2002 highlighting open issues at the time, such as agronomic and ecological principles for optimized recommendations for inputs [21]. Due to reported excessive use of agro-chemicals plus lagging governmental legislation for example in Mexico and questionable terrain suitability in many traditional agricultural enclaves, the contribution of PA to solve 3F problems remains debatable. However, it has been presented here as an example of advanced ICT proposals to be discussed for other regions in Latin America.

Other examples of the positive impact of ICT developments on 3Fs are apps accessible by mobile phones, particularly when these are able to access comprehensive databases and extract information of interest to users in agricultural environments. In what follows we include two of them developed in Mexico: the RNEAA and the CapCarb-Huatusco systems.

RNEAA is a downloadable app offering friendly access to the National Network of Automated Agro-meteorological stations developed by a group of researchers in INIFAP, the National Institute of Forestry, Agriculture and Livestock Research [22, 23]. In Mexico, big databases of meteorological and weather information are collected, some feed RNEAA which provides short-term meteorological forecasts, forecast models for agricultural pests and plant illnesses as well as national thematic maps of main meteorological variables. This comprehensive app relies on data-takes every 15 min about temperature, relative humidity, precipitation, solar radiation, and wind speed and direction. Every hour users can visualize the status of meteorological conditions over the country and over the oceanic platform. With this, information researchers are able to follow and even prevent in the short, medium and long terms conditions affecting crops.

CapCarb-Huatusco systems are among other examples that rely on a multi-platform application. A first version of the system has been developed to address data gathering of carbon sequestration in coffee agro-ecosystems in Mexico. The system works in mobile and web platforms to calculate plant carbon capture, responding to information needs of users and social actors interested in the sustainability of crops within agro-ecological systems. It systematizes data collected in situ and in real time to generate estimates of carbon sequestration in coffee agro-eco-systems. The first version is already used in experimental fashion in the State of Veracruz, Mexico. Registration of vegetation, soil and basic indicators of the productive structure of selected *coffea arabica* plots are input to the system and are the basis for a diagnosis of plot productive structure. A process workflow starts with the design of system components: general system architecture, conceptual model of the database, user profiles/interfaces and internal navigation. A main component in the CapCarb-Hautusco system, is the comprehensive data and information storage module, which includes geolocation of sites and plots, technical, documents and maps. An open-source data base management system is needed to allow access to all required components. CapCarb-Huatusco aims at flexibility through the possibility of updating algorithms as required, for example to estimate biomass and carbon sequestration. CapCarb-Huatusco illustrates the relevance of user, researchers and developers involved in all stages of system development [24].

Among other relevant ICT developments for agriculture is Big Data and Analytics. The concept of Big Data arises as a generic label for data in domains or organizations, that is continuously becoming massive, complex and in certain cases intractable. A variety of remote sensors onboard satellite platforms and drones gather Earth data continuously at different geographical scales. Other huge volumes of digital data are collected at farm level for precision agriculture and government statistics. All these archives make for Big Agricultural Data [25]. Analytics focuses on the knowledge management process which can certainly be applied to such massive data in the agricultural sector [26].

(b) **Productive potential of crops in Mexico.** The productive potential is the spatial geographical delimitation of the areas where it is feasible to carry out, with the greatest probability of success, the production of different agricultural species without deterioration of the environment [50]. As part of the strategies for modernizing the countryside and reconversion of agriculture in Mexico, the Mexican Ministry of Agriculture, through National Institute of Research for Forestry Agricultural and Livestock (*acronim in spanish* INIFAP), has been developing studies since 1992 that would allow the identification of areas/zones of the country where it can be developed with the greatest probability of success. In the studies of the productive potential, it is important to determine the agroecological requirements of the crops, for this, the ECOCROP world database is used [27].

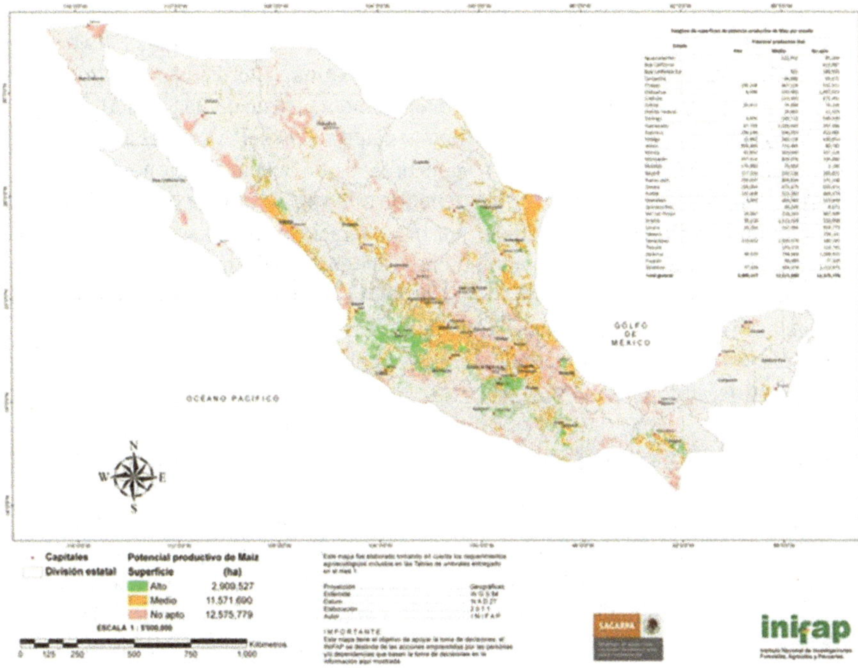

Fig. 1 Areas with productive potential for maize crop in Mexico

This database contains information on the agroecological requirements of 2568 species of plants of economic and social importance. In order to improve the quality of the studies of productive potential, the databases of climate and soil and crop requirements were improved to determine areas with productive potential of basic crops, fruit trees and vegetables in various regions of Mexico [28]. Studies have been carried out at the national level of the productive potential of 55 agricultural species of socio-economic importance in Mexico, where the potential zones of basic crops, horticultural, industrial, fruit trees, ornamentals, pastures and forage plants, bioenergetics crops and other species were obtained. In most cases, they are studies that show a broad perspective of the areas with the highest probability of success for each crop, which are used as an auxiliary tool and aid in the planning and reconversion of crops. In Fig. 1, the cartography of the productive potential for corn in Mexico is shown [29].

(c) **Phytosanitary alert systems in agriculture: the case of Mexico.** Currently food production systems, mainly based on agricultural production processes, face the challenge of producing more and better food and at the same time reducing the environmental impact associated with this process. Science and technology have contributed to improving agriculture and food production by designing new ways of producing that use soil, water, and input resources more efficiently while increasing production and eliminating threats to human health and the environment. Pest's management is one of the issues that require constant innovation due to the changing nature of the interactions between the harmful organism, the crop and the climate. Among the different technologies that have been developed for pest management, Phytosanitary Alert Systems (PAS) constitute one of the most innovative approaches. PAS are associated with early notification systems about the appearance of a pest in a region, which are of great importance to prevent the introduction and spread of new pests considering the intense commercial exchange that has characterized the world in the recent years [30]. In 2009, Mexico implemented the National Phytosanitary Epidemiological Surveillance System, SINAVEF [31] and since 2000 participates together with the United States and Canada in the PAS of the North American Organization for the Protection of Plants [32]. Technological advances in data acquisition and communication are of great importance to early detection and notification of new pests in a region or a country. Beltrán-Peña et al. [33] proposed a methodology to detect the avocado *sunblotch viroid* using satellite spectral imaging.

PAS also include technologies that integrate data on the pest, crop and weather conditions and complement their analysis with analytical tools such as simulation models to forecast and inform producers in a timely manner about possible pest outbreaks and help them prevent damages to production. This type of model allows predicting when the critical development stage of the organism occurs for the application of preventive measures to avoid the presence of an epidemic outbreak [34, 35]. In Mexico, efforts to develop alert systems based on forecasting models began in 2003 and to date, important advances have been published in monitoring pests such as the fall armyworm *Spodoptera frugiperda* [30] and the Central American Locust *Schistocerca piceifrons piceifrons* [36]. A national phytosanitary alert system based on forecasting models is currently under development to support surveillance of the most important pests in the country.

(d) **Inspection and risk analysis of agricultural products.** Developing and implementing sampling schemes that comply with phytosanitary import requirements is essential to guarantee access to agricultural products, which minimize the probability of acceptance of shipments that do not comply with international standards for phytosanitary measures. In this sense, Mexican researchers have designed sampling schemes for the inspection of imported seeds according to the specifications of the International Standard for Phytosanitary Measures (ISPM). It is important to guarantee that the inspections carried out at ports, airports and borders can be detected with a certain level of confidence, quarantining pests that affect crops of economic importance. Manuals have been generated for the inspection of grains, propagating plant material, fruits and vegetables and dehydrated products [37].

In addition, seed sampling schemes of imported products have been carried out by sampling CSP-3 skipped batches. Training is continuous for inspectors, characterized by updating the sampling schemes with the application of the hypergeometric distribution, according to NIMF 31 and NIMF 32. Other member countries of the International Regional Organization for Agricultural Health (OIRSA) such as Nicaragua and Panama have been trained following the Mexican model in Phytosanitary Inspection and sampling; risk-based for surveillance, certification and inspection. In addition, as part of the development of the inspection in Mexico and derived from the COVID 19 pandemic, Information and Communication Technologies (ICT) are applied to share the sampling schemes through various technological supports, such as the *Shyni package* from R [38], which offers the facility to interact with the user in real time through the development of an application so that inspectors from their smartphone determine the size of the sample to be inspected, according to the size of the lot, level of stipulated reliability, the prevalence of the pest of interest and based on risk with the hypergeometric distribution, according to the North American Plant Protection Organization [32], where Mexico is a member.

On the other hand, the National Center for Phytosanitary Reference (*acronim in spanish* CNRF), leader in the development of pest risk analysis in Mexico, is in an update phase with the complement of risk analysis, where the risk matrices will be parameterized to incorporate, uncertainty, probability and impact to assess the risk of various dangers that may occur during the introduction of agricultural products for import or export, which are associated with the presence of pests such as fungi, insects, viruses, bacteria and nematodes, among others. This implies the incorporation of discrete and continuous statistical distributions to model the probability of different risk events, such as the frequency of historical interceptions of the aforementioned pests (discrete event) and the economic impact incurred when a risk event occurs (continuous variable). For such risk analyses, a very versatile tool is used [39]. Mexico's leadership, in the inspection of agricultural products and attentive to international phytosanitary provisions, has allowed it to compete in the international market for agricultural products and to be a model for other countries such as those formed by OIRSA.

(e) **Seed quality in Agriculture.** As an important input in food production, seeds play an essential role. Latin America is a region that has great environmental, cultural and social wealth, with variable agricultural systems according to its production conditions, crops, and cultural, economic and political factors. This wide diversity of factors influences the levels of agricultural productivity and the competitiveness of the region. The seed sector is different in each country and has its own characteristics, dynamics and norms according to its local contexts. Even within a country there are differences between the seed sectors between crops and/or regions [40].

In the seed production systems in Latin America, the formal and informal sectors coexist depending on the crop and the country. In Colombia and Peru, farmers resort to sectors according to their particular needs, this is mainly due to the development characteristics of the systems and to the fact that governments must establish clear measures for the proper functioning and recognition of both. In Brazil, the seed production system is characterized by the association between private and public-private companies, guaranteeing the strengthening of research, training and development of this industry. In Guatemala, seed production systems are in a state of emerging development [40].

In Mexico, the purpose of seed production is to link grain and seed production with genetic improvement, with emphasis on aspects related to productivity and postharvest handling. Research work on seeds is focused on designing and evaluating procedures for the description and conservation of genetic identity, on field control of genetic, physical and sanitary quality, on field production of different categories, on establishment and management from production batches; to certification and production standards, to control of pests and diseases in the field and to the use and evaluation of techniques that increase the production index in recalcitrant such as tissue culture. In grains, it will focus on field production, regulations and nutritional quality, in addition to its processing and industrialization.

(f) **Beneficial microorganisms in agriculture.** Several crops such as corn, beans, wheat, rice, oats, tomatoes and chili are affected in their production by a wide variety of *phytopathogenic* organisms that affect plant health [41]. Among these microorganisms are *phytoparasites*, bacteria, viruses, insects, protozoa, mites, fungi, nematodes, which cause diverse symptoms that depend on each organism (genus and species) affecting different parts of the plant in agriculture [42]. The conventional type of control of these *phytopathogenic* organisms is the use of mainly organophosphate chemicals (*e.g.* organophosphates: fenamiphos, ethoprofos, cadusafos; carbamates: aldicarb, carbofuran, oxamyl) and fumigant nematocidal. However, the frequent and indiscriminate use of these products has generated the problem of resistance to these products, in addition to the damage to beneficial organisms (dung beetles, worms, among others), microbial consortiums present in the soil and contamination of aquifers [43]. For this reason, it is important to look for alternative, sustainable and environmentally friendly control methods. In this context biocontrol is an ecological method that uses natural antagonists such as: parasitoids, nematode predators of other nematodes, bacteria, entomopathogenic nematodes, insects, viruses, tardigrades, nematophagous mites, micro and macro fungi [44]. Within this repertory of beneficial organisms for agriculture, studies have been carried out on the bacterium *S. marcescens* that produces chitinases, which hydrolyze the ß-1, 4 glycosidic bonds of chitin [45, 46]. This bacterium has a complete chitinolytic system formed by three classes of enzymes: (1) endochitinases, (2) exochitinases and (3) chitobiases and a protein that binds to chitin. The use of beneficial microorganisms in agriculture through biocontrol is a compatible approach to agricultural production, additionally a tool for producers for the development of a sustainable and regenerative food system, as it would reduce the use of chemicals such as pesticides, public health risks and negative impact on the environment.

3 Food Demand in Latin America by 2050

The Latin American and Caribbean region depends on agriculture as the basis of its economy in terms of work, internal trade, and the generation of foreign exchange through exports [47]. As global trends affect food security and the sustainability of food and agricultural systems, a number of uncertainties arise. Today's food and agricultural systems are not capable of meeting the needs of a population that is estimated to exceed 9.7 billion people by 2050. Therefore, it is necessary to increase production in the context of climate change [10]. The trends point to specific challenges facing food and agriculture to achieve hunger eradication and food security by 2050. The challenges identified are relevant to achieving food security, improving rural livelihoods, and make agriculture, fisheries, forestry and natural resources more resilient, productive and sustainable [10].

In Latin America and the Caribbean, substantial improvements are being made in the conservation and use of resources to meet the demand for food in the current context, and agricultural research institutions such as INIFAP of Mexico, CATIE of Costa Rica, EMBRAPA of Brazil and INTA of Argentina among others, are developing new methods and techniques to produce more and better food through the efficient use of soil and water resources, genotype, management taking into account the behaviour of the present and future climate by 2050. Any increase in agricultural production is based mainly in the conservation and efficient use of natural resources. Biological control of pests and diseases to enhance agricultural production and food safety.

Regional trends in agriculture require finding new ways to produce food, to meet the needs of a constantly growing population, favouring the ecological balance and the efficient use of resources. In this sense, the FAO mentions ten challenges that impede food security for all and the sustainability of agriculture [9], and indicates that it is necessary to introduce fundamental changes in the agricultural systems of the world and particularly of the countries of Latin America, rural economies and the management of natural resources. To develop the maximum potential of food and agriculture and guarantee a healthy future for all in 2050, it is necessary to adopt innovative but sustainable agricultural practices, where traditional and ancestral knowledge must be privileged and technological developments that increase the productivity of the agricultural systems to be produced, preferably without the use of pesticides that cause damage to ecosystems.

The immediate challenge for the Latin American countries is to develop, manage, implement and promote agricultural systems with the following characteristics: economically profitable, biologically efficient and ecologically sustainable. Through systems of this type, it is expected to achieve certain goals such as: increasing food production to supply the population's needs at fair prices; generation and saving of

Table 1 Increase of agricultural production necessary to supply anticipated demand, 2005/2007–2050 (%)

Region	2005/ 2007	2012/ 2050	2005/ 2007–2012	2013–2050
Worldwide:				
According to AT2050 [48]	100	159.6	14.8	44.8
Population projections [49]	100	163.4	14.8	48.6
Sub-Saharan Africa and South Asia:				
According to AT2050	100	224.9	20.0	104.9
Population projections	100	232.4	20.0	112.4
Rest of the World:				
According to AT2050	100	144.9	13.8	31.2
Population projection	100	147.9	13.8	34.2

foreign exchange through the export of agricultural surpluses; increase the efficiency of the use of natural resources in the region; reduce damage to the environment by conducting sustainable and smart agriculture to provide a more secure future for future generations.

To meet the demand for food in 2050, FAO estimates that agriculture will have to produce almost 50% more food, fodder and biofuel than it did in 2012. In sub-Saharan Africa and South Asia, agricultural production should multiply for more than two to meet the growing demand, while in the rest of the world where Latin America and the Caribbean is located, growth forecasts would be around a third above current levels (Table 1) [48]. This assessment takes into account recent United Nations projections that the world population will reach 9.7 billion people by 2050.

4 Conclusions

The current situation of Agriculture in Latin America and the Caribbean, is very diverse and with different levels of technification. From that where it is still practiced in hillside areas with massive labor and rudimentary techniques with unit yields below the world average, to intensive agriculture (Digital Agriculture) characterized by mechanization with information technologies, where yields are recorded higher than the world average. At both levels of modernization, climate change is causing direct impacts on this sector. However, research institutions and the government sector of these countries have implemented mitigation and adaptation measures to climate change in agriculture, applying new knowledge and techniques (some of them are described in this chapter), to increase the production and productivity of crops in this region of the world. It is clear that, to meet the demand for food in 2050, it will be necessary to produce more than 50% of what is currently produced; with the need to increase the agricultural area of at least, 2% compared to 2012.

It is recommended that the governments of the countries of Latin America and the Caribbean continue to promote scientific and technological research to mitigate the effects of climate change on agriculture and that local and regional mitigation measures consider the various production systems, the level of knowledge of farmers, favoring the use of native seeds and organic agriculture, particularly in marginalized regions.

References

1. CEPAL (2020). América Latina y el Caribe: Estimaciones y proyecciones de población. Comisión Económica para América Latina y el Caribe. Naciones Unidas.
2. Alston, J. M., & Pardey, P.G. (2014). Agriculture in the Global Economy. *The Journal of Economic Perspectives* 28 (1), 121–146.
3. Howden, S. M., Soussana, J.F., Tubiello, F.N., Chhetri, N., Dunlop, M. & Meinke, H. (2007), "Adapting Agriculture to Climate Change", *Proceedings of the National Academy of Sciences* 104 (50), 691–696
4. Fischer, G., Shah, M., Tubiello, F. & Van Velhuizen, H. (2005), "Socio-economic and Climate Change Impacts on Agriculture: An Integrated Assessment, 1990-2080", *Philosophical Transactions of the Royal Society B: Biological Sciences* 360, (1463), 2067–2083.
5. Mendelsohn, R. (2009). The Impact of Climate Change on Agriculture in Developing Countries. *Journal of Natural Resources Policy Research* 1 (1), pp. 5–19.
6. Lopez, F. A.J. & Hernández C. D. (2016). Climate Change and Agriculture: A Review of the Literature with Emphasis on Latin America. *El Trimestre Económico* 83(4), 332, 459–496.
7. Magrin, G., Marengo, J., Boulanger, J., Buckeridge, M., Castellanos, E. & Vicuña, S. (2014), "Central and South America", In: V. Barros, C. Field, D. Dokken, M. Mastrandrea, K. Mach, T. Bilir & L. White (eds.), Climate Change 2014: Impacts, Adaptation, and Vulnerability. Part B: Regional Aspects. Contribution of Working Group II to the Fifth Assessment Report of the Intergovernmental Panel on Climate Change, Cambridge University Press/Cambridge, UK/ Nueva York, 1499–1566.
8. Soria-Ruiz, J., Contreras, M., Piña, J., Torres, H. (2009). Metodología para la evaluación de daños en la agricultura causados por huracanes. INIFAP-CIRSE. Publicación Especial 1, 51p. Mérida, Yuc., México. ISBN: 978-607-425-238-5.
9. IICA (2019). Perspectivas de la agricultura y del desarrollo rural en las Américas: una mirada hacia América Latina y el Caribe 2019-2020./CEPAL, FAO, IICA. – San José, C.R. 144 p.
10. FAO (2017). El futuro de la Alimentación y la Agricultura, Tendencias y desafíos. Organización de las Naciones Unidas para la Agricultura y la Alimentación. Versión resumida. Roma.
11. Quiroga-Parra, D.J., Torrent-Sellens, J. & Murcia-Zorrilla, C. (2017). Information technology in Latin America, its impact on productivity: A comparative analysis with developed countries. 84 (200), 281–290 (SP) https://doi.org/10.15446/dyna.v84n200.60632
12. COLPOS (2020). Colegio de Posgraduados - Agroecology for sustainable development. Research theme on local rural and indigenous knowledge. Accessed 23/08/2020: http://www.investigacion.colpospuebla.mx/about-co.html
13. Hofman, A. (1990). Information and communication technologies and their impact in the economic growth of Latin America. https://doi.org/10.1016/j.telpol.2016.02.002.

14. Barraclough, S. and Utting, P. (1987). Food Security Trends and Prospects in Latin America. Working Paper #99 - August 1987. (Accessed: 7/20/2020). https://kellogg.nd.edu/sites/default/files/old_files/documents/099_0.pdf.
15. Espinosa-Cristia, J.F., Feregrino, J. and Isla, P. (2019). Emerging, and old, dilemmas for food security in Latin America. *Journal of Public Affairs* 19(3): e1999.
16. Setboonsarng, S., Jun, S. & Lucia, V. (2009). Food safety and ICT traceability systems: Lessons from Japan for developing countries, ADBI Working Paper, No. 139, Asian Development Bank Institute (ADBI), Tokyo. Available at: https://www.adb.org/sites/default/files/publication/155994/adbi-wp139.pdf.
17. Carolan, M. (2017). Publicising Food: Big Data, Precision Agriculture, and Co-Experimental Techniques of Addition. *Sociologia Ruralis* 57 (2), 135–154. Available: https://onlinelibrary.wiley.com/doi/epdf/10.1111/soru.12120
18. Hofman, A., Aravena, C. and Aliaga, V. (2016). Information and communication technologies and their impact in the economic growth of Latin America. *Telecommunications Policy* 40, 485–501. https://doi.org/10.1016/j.telpol.2016.02.002.
19. Leach, R. J. (1999). Introduction to Software Engineering. CRC Press. ISBN 0849314453, 9780849314452, p 448.
20. Sabin, M., Viola, B., Impagliazzo, J., Angles, R., Curiel, M., Leger, P., Murillo, J., Hernán, N., Pow-Sang, J.A. and Trejos, I. (2016). Latin American Perspectives to Internationalize Undergraduate Information Technology Education. https://doi.org/10.1145/3024906.3029847.
21. Zhang, N., Wang, M. and Wang, N. (2002). Precision agriculture—a worldwide overview. Computers and Electronics in Agriculture. Volume 36, Issues 2–3. November, 113–132. https://doi.org/10.1016/S0168-1699(02)00096-0.
22. INIFAP (2020a). Databases converted into tools for agricultural development (blog: https://www.gob.mx/inifap/articulos/convierten-bases-de-datos-en-herramientas-para-el-desarrollo-del-campo?idiom=es)
23. INIFAP (2020b). National Modeling and Remote Sensors Laboratory (Laboratorio Nacional de Modelaje y Sensores Remotos). Mexico. https://clima.inifap.gob.mx/lnmysr.
24. Marquez De La Cruz, S.E. (2020). CapCarb-Huatusco: a computational system to estimate carbon sequestration in a *Coffea arabica* agro-ecosystems: Case of Huatusco, Veracruz, Mexico. Master's thesis, Colegio de Postgraduados-Montecillo, Mexico. (SP).
25. Kamilaris, A., Kartakoullis, A. and Prenafeta-Boldú, F.X. (2017). A review on the practice of big data analysis in agriculture. *Computers and Electronics in Agriculture* 143, 23–37, https://doi.org/10.1016/j.compag.2017.09.037
26. Vinaja, R. (2020) Data Analytics Applications in Latin America and Emerging Economies, Book Review. Journal of Global Information Technology Management, 23:1, 79–81, https://doi.org/10.1080/1097198X.2019.1701728
27. FAO (2020). Base de datos de requerimientos agroecológicos para el desarrollo de cultivos. http://ecocrop.fao.org/ecocrop/srv/en/home.
28. Medina G.G., Rumayor, A., Cabañas, B., Luna, M., Ruiz, J., Gallegos, C., Madero, J., Gutiérrez, R., Rubio, S. & Bravo, A. (2003). Potencial productivo de especies agrícolas en el estado de Zacatecas. INIFAP-CIRNOC, Campo Experimental Zacatecas. México. 157 p.
29. Díaz, G., Guajardo, R., Medina, G., Sánchez, I., Soria-Ruiz, J., Vázquez, J., Quijano, J., Legorreta, F. & Ruiz, A. (2012). Potencial productivo de 50 especies agrícolas de importancia socio-económica en México. INIFAP- CIRGOC. Publicación especial No. 8. ISBN: 978-607-425-766-3.
30. Yáñez-López R., Vázquez-Ortega, A., Arreguín Centeno, A., Soria-Ruíz, J. and Quijano-Carranza J.A. (2019). Alert system against the fall armyworm Spodoptera frugiperda (J. E. Smith) (Insecta: Lepidoptera: Noctuidae). *Revista Mexicana de Ciencias Agrícolas* 10(2), 405–16.
31. Galindo M. M.G., Olvera, L.A. and Contreras, C. (2015). Sistema Nacional de Vigilancia Epidemiológica Fitosanitario. Caso México. In Jaramillo G., J.L. (Ed.). 2015. Memorias Congreso Colombiano de Entomología 42. SOCOLEN, Medellín, Colombia, 177–187.

32. NAPPO (2020). Northamerican Plant Protection Organization. https://www.nappo.org/spanish/nosotros/introduccion. Accessed: 08/18/2020.
33. Beltrán-Peña, H., Soria-Ruiz, J., Teliz-Ortiz, D., Ochoa-Martínez, D., Nava-Díaz, C. & Ochoa-Ascencio, S. (2014). Detección satelital y molecular del viroide de la mancha de sol del aguacate (Avocado Sunblotch Viroid, ASBVd). *Revista Fitotecnia Mexicana* Vol. 37 (1), 21–29.
34. Frolov A. (2019). Patterns of pest population dynamics and phytosanitary forecast. *Plant Protection News*, No 3 (2019): 3(101).
35. Wieser, A., Reuss, F., Niamir, L., Muller, L., O'Hara, R.B. and Pfenninger, M. (2019). Modelling seasonal dynamics, population stability, and pest control in *Aedes japonicus japonicus* (Diptera: Culicidae). *Parasites Vectors* 12, 142. https://doi.org/10.1186/s13071-019-3366-2
36. Hernández-Zul, M. I., Quijano-Carranza, J.A., Yáñez-López, R., Ocampo-Velázquez, R.V., Torres-Pacheco, I., Guevara-González, R.G. & Castro-Ramírez A.E. (2013). Dynamic Simulation Model of Central American Locust Schistocerca piceifrons (Orthoptera: Acrididae). *Florida Entomologist* 96(4), 1274–1283
37. Ramírez-Guzmán, M.E. & López-Tirado, Q. (2007). Determinación de metodologías de muestreo y tamaños de muestra en embarques de importación para realizar análisis de laboratorio para granos. https://www.researchgate.net/profile/M_E_Ramirez_Guzman/research. (Accessed: 08/18/2020).
38. RStudio Team (2015). RStudio: Integrated Development for R. RStudio, Inc., Boston, MA URL http://www.rstudio.com/.
39. @Risk (2020). Análisis de Riesgo para Excel con simulación Monte Carlo. https://www.palisade-lta.com/productos.asp. (Accessed: 08/18/2020).
40. Arenas, W.C., Cardozo, C.I. & Baena, M. (2015). Analysis of seed systems in Latin American countries. *Acta Agron.* 64 (3), 239–245.
41. Servicio de Información Agroalimentaria y Pesquera (2020). 1306–1315. http://infosiap.siap.gob.mx:8080/agricola_siap_gobmx/ResumenDelegacion.do.
42. Aguilar-Marcelino, L., Mendoza, P., Al-ani, L., Lopez, M., Gómez, O., Villar-Luna, E. and Reyes-Guerrero, D. (2020). Using molecular techniques applied to beneficial microorganism as biotechnological tools for controlling agricultural plants pathogens and pest. *In:* Molecular aspects of plant beneficial microbes in agriculture. Vivek Sharma, Richa Salwan, Laith Khalil Tawfeeq Al-Ani (Eds). London, United Kingdom. pp: 333–349.
43. Heap, I. (2014). Global perspective of herbicide-resistant weeds. Pest Manag. Sci. 70 (9). https://doi.org/10.1002/ps.3696
44. Al-ani, L., Aguilar-Marcelino, L., Fiorotti, J., Sharma, S., Sharif, M., Furtado, E., Wijayawardene, N. & Herrera-Estrella, A. (2020). Biological control agents and their importance for plant health. In: J. S. Singh & S. Raj Vimal (Eds.), *Microbial services in restoration ecology* (pp. 13–28). Elsevier.
45. Chun Su, Liu Y, Sun Y, Li Z. (2017). Complete genome sequence of *Serratia* sp. YD25 (KCTC 42987) presenting strong antagonistic activities to various pathogenic fungi and bacteria. Journal of Biotechnology 245:9–13. https://doi.org/10.1016/j.jbiotec.2017.01.011
46. Méndez-Santiago, E.W., Gómez-Rodríguez, O., Sánchez-Cruz, R., Folch-Mallol, J.L., Hernández-Velázquez, V.M., Villar-Luna, E., Aguilar-Marcelino, L., Wong-Villarreal, A. (2020). *Serratia* sp. and endophyte of *Mimosa pudica* nodules with nematicidal antifungal activity and growth-promoting characteritics. Archives of Microbiology. https://doi.org/10.1007/s00203-020-02051-2.
47. Wadsworth, J. (1997). Análisis de sistemas de producción animal. Tomo 1: Las bases conceptuales. Organización de las Naciones Unidas para la Agricultura y la Alimentación. Roma, © FAO.

48. FAO (2015). Estudios de perspectivas mundiales, datos basados en ONU (Accessed at: https://esa.un.org/unpd/wpp).
49. Alexandratos, N. and Bruinsma, J. (2012). World agriculture towards 2030/2050: the 2012 revision. ESA Working Paper No. 12-03. Roma.
50. Medina, G.G., R. Gutiérrez Luna, R., Echavarría Chairez, F., Amador M.D. and Ariel Ruiz Corral, J. (2009). Estimación de la producción de forraje con imágenes de satélite en los pastizales de Zacatecas. *Técnica pecuaria en México* 47(2):135–144.
51. Braun, J., M. Braun, M. and Diaz, J (2000). Economía Chilena 1810–1995. Estadísticas Históricas. Documento de Trabajo No. 187. Universidad Católica de Chile. ISSN electrónica 0717-7593.

The Development Opportunities of Agri-Food Farms with Digital Transformation

Alessandro Scuderi, Giuseppe Timpanaro, Luisa Sturiale, Giovanni La Via, and Biagio Pecorino

1 Introduction

The agriculture and agri-food sectors are facing multiple challenges, with the global population projected to grow over 9.6 billion until 2050 resulting in a significant increase in the demand for food. At the same time the availability of natural resources is becoming increasingly constrained [1]. The agri-food chain remains critical for livelihoods and employment, hence achieving the United Nations Sustainable Development Goal of a 'world with zero hunger' by 2030 will require more productive, efficient, sustainable, inclusive, transparent, and resilient food production systems [2]. This requires an urgent transformation of the agri-food chain towards digitalization. The digital innovations (the so-called Industry 4.0) is seeing several sectors rapidly transformed by 'disruptive' digital technologies such as Blockchain, Internet of Things, Artificial Intelligence and Immerse Reality. In the agriculture and agri-food sectors, the spread of mobile technologies, remote-sensing services, and distributed computing are already improving smallholders' access to information, inputs-output flows, and markets [3, 4].

Digital transformation is a process that influences every aspect of human society. It is a transformation, determined by new technologies, that not only enhances traditional processes of innovation and development, but creates new forms of innovation characterized by clear and rapid changes, and affects every segment of

A. Scuderi · G. Timpanaro · G. La Via · B. Pecorino
Department of Agri-Food and Environmental Systems and Management (Di3A), University of Catania, Catania, Italy
e-mail: alessandro.scuderi@unict.it

L. Sturiale (✉)
Department of Civil Engineering and Architecture (DICAR), University of Catania, Catania, Italy
e-mail: luisa.sturiale@unict.it

D. D. Bochtis et al. (eds.), *Information and Communication Technologies for Agriculture—Theme IV: Actions*, Springer Optimization and Its Applications 185,
https://doi.org/10.1007/978-3-030-84156-0_8

society, such as the economy, communication, governance, information flow, art, medicine, and science [5–7]. From the economic point of view, digital transformation can be defined as the process that redesigns and makes the company's overall offer more competitive, through the transformation of production processes, analysis, and directly listening to market needs using digital technologies.

The definition highlights the importance of the innovative aspect of digital transformation linked to the originality of the transformation. In order to understand the digital transformation process, it is necessary to analyze some enabling technologies, distinct in product-service and process innovations, which assume a strategic economic significance [8–10]. In particular, it is useful to give some hints on the key concepts: Internet of Things and Big Data.

"Internet of Things" (IoT) is a neologism referring to the extension of the Internet to the world of objects and concrete places, equipped with a more or less permanent connection to the Internet as well as sensors and other devices capable of monitoring and recording people's actions and habits. The connection of these objects (IoT devices) to the Internet allows the exchange, storage, sharing, processing of huge flows of information and data [11]. The term "Big Data" refers to the set of data with dimensions that go beyond the capacity of commonly used software tools. Digital technologies have multiplied the available data at an exponential rate, generated by sensors, social media, transactions, smartphones and other sources. Big Data can represent a real asset for companies, whose potential can only be expressed through their intelligent use [12].

The digital transformation is proceeding at an increasing pace but in a diversified manner in the individual countries [13]. As far as the European Union (EU) is concerned, a picture of the situation of this phenomenon can be taken from the Digital Economy and Society Index (DESI). The DESI is a composite index that summarizes relevant indicators on Europe's digital performance and tracks the progress of EU Member States in digital competitiveness. The five dimensions of the DESI are: connectivity; human capital; use of internet; integration of digital technology; and digital public services. Italy still has a large gap to catch up, in fact, it is in 24th place in the ranking of the 28 EU Member States. On the other side, the first places are kept by Finland, Sweden, Holland, Denmark.

In general, in Italy it seems that companies are slow to understand the potential of the network: 40% of entrepreneurs declare that it is not useful to their activity. Many entrepreneurs are still not aware of the potential offered by the network for the promotion of products, for business turnover thanks to e-commerce, and interaction with customers with social media [14]. The data for the recent few years show a slow improvement for Italy, but the distance from the European average is still evident, an alarming situation especially if we consider the growing importance of the digital economy, especially in the short-term future [15].

The Chapter aims to provide a first contribution to the perception that Italian agricultural operators have about the opportunities and limits derived from the adoption of smart agri-food. The first results, obtained from a multicriteria analysis approach (MCA), will be presented to define possible future scenarios deriving from the implementation of digital transformation.

2 Materials and Methods

2.1 Digital Transformation in the Italian Agri-Food System

The agri-food sector has undergone a series of revolutions that have driven efficiency, yield and profitability to previously unattainable levels. Market forecasts for the next decade suggest a 'digital agricultural revolution' will be the newest shift which could help ensure agriculture to meet the needs of the global population into the future. Digitalization will change every part of the agri-food chain. Management of resources throughout the system can become highly optimized, individualized, intelligent and anticipatory. It will function in real time in a hyper-connected way, driven by data. Value chains will become traceable and coordinated at the most detailed level whilst different fields, crops, and animals can be accurately managed to their own optimal prescriptions. Digital agriculture will create systems that are highly productive, anticipatory and adaptable to changes such as those caused by climate change. This, in turn, could lead to greater food security, profitability and sustainability [16].

In the context of the Sustainable Development Goals, digital agriculture has the potential to deliver economic benefits through increased agricultural productivity, cost efficiency and market opportunities, social and cultural benefits through increased communication and inclusivity, and environmental benefits through optimized resource use as well as adaptation to climate change [17].

The potential benefits of digitalizing the agri-food sector are convincing but it will require major transformations of farming systems, rural economies, communities and natural resource management. This will be a challenge and requires a systematic and holistic approach to achieve the full potential benefits [18]. The conditions for the digital technologies and therefore for digital transformation of the agri-food chain are the infrastructure for connectivity and the institutional support.

The introduction of digital technologies in rural areas can be a challenge for development of the rural populations. There is a lack of infrastructure, including basic information technologies infrastructure, particularly in very remote rural communities. The cost associated with information technologies infrastructure presents a major challenge in rural areas where rates of poverty are often higher than urban areas, especially in least-developed countries.

Digital technologies are rapidly transforming our economies and societies. Their adoption is driving down information and transaction costs, improving efficiency, creating new jobs, generating new income streams, and saving resources. At the same time, they can be disruptive, modifying or displacing activities and products [19]. Digital technologies can help agriculture meet the global challenges it faces. These include increasing the production of sufficient, safe and nutritious food for a growing population to ensure food security; generating jobs, improving incomes, reducing poverty and promoting rural economic growth; and sustainably managing natural resources [20, 21].

Some digital technologies accelerate the evolution of agricultural and food value chains. Other technologies significantly affect the contribution of labor, capital and other inputs to the production, processing and marketing of food. Thus, the adoption of digital technologies can result in changes in relative prices, disrupting markets. Sensors, satellites, robots, and drones are examples of digital technologies that can revolutionize farming and value chains. Sensors and satellites provide information on soil conditions, weather and temperature, or crop growth. They enable farmers to achieve better yields by optimizing farm management, reducing the use of fertilizers, pesticides and water, and also contributing to better and more sustainable outcomes. The Internet of Things that connects robots, drones, and vehicles to the internet can make labor intensive tasks, such as monitoring plant health or sowing crops, more cost effective. These technologies also generate large amounts of data that can be combined with other information, stored and analyzed to support decision making [22, 23]. Such Big Data can contain high variety of information assets which can be processed by new methods of analysis, such as artificial intelligence, to assess possible outcomes based on a range of actions and conditions to help guide future interventions. Thus, agriculture can become a knowledge intensive activity.

Farmers assess the weather, soil nutrient and moisture levels, plant and livestock appearance, the presence of parasites, market prices, and many more variables before they make decisions on farm practices and production. Technological improvements have greatly facilitated these decision-making processes [24]. Though the access to technology and the rate of adoption differ greatly across the world and also within countries, technology can be present at every stage of farming, marketing and processing.

In the followings, a number of definitions on entities around digital agriculture are listed:

- Information and Communications Technology (ICT) refers to the integration of telecommunications, computers and the necessary systems that enable users to access, store, share and use information.
- Digital technology is an all-encompassing term to refer to computerized tools that generate, store, and use data for a variety of purposes.
- Digital platforms are virtual hubs for trading goods and services (e-commerce).
- Internet of Things (IoT) is a term coined to refer to the collection of internet enabled devices that capture information from the real world. The information collected is processed with the help of a software application (app).
- Distributed ledger technology (DLT), is in essence, a decentralized, consensus-based record keeping system.
- Precision agriculture (PA) is a whole farm management approach having as a prerequisite the implementation of information technology, satellite positioning (GNSS) data, remote sensing and proximal data gathering.
- Artificial intelligence (AI) refers to software systems that can make decisions which normally require a human level or superior of expertise, often using real-time data.

- Big Data is an umbrella term referring to the large amounts of digital data continually generated by the global population as a byproduct of everyday interactions with digital products or services.

Digital transformation in the agri-food sector plays a crucial role in our society and to counteract the critical factors of globalization and the growing environmental impact [6, 25]. In Italy, the market growth potential of "Agriculture 4.0" and "Farming 4.0" solutions is very high, but the adoption of technologies such as robots and precision farming sensors is still at low levels [26–28]. In this context, "Agriculture 4.0" solutions are integrated with "Farming 4.0" solutions, according to an approach based on the integration of various ICT/geo-space technologies [29]. That is, reliable remote monitoring is possible through space-time and spectral measurements, able to monitor the phenomena at the level of individual sites from various altimetric positions.

Similarly, the Blockchain technology applied to the agri-food supply chain makes it possible to guarantee a transparent, safe and shared environment for the traceability of the components and processing processes of agri-food products offered to the consumer [30–33]. For example, the Italian Food chain makes it possible to securely collect, record, analyze, validate and certify data, information and documentation at every stage of the supply chain, through the open functionalities of blockchain, through the use of the "smart contract" concept [35, 52].

In summary, systems and technologies such as GIS/geo-spatial infrastructures, fixed and mobile broadband networks, Internet of Things, Artificial Intelligence, Blockchain, Augmented and Virtual Reality, etc. are available. These make it possible the provision of digital services through intelligent platforms for "green & sustainable development" applications (Precision Farming/Farming 4.0, food chain tracking, e-health, etc.) [36], using case by case the most appropriate combinations of these technologies. With the availability of advanced skills and technologies available "as a service" in the Cloud, and the support of researchers and experts in the various "verticals", it is possible to implement initiatives (market-driven) for the provision of "digitized" value-added services in the field of "green" development [37]. It will be necessary to guarantee users the transparency of the process, i.e., the mix of advanced technologies used to generate the value of the chain (Fig. 1).

The Digital Transformation Impact on agriculture is [39]:

- Big Data: contribute in the decision-making process to increase efficiency in crop planning, intelligent irrigation systems development, pest control, weather alerts implementation.
- Biotechnology Genetic: modelling to increase the production while decreasing the volatility of the yield and the usage of pesticides; laboratory-grown meat.
- Synthetic biology: crops with higher nutritious values with less resources, and resistant to more variable climates.
- Internet of Things (IoT): IoT to collect and publish information on the production processes and the farm.
- Automation and Robotization: increased productivity by reducing the need for human workforce.

Fig. 1 The elements of digital transformation. Source: FAO [38]

- Artificial Intelligence (AI): contribute in agricultural robotics, soil and crop monitoring, and predictive analytics.
- Global Navigation Satellite System (GNSS): improve crop yield and reduce environmental impact through the application of for example farm machinery guidance, automatic steering, variable rate applications, yield and soil condition accurate geo-referenced monitoring.
- Drones: field and crop analysis and monitoring, variable rate applications, e.g. crop spraying and irrigation.
- Blockchain: enhance transparency, accountability and efficiency in agricultural insurance, land registration, and agricultural supply chains.
- Augmented Reality: optimization of the farming process.

There is an enormous potential for growth and market development in the agri-food sector (Fig. 2). In fact, only 2% of the Italian agricultural area uses robots and precision farming sensors, which are not uniformly distributed in the various regions of the country [40, 41]. Digital agriculture (ICT-assisted) varies between minus 1% and 4–5%, compared to 40–70% in China, Israel, and the USA. The most frequent solutions are systems that can be used transversally in several agricultural sectors, followed by those aimed at the cereal, fruit and vegetable and wine sectors. The focus on the Internet of farming is growing, albeit very slowly [27].

Digital entrepreneurship involves the transformation of existing businesses through novel digital technologies and the creation of new innovative enterprises characterized by: the use of digital technologies to improve business operations, the invention of digital business models and engaging with customers and stakeholders through digital channels [42].

Globally, there is an increasing number of initiatives to foster digital entrepreneurial activity related to the creation, development of 'digital start-up', including the agriculture and food sector [43]. Modern day farmers may be particularly suited to entrepreneurial activities. These days, farmers often design business plans, scout for funding, make use of farming enterprise 'incubators', and attend scientific conferences [16]. Youth farmers in particular are also more likely to take risks in their farm management.

An application of digital transformation is predictive analysis, which uses historical data to predict future events. Usually, historical data is used to build a mathematical model able to detect the most important trends. This predictive model is then applied to current data to predict future events or suggest measures to be taken to achieve optimal results [44]. Predictive analysis is often referred to as big data

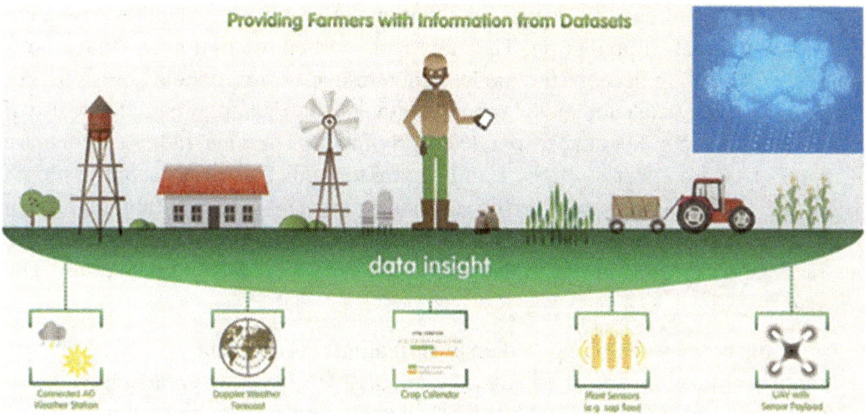

Fig. 2 Open data for agriculture and nutrition [38]

analysis. Engineering data, for example, is derived from sensors, instruments and connected systems around the world. Data from a company's business systems may include transaction data, sales results, customer complaints, and marketing information. Companies are increasingly making data-based decisions from this valuable source of information.

As competition increases, companies are looking for scope to introduce products and services into saturated markets. Data-based predictive models can help companies solve long-term problems in new ways. Predictive analysis is a process of using data analysis to make data-based predictions. This process uses data along with analysis, statistics and machine learning techniques to create a model for predicting future events [12, 45].

To exploit the value of big data, companies apply algorithms to large data sets using tools such as Hadoop and Spark. Data sources can consist of transactional databases, equipment log files, images, video, audio, sensors or other types of data. Innovation often comes from combining data from different sources. The term "predictive analysis" describes the application of a machine learning technique to create a quantitative forecast of the future. Often, supervised machine learning techniques are used to predict a future value [39]. Predictive analysis starts with a business objective: to use data to reduce waste, save time or cut costs. The process uses heterogeneous, often large data sets in models that can generate clear, immediately usable results in order to more easily achieve that goal, such as reducing material waste and inventory, and to obtain a finished product that meets specifications.

Predictive modelling uses mathematical and numerical methods to predict an event or result. A mathematical approach uses a model based on equations that describes the phenomenon under consideration. The model is used to predict a result in a given future state or time instant as a function of input changes. The model's parameters help to explain how the model's inputs affect the result. The numerical predictive modelling approach differs from the mathematical approach in that it uses models that are not easy to explain in equation form and often require simulation techniques to create a prediction. This approach is often referred to as "black box" predictive modelling because the model structure does not provide a description of the factors that associate the model input with the result. Examples include the use of neural networks to predict the corporate origin of a glass of wine, rather than the use of models based on decision trees with bagging to predict a debtor's credit rating.

Predictive modeling is often performed using curve and surface fitting, regression of historical series or machine learning approaches. Regardless of the approach used, the process of creating a predictive model is the same as other methods. The steps are:

1. Clean the data by removing outliers and treating missing data.
2. Identify a parametric or non-parametric predictive modelling approach to use
3. Pre-process the data in a form appropriate to the chosen modeling algorithm
4. Specify a subset of the data to be used to train the model
5. Training, or estimating, model parameters from the training data set
6. Carry out model performance or fitting tests to verify the suitability of the model
7. Validating the accuracy of predictive modeling on data not used to calibrate the model

2.2 *Methodology*

The present study analyzes the digital transformation to identify new approaches and opportunities in the agri-food sector that can be used to develop guidelines, to enhance production, consumer protection, and to analyze the value chain. The proposed approach is based on integrating participatory planning and the novel approach to imprecise assessment and decision environments as a possible methodological structure to acquire and evaluate the "complex" information collected on possible alternative scenarios in relation to digital transformation [46, 47].

The aim is to develop a methodological structure using suitable tools to acquire firstly, and process secondly, qualitative and quantitative information concerning the possible alternative scenarios of the problem under study. The opinions were collected through specific focus groups with local stakeholders, operators, consumers, and producers interested in the issue in question.

The proposed model is based on:

- The individualization of stakeholders involved (ten questionnaires);
- The definition of the alternative scenarios (definition of the three hypotheses of scenario: Farm, Chain and Consumers).

The model used focus groups as a social research methodology, aiming to acquire information on the opinions of stakeholders regarding a variety of scenarios for future development [47]. The matrices of impact and equity constitute the basis for the use of the discrete multicriteria evaluation NAIADE model (Novel Approach to Imprecise Assessment and Decision Environment) [48], which is able to manage

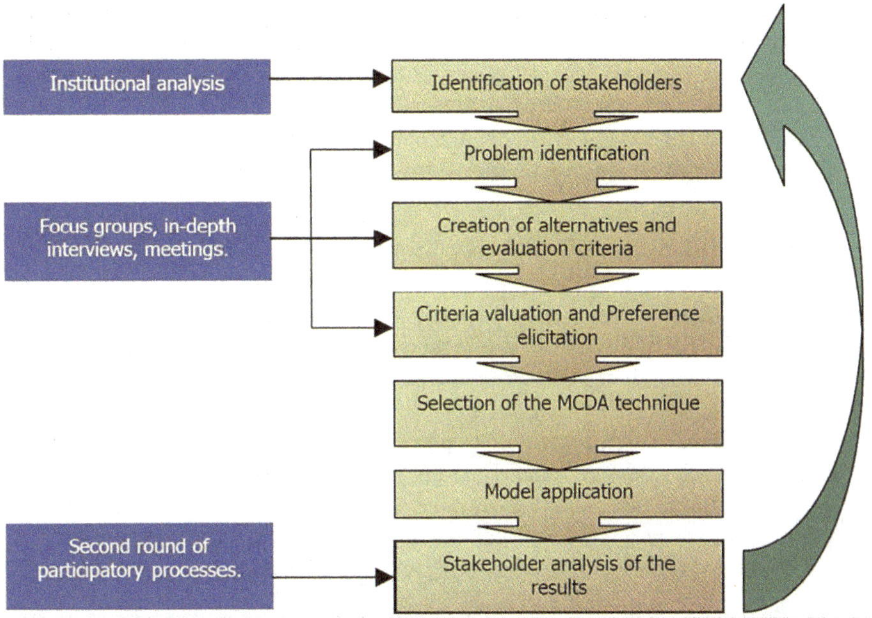

Fig. 3 The theoretic structure of the NAIADE model. Source: Munda [48]

qualitative and quantitative data in order to evaluate the measures of intervention (Fig. 3). This instrument supports the classification of the alternative scenarios proposed on the basis of determined decisional criteria and considerations of possible "alliances" and "conflicts" between the groups of stakeholders for the proposed scenarios, thus measuring their acceptability [49].

The objective of this study is to analyze the principal priorities, using as its methodology the model of digital transformation in the agri-food sector. The evaluation through the focus groups was divided into three phases, referring in this specific case to the potential repercussions. The questionnaire used for the interviews was designed to explore the perception of traceability issues in the citrus-supply-chain context and to evaluate the real needs of actors in the supply chain. It comprised ten questions aiming to collect information and opinions useful for the research related to three hypotheses proposed (Farm, Chain and Consumers):

- Scenario Farm: application of digital transformation for the valorization of the agricultural productions on the basis of the quality of the product.
- Scenario Chain: application of digital transformation in order to gain control information and prices along the chain.
- Scenario Consumer: application of the digital transformation is aimed at protecting the health of the consumer.

The input of the NAIADE method is constituted by the impact matrix (criteria/alternative matrix), including scores that can take the following forms: crisp numbers; stochastic elements; fuzzy elements; and linguistic elements (such as "very poor", "poor", "good", "very good", and "excellent"). To compare alternative scenarios, the concept of distance is introduced. In the presence of crisp numbers, the distance between two alternative scenarios with respect to a given evaluation criterion is calculated by subtracting the respective crisp numbers.

The classification of alternative scenarios is based on data from the impact matrix, used for:

- comparison of each single pair of alternatives for all the evaluation criteria considered;
- calculation of a credibility index for each of the aforementioned comparisons that measures the credibility of one preference relation, e.g. alternative scenario (a) is better/worse, etc. than alternative scenario (b) (preference relationships were used);
- aggregation of the credibility indices produced during the previous stage leading to a preference intensity index [* (a, b)] of an alternative (a) with respect to another (b) for all the evaluation criteria, associated the concept of entropy [H * (a, b)] as an indication of the variation in the credibility indices; and classification of alternative scenarios on the basis of previous information.

The final classification of the alternatives is the result (intersection) of two different classifications: the classification "+" (a) (based on the "best" and "decidedly better" preference relationships); and the classification "–" (b) (based on the "worst" and "decidedly worse" preference relationships). In relation to the objective of the present study, the analysis will be applied to the main priorities, for the assessment of the scenario that benefits most from digital transformation implementation in the agri-food sector.

3 Results and Discussion

The results of the present study provide a further multidisciplinary contribution to research on the management of digital transformation. Specifically, the analysis was conducted to address the research question:

What are the opportunities that Digital transformation for the agri-food sector?

The evaluation criteria that were used is technology, communication, data, internet of things, automation and networking. These criteria were defined on the basis of the purpose and objectives of the evaluation of the analyzed case, which can be considered representative of agri-food sector. The scenario Chain was revealed to be the best option for sharing, closely followed by scenario Farm and scenario Consumer, but all three hypotheses had positive evaluations (Table 1).

This provided the views of interested parties on the three suggested hypotheses. The selection of interested parties was based on their potential to assess the major advantages for agri-food sector. A total of six groups of stakeholders were involved: producers; trade associations; dealers; consumer associations; institutions and scientific associations. It is important to underline that the opinions of the interested parties in the NAIADE model can only be of a qualitative type, i.e. linguistic expressions: bad; poor; medium; good; very good; and excellent. The results show that a large number of stakeholders and groups of selected operators agreed with the assessment of the three hypotheses. The results of the multi-criteria analysis revealed that the scenario Chain was the predominant hypothesis, closely followed by scenario Chain, while scenario Consumer acquired only a lower rating (Table 2).

The results obtained through the analysis of the single answers were used to examine possible alliances or conflicts between the opinions of the interested parties regarding the decision on which hypothesis to adopt. The results in Table 3 show that a large number of interested parties, in addition to agreeing on the classification of the different hypotheses to be applied, agreed with scenario Chain, while noting that there were also significant consequences for the Farm and Consumer scenarios.

The results include different perspectives of digital transformation, the different groups involved the perception and acceptability of the proposed alternatives, which can lead to improving strategic decisions and creating innovative ideas and new solutions to enhance and protect, based on the possibilities offered by this participatory processes (Fig. 4).

The results obtained from this model, developed through the integration of a participatory tool and a multicriteria analysis, become strategic for investment choices in the agri-food system, particularly in relation to the current situation in which the supply chain, the farm and the consumer try to define their role through digital transformation.

Some specific priorities for future work are:

- Facilitate the collection of better data about digital technologies and digitalization at the regional and population level, particularly to show differentiated information about urban and rural areas;
- Create of sustainable business models that provides viable digital solutions for inclusion of small-scale farmers in the digital agriculture transformation process;
- Creation of an index to consider the development of digital agriculture in the context of cultural, educational and institutional dimensions of a given country, both in terms of the availability of basic conditions and enablers for digitalization

Table 1 Objectives and evaluation criteria of digital transformation for agri-food sector

Evaluation criteria	Scenario FARM "P"	Scenario CONSUMER "C"	Scenario CHAIN "M"
Technology	Excellent	Good	Excellent
Communication	Very good	Excellent	Excellent
Data	Good	Excellent	Very good
Internet of things	Poor	Very good	Excellent
Automation	Very good	Good	Very good
Networking	Good	Excellent	Very good

Source: our elaborations

Table 2 Classification of the scenarios at the highest consensus level

Groups and stakeholders		Scenario FARM "P"	Scenario CONSUMER "C"	Scenario CHAIN "M"
A1	Producers	0.7387	0.6311	0.8470
A2	Trade associations	0.6732	0.7334	0.6213
A3	Dealers	0.5764	0.6218	0.6138
A4	Consumer associations	0.5216	0.6392	0.8723
A5	Institutions	0.6283	0.5357	0.8231
A6	Scientific association	0.8329	0.7342	0.6379

Source: our elaborations

Table 3 Consensus and related prioritization of scenarios

	Consent levels			
Scenario classification	0.7525	0.7603	0.7323	0.7863
	M	M	C	M
	P	P	M	P
	C	C	P	C
Groups of "alliances" At each level of consensus	All groups	All groups except A3	All groups except A3 and A4	All groups except A2 and A6

Source: our elaborations

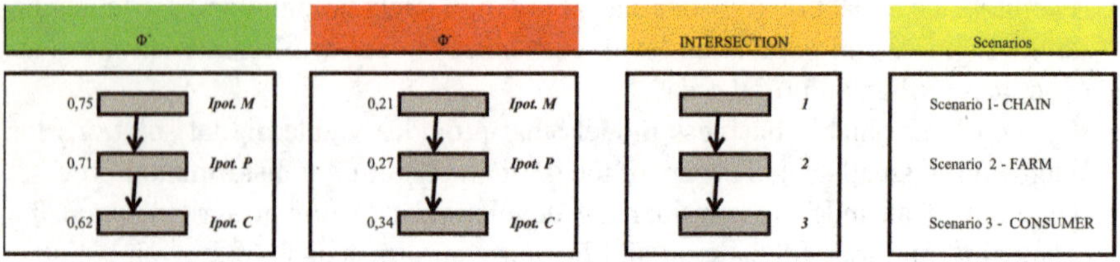

Fig. 4 Classification of alternative hypotheses and multicriteria assessment. Source: our elaborations

and the potential economic, social and environmental impacts of the process. This could involve further development of a Digital Agriculture Readiness Index, expanding on previous work by the FAO Regional Office for Europe and Central Asia in 2015. Such an index would help provide context for the development of future digital agriculture strategies for the FAO member countries, which starts with sensitizing countries to the concept of digital agriculture and the importance of digital technologies for the agri-food sector and continues with steps towards the digital agriculture transformation process.

4 Conclusion

The digitalization of agriculture will cause a significant shift in farming and food production over the coming years. Potential environmental, economic and social benefits are significant, but there are also associated challenges. Disparities in access to digital technologies and services mean there is a risk of a digital divide. Smallholder famers and others in rural areas are particularly at risk of being left behind, not only in terms of e-literacy and access to digital resources but also in terms of productivity and aspects of economic and social integration.

Simply introducing technologies is not enough to generate results. Social, economic and policy systems will need to provide the basic conditions and enablers for digital transformation. The "Law of Disruption" [50] states that technology changes exponentially, but economic and social systems change progressively and have trouble keeping up. Work is especially needed to ensure the necessary conditions for digital transformation are created in rural areas.

The Italian agri-food sector has begun to understand that digital innovation is a strategic lever, able to guarantee greater competitiveness to the entire supply chain, from production in the field to food distribution [51].

Digital transformation is fundamental to improve the competitiveness of the agri-food sector not only for economic needs but also for social and environmental ones [34]. The remuneration of all phases of the agri-food chain includes correct economic and contractual relations between all actors: agricultural producers, processing and distribution industry; greater cooperation and transparency, adoption of product and process innovations. This condition is essential to allow the improvement of quality, social and environmental standards, also in the logic of improving the efficiency of production, innovation and marketing processes. The success of agricultural enterprises increasingly depends on the ability to collect and enhance the large amount of data that will be generated, especially to achieve cost control and increase the quality of production. It should be noted, however, that there is still little clarity among those involved in the sector on how to exploit these opportunities. It is necessary to invest in the creation of skills, in a sector characterized by a level of 'corporate' culture and operational processes based more on the transfer of generational skills and knowledge than on innovation and optimization of production processes.

The food economy should therefore constitute a resource capable of responding to the most urgent and immediate needs of the planet [53], regulating the production of this primary resource, encouraging innovative and environmentally friendly production techniques, but above all ensuring a fair distribution of the resources produced through the aid of digital transformation.

References

1. UN DESA. (2017) The sustainable development goals report.
2. FAO. (2017) The state of food security and nutrition in the world. Rome.
3. Iakovou, E.; Bochtis, D.; Vlachos, D.; Aidonis, D. *Supply Chain Management for Sustainable Food Networks*; 2015; ISBN 9781118937495.
4. Townsend, R., Lampietti, J., Treguer, D., Schroeder, K., Haile, M., Juergenliemk, A., Hasiner, E., Horst, A. and Hakobyan, A. (2019) The future of food: Harnessing digital technologies to improve food system outcomes. World Bank.
5. Foti, V.T., Scuderi, A., Stella, G., Sturiale, L., Timpanaro, G., Trovato, M.R. (2018) The integration of agriculture in the politics of social regeneration of degraded urban areas. Green Energy and Technology: 99–111.
6. Weiss C.R. and Wittkopp A. (2005) Retailer concentration and product innovation in food manufacturing, European Review of Agricultural Economics, 32 (2):219–244.
7. Zheng, Z., Xie, S., Dai, H., Chen, X. and Wang, H. (2016) Blockchain challenges and opportunities: A survey. International Journal of Web and Grid Services, 14(4): 352–375.
8. Janvier-James, A.M. (2012) A new introduction to supply chains and supply chain management: Definitions and theories perspective. International Business Research, 5(1).
9. Sturiale, L. and Scuderi, A. (2011) Information and communication technology (ICT) and adjustment of the marketing strategy in the agrifood system in Italy. CEUR Workshop Proceedings, 1152: 77–87.
10. West, D.M. (2018) What is artificial intelligence? https://www.brookings.edu/research/what-is-artificialintelligence.
11. FAO (2018) The future of food and agriculture – Alternative pathways to 2050. Supplementary material. Rome. 64 pp. Licence: CC BY-NC-SA 3.0 IGO.
12. Wolfert, Sjaak, Lan Ge, Cor Verdouw, Marc-Jeroen Bogaardt. (2017) Big data in smart farming – a review.153: 69–80, https://doi.org/10.1016/j.agsy.2017.01.023.
13. Spielman, D. J. (2006) A critique of innovation systems perspectives on agricultural research in developing countries. Innovation Strategy Today, 2 (1):41–54.
14. Cagnina M.R., Cosmina M., Gallenti, G., Marangon, F., Nassivera, F. and Troiano, S. (2018) The role of information in consumers' behavior: A survey on the counterfeit food products. Economia Agro-Alimentare, 20(2):221–231.
15. OECD (2019). Measuring the digital transformation. A road map for the future. OECD.
16. OECD (2018) Agricultural Policy Monitoring and Evaluation 2018. Paris. Available online at https://www.oecd-ilibrary.org/agriculture-and-food/agricultural-policy-monitoring-and-evaluation-2018_agr_pol-2018-en.
17. Kayumova, M. (2017) The role of ICT regulations in agribusiness and rural development. World Bank. https://openknowledge.worldbank.org/bitstream/handle/10986/29041/121932-WP-ICT
18. Divanbeigi, R. and Saliola, F. (2017) Regulatory Constraints to Agricultural Productivity. Policy Research Working Paper No. 8199. World Bank.
19. Marinoudi, V., Sørensen, C.G., Pearson, S., and Bochtis, D. Robotics and labour in agriculture. A context consideration. Biosyst. Eng. 2019, 184, 111–121.
20. Bloom, N., Garicano, L., Sadun, R. and Van Reenen, J. (2014) The Distinct Effects of Information Technology and Communication Technology on Firm Organization. Management Science, 60(12): 2859–2885.
21. ITU (2019) Measuring digital development: Facts and figures. https://www.itu.int/en/ITU-D/Statistics/Documents/facts/FactsFigures2019.pdf
22. Nakasone, E., Torero, M. and Minten, B. (2014) The Power of Information: The ICT Revolution in Agricultural Development. Annual Review of Resource Economics, 6(1): 533–550.
23. Tinsley, E. and Agapitova, N. (2018) Private Sector Solutions to Helping Small holders Succeed. Social Enterprise Business Models in the Agriculture Sector. World Bank.

24. Durié, I. (2020) Digital Technology and Agricultural Markets.Background paper for The State of Agricultural Commodity Markets2020. Rome, FAO.
25. Ge, L. and Brewster, C.A. (2016) Informational institutions in the agri-food sector: Meta-information and meta-governance of environmental sustainability. Current Opinion in Environmental Sustainability, 18:73–81.
26. FAO (2016) Information and Communication Technology (ICT) in Agriculture – A Report to the G20 Agricultural Deputies. Rome, FAO.
27. Osservatori.net (2019). Smart Agri-food: boom dell'agricoltura 4.0. (Available on www.osservatori.net on 15 march 2020).
28. Vecchio, Y.; Agnusdei, G.P.; Miglietta, P.P. and Capitanio, F. (2020) Adoption of Precision Farming Tools: The Case of Italian Farmers. Int. J. Environ. Res. Public Health, 17, 869.
29. Lee, H., Mendelson, H., Rammohan, S. and Srivastava, A. (2017) Technology in agribusiness: Opportunities to drive value. White paper, Stanford Graduate School of Business.
30. Kempe, M., Sachs, C. and Skoog, H.(2017) Blockchain use cases for food traceability and control: A study to identify the potential benefits from using blockchain technology for food traceability and control. Axfoundation, SKL Kommentus, Swedish Country Councils and Regions, Martin & Servera, and Kairos Future.
31. Krishnan, H. and Winter, R.A. (2012) The economic foundations of supply chain contracting. Foundations and Trends in Technology, Information and Operations Management, 5:147–309.
32. Scuderi, A., Sturiale, L. and Timpanaro, G. (2018) Economic evaluation of innovative investments in agri-food chain. Quality-Access to Success, 19 (51):482–488.
33. Tapscott, D. (2016) Blockchain Revolution, Random House, New York, NY.
34. Scuderi, A., Foti, V.T., Timpanaro, G., 2019. The supply chain value of pod and pgi food products through the application of blockchain. Quality - Access to Success, 20:580–587.
35. Timpanaro, G., Foti, V.T., Scuderi, A., Schippa, G. and Branca, F. (2018) New food supply chain systems based on a proximity model: The case of an alternative food network in the Catania urban area. Acta Horticulturae, 1215:213–217.
36. Abeyratne, S.A. and Monfared, R.P. (2016) Blockchain ready manufacturing supply chain using distributed ledger. International Journal of Research in Engineering and Technology, 5 (9):1–10.
37. Chen, L. and Lee, H.L.(2016) Sourcing under supplier responsibility risk: The effects of certification, audit, and contingency payment. Management Science, 63: 2795–2812.
38. FAO (2019) Digital Technologies in agriculture and rural areas. Briefing paper. By Nikola M. Trendov, Samuel Varas and Meng Zeng. Food and agriculture Organization of the United Nations. Rome.
39. FAO (2020) Agricultural markets and sustainable development: global value chains, smallholder farmers and digital innovations. Food and Agriculture Organization of the United Nations. Rome, 2020.
40. De Molli, V., De Biasio, A., Lovati, S., Andrisani, N., Barchiesi, R., Brioschi, B. and Brugora, A. (2017) Sfide e priorità per il settore alimentare oggi. Strumenti e approcci per la competitività, The European House – Ambosetti, Retrieved from https://www.ambrosetti.eu/ricerche-e-presentazioni/settore-alimentare-oggi/
41. De Paulis, G. (2015) Food marketing: web e social. Strategie di business online per avere successo nell'agroalimentare. FrancoAngeli, Milano, 2015.
42. European Commission. (2013) Powering european public sector innovation: Towards a new architecture. Available online at https://ec.europa.eu/digital-single-market/en/news/powering-european-public-sectorinnovation-towards-new-architecture
43. Kurt Salmon (2017), The Future of food: New Realities for the Industry.
44. Jin, X.; Yang, N.; Wang, X.; Bai, Y.; Su, T. and Kong, J. (2019) Integrated predictor based on decomposition mechanism for PM2. 5 long-term prediction. Appl. Sci., 9, 4533.
45. Sonka S. (2016) Big data: Fueling the next evolution of agricultural innovation. 4 (1): 114–136.

46. Munaretto, S., Siciliano, G. and Turvani, M. (2014) Integrating adaptive governance and participatory multicriteria methods: A framework for climate adaptation governance. Ecology and Society, 19.
47. Scuderi, A. and Sturiale, L. (2016). Multicriteria evaluation model to face phytosanitary emergencies: The case of citrus fruits farming in Italy. Agricultural Economics, 62:205–214.
48. Munda, G. (2006). A NAIADE based approach for sustainability benchmarking. International Journal of Environmental Technology and Management, 6:65–78.
49. Sturiale, L, Scuderi, A. (2018) The evaluation of green investments in urban areas: a proposal of an eco-social-green model of the city. Sustainability, 10 (12): 4541.
50. Downes, L. (2009) The Law of Disruption. Hachette. United Kingdom.
51. Sturiale, L. and Scuderi, A. (2016) The digital economy: new e-business strategies for food Italian system. International Journal of Electronic marketing and Retailing, 7 (4):287–310.
52. Scuderi, A. and Sturiale, L. (2019) Evaluation of social media strategy for green urban planning in metropolitan cities. Smart Innovation, Systems and Technologies, 100:76–84.
53. Weis, T. (2007) The Global Food Economy: The Battle for the Future of Farming. Ed. Zed Books.

Precision Agriculture's Economic Benefits in Greece: An Exploratory Statistical Analysis

Athanasios Falaras and Stratos Moschidis

1 Introduction

World population is expected to be around ten billion people in 2050, while the climate change situation does not allow to increase arable lands [1]. So, a new challenge rises. More people must be fed, therefore there will be a demand to increase global food production while retaining quantity of arable lands steady. Then modern farmers have to focus on how to effectively utilize arable land they use.

The purpose of this study is finding how Greek districts are bundled on crop type (arable drops, horticulture, permanent crops), arable land size, improved agricultural field actions/processes (irrigation, fertilization and pesticide use, plant disease prevention, harvesting, sowing, ploughing, agricultural vehicle use) and specific economic benefits (reduced resource spending, increased labor productivity, improved use of commodities and equipment, increased product quantity and quality, increased income, improved environment protection). This grouping would make it possible to have a wide view of Greece's agriculture sector and see how improved agricultural processes benefit farmers.

To achieve the goals that have been set, an exploratory study took place on almost all districts of Greece. This research could be the start of a series of studies, focusing on agricultural economics in Greece. It could also be a useful farmers' guide. Every farmer has specific investing restrictions on improving agricultural processes due to limited budget. Consequently, every farmer could choose an optimized solution in terms of the agricultural processes she/he should focus, to improve farm sustainability.

A. Falaras (✉) · S. Moschidis
University of Macedonia, Thessaloniki, Greece
e-mail: afalaras@uom.edu.gr; smos@uom.edu.gr

D. D. Bochtis et al. (eds.), *Information and Communication Technologies for Agriculture—Theme IV: Actions*, Springer Optimization and Its Applications 185,
https://doi.org/10.1007/978-3-030-84156-0_9

To improve agricultural actions, Precision Agriculture (PA) practices and related technologies are the timeliest way providing also a sustainable solution to agricultural development. PA could be defined as

> "a whole-farm management approach using information technology, satellite positioning (GNSS) data, remote sensing, and proximal data gathering. These technologies have the goal of optimizing returns on inputs whilst potentially reducing environmental impacts" [2].

As a concept, PA emerged around 1990, when Global Positioning System (GPS) and sensor technologies were made widely available at civil economy, while important advances took place in soil sampling, statistics, and computing power [3, 4]. Sensors started to be placed on agricultural equipment to measure specific yield and soil properties, aiming at maximizing outputs and reducing inputs [5, 6].

Till the appearance of PA, a farm field was considered as a uniform entity. The decisions a farmer made considering various agricultural actions, applied on the whole field [7]. PA offers a new perspective on agricultural fields' management. Therefore a farm field is divided to management zones (MZ), which are areas with similar characteristics under a specific criterion or a set of criteria [8–10]. Based on this approach, a farmer can differentiate decisions regarding a field by optimizing inputs on each individual MZ.

2 Literature Review

PA technologies are categorized into data acquisition, data analysis and evaluation, and precision application technologies [11]. The literature of this chapter will focus on precision application technologies, which are the most directly affecting agricultural actions and there are references to Unmanned Aerial Vehicles (UAV) and weather stations, which are some of the most trending equipment these times. It is interesting to see PA's application in practice, through the presented literature and seek benefits that come out as a result.

One of the most important PA technologies is the variable rate application (VR). It is based on the MZ philosophy and its application varies depending on the agricultural process that takes place on the field. There are two VR methods. The first one is called "map-based" and is based on (off-line) historical data. Information are extracted from a GIS so as to control a set of farm processes to optimize inputs. The second one is called "sensor-based". In this method sensors are used to detect soil properties and (on-line) adjust inputs [3]. VR is distinguished into the variable rate nutrient application (VRNA), the variable rate irrigation (VRI), the variable rate pesticide application (VRPA) and the variable rate planting/seeding (VRP/VRS).

Extended fertilizer use harms environment, damages soil quality and pollutes water. Optimized fertilizer use can achieve a considerable chemicals use reduction [6]. VRNA helps farmers optimize fertilizer application by measuring crop nutrient status and adjusting fertilizer rate accordingly [12]. Consequently, yield quantity can be increased, and crop quality can be improved. Therefore, economic gain is increased and, in parallel, agricultural activities are less hazardous for the environment [13]. A study on apples in Greece applied homogeneous and variable rate fertilization on orchard [14]. The results showed significant nitrogen inputs reduction on VRNA case. Yield quantity was a little smaller, but the farmer's profit increased, and the quality of apples improved. The same comparison was conducted by another research team in Italy during 2005–2008 [7]. The results showed that it is possible to identify the N ideal fertilization rate so as to maximize yield production and economic profit while affect the environment as least as possible. In summary, VRNA's applications of Nitrogen (N) are considered profitable by 72% of studies in corn and 20% in wheat. Respective applications on Phosphorus (P) and Potassium (K) are profitable by 60% of studies in corn [15].

Irrigation is one of the most critical actions on agriculture, since the invention of agriculture. Its optimization can lead to substantial water saving [6]. VRI (Variable Rate Irrigation) achieves micro-irrigation across the farm according to every MZ's requirements [13]. It increases crop yield, water efficiency, while it maintains soil temperature and might result in less pesticide use. By testing the application on corn, the results indicated higher yield output and improved water efficiency [16]. The HydroSence project studied the VRI on cotton in Greece. The results showed 5–34% savings in water consumption, while yield output increased from 18 to 31% [13]. VRI was also tested on New Zealand on a maize farm. MZs were created based on electrical conductivity and soil water availability. The results showed water savings of around 26.3% [17].

VRPA can reduce the quantity of pesticide use. Reduced pesticide use improves the final quality of agricultural products [13]. Water and ground contamination is also reduced and the biodiversity is less affected than with conventional pesticide use. So there are several environmental benefits [18]. In a study on maize-based cropping systems within Europe, it was evaluated that VRPA can result in net profit in about 3–4 years [19].

VRP/VRS is a technology also based on the MZ philosophy on which farmers can selectively plant/seed depending to soil potential. The main benefit is increased yield quantity [13, 20, 21].

Another important aspect of farming is vehicle use in agriculture. PA also applies on this action. There are two matters that PA can improve on an agricultural vehicle, namely navigation and route optimization. A possible solution for improved navigation comes through the Multi Global Navigation Satellite System (GNSS) precise point positioning (PPP) which provides high accuracy positioning, with higher flexibility and potentially lower capital and running costs [22]. Route optimization is possible by use of real–time extend GNSS which optimize farmers' routes by avoiding missed-area and overlaps in area coverage operations [23].

Vehicle use optimization technologies can be used during seeding, tillage, planting, weeding, harvesting and autonomous vehicles enabling [13]. The use of machine guidance on corn and soybean crops on Kentucky resulted to 2.4% cost savings in seeding, 2.2% in fertilizing, and 10.4% in planting operations [24]. It can also reduce working hours by 6.04% and fuel consumption by 6.32% [25]. A study in Alabama demonstrated that the average net returns could be calculated between

83 and 612 € ha^{-1} [26]. Machine guidance has also a positive effect on fertilizer and pesticide costs by reducing them 3–5% [27].

Control traffic farming (CTF) is a system which optimizes driving patterns, operations and input applications. This system was implemented in Denmark on cereal farms, in areas where the arable land was greater than 300 ha. Its application resulted in fuel cost reduction by 25–27%. Application of CTF in Australia successfully cut machinery cost by 75%, while crop yields rose in different levels depending on crop [13, 27].

Precision physical weeding (PPW) technology offers optimized herbicide application and weed control according to observed weed density without damaging crops [12]. Its function is based on a continuous ground-based image analysis system that locates crop row on the field. It benefits the environment due to reduced pesticide and fuel consumption, and also due to the fact that a tractor using a weeding implement will tackle with lower draught forces in comparison with the mainstream methods [13, 28].

UAV is defined as a device which can be remotely controlled. There have been various UAVs classifications. A first one is based on aerodynamic features. A second one on autonomy level. A third one is based on physical characteristics (size, weight). Finally, a fourth one is based on power source [29].

UAV utilizes optimal sensors and uses various cameras to capture Red Green Blue (RGB) images which are useful to extract the Normalized Vegetation Index (NDVI) [4, 30]. NDVI values can provide useful data regarding crop disease, water stress, pest infestations, nutrient deficiencies, and other relevant conditions affecting crop productivity. UAVs can be applied on harvesting, spraying and yield estimation, weed detection, disease outbreaks, and insect infestations prediction [29, 31–33]. There had been plenty UAV applications. One of them took place in a vineyard in Spain. An UAV collected data via thermal sensors, which were used to extract information about the vineyard water status [29]. Another application took place on sweet potatoes, grapes asparagus and sugar crops in Peru. A UAV equipped with optical sensors was used to calculate the NDVI [29]. The results showed that the targeted use of pesticides results in pesticide quantity, economic expenses and ecological impact reduction. As a result, health crop status can be successfully estimated and plant diseases can be prevented. A similar application took place in a pomegranate orchard in Greece. The researchers concluded that UAVs are useful for supporting irrigation systems, accompanying maintenance systems and discriminating MZs [29].

A weather station is a system which is used to collect a series of data to monitor field weather. It can record gust speed, wind speed, leaf wetness, soil moisture, Photosynthetically Active Radiation (PAR), soil temperature, air pressure, air temperature, rain, relative humidity, dew point, solar radiation and electrical conductivity. A weather station is a prerequisite for applications such as control irrigation and diseases forecasting [15, 34, 35].

This study was carried out within the context of innovation. Innovation is an important aspect of economic development [36].

> "An innovation is the implementation of a new or significantly improved product (good or service), or process, a new marketing method, or a new organizational method in business practices, workplace organization or external relations" [37].

It can be classified as product, process, organizational and marketing innovation [37]. PA can definitely be regarded as a way to innovate. As the presented literature has already demonstrated, PA improves agricultural processes. Hence, PA technologies are regarded as process innovations.

> "A process innovation is the implementation of a new or significantly improved production or delivery method. This includes significant changes in techniques, equipment and/or software" [37].

3 Research Methodology

3.1 Data Collection

Aiming to investigate the range of innovation diffusion on agriculture in Greece and see the differentiation of Greek districts, a research was conducted among farmers in Greece. So as to successfully represent the agricultural population, Greece was divided to districts, according to the Association of Greek Regions (EN.P.E) [38] and there were certain answer goals on each district according to the farmers' population which is online published by Hellenic Statistical Authority [39]. A questionnaire was chosen as the data collection method and the percentage sampling was selected as the sampling method. The research was conducted specifically on Greece mainland and Crete district.

The questionnaire was created as a Google form. It included 16 questions, where each question represents a variable. The initial thought regarding the questions was to include directly PA technologies as variables. However, Greek farmers are generally low educated and not informed about PA [40]. Thus, on a second thought it was decided that PA technologies would be indirectly implied by variables which express actions, e.g. actions that a farmer performs on farm soil, so as to make the questions' content simpler.

Most questions were divided in two categories. The first one refers to actions that take place on a field (irrigation, fertilization and pesticide use, plant disease prevention, harvesting, sowing, ploughing, agricultural vehicle use) and the second one refers to the economic benefits that a farmer could have by improving farm actions, reduced resource spending, increased labor productivity, improved use of commodities and equipment, increased product quantity and quality, increased income, improved environment protection. Every question category followed a specific formulation based on the Community Innovation Survey (CIS) questionnaire, which is used by Eurostat about innovation at industry and services sectors every 3 years (European Commission, 2018). The question formulations are presented below:

- Did you introduce new or improved processes of "variable_name" during the last 3 years?
- Did you have "variable_name" during the last economic year in comparison with 3 years ago?

The possible answers for these question formulations were "yes" or "no".

There were also three other questions. The first one regarded the Greek district in which each respondent resided. The possible answers were the districts of Eastern Macedonia and Thrace, Central Macedonia, Western Macedonia, Epirus, Thessaly, Western Greece, Central Greece, Peloponnese and Crete. The second one concerned the crop category. Based on HELSTAT's categorization there are three main crop categories [41]:

- Arable crops (wheat, legumes, industry plants, aromatic plants, fodder plants, melons, potatoes)
- Horticulture
- Permanent crops (vineyards, orchards, citrus, fruit trees, dry fruits, olive groves)

The third one referred to the number of hectares each farmer cultivates. This question had four possible answers. The first category included farmers cultivating up to 8 hectares, the second one from 8 to 20 hectares, the third one from 20 to 40 hectares and the last one regarded arable land above 40 hectares. There are studies that do not regard professional farmers, those who utilize less than 8 hectares [42]. But it is regarded interesting to search if this farmers' category differentiates among other farmers.

All questions on the Google form questionnaire were chosen as closed, to make them as simple and less time-consuming as possible and mandatory so as to avoid missing values. Missing values is a serious issue for nominal data, hence it had to be avoided. An exception was made for the crop type question, in contingency that a farmer might cultivate multiple crop types. On the questionnaire beginning, a brief description of the survey purpose was mentioned and contact details where included in case of further clarifications needed. It was also stated that answers were completely anonymous. Before the process innovation questions a brief description of innovation and process innovation definition was quoted.

The answer collection strategy was the indirect connection with farmers via people, who have regular communication and trust them. Such people are agricultural consultants, agricultural cooperatives, agronomy shop owners, and accountants. These middlemen could carry forward an email in which the research description, contact details and the questionnaire link on Google forms were included. On special occasions a printed form of the questionnaire was used and, in a few cases, farmers called back asking to answer the questionnaire as a phone call interview. The research started on July 2019 and ended on December of the same year.

Like every research has its difficulties and setbacks, so this research does. The main problem acquiring answers was that in Greece there is not an official database to include farmers' names and contact details. Also Greek farmers were considered e-illiterate due to their low education level [40]. As a first step there was an effort to directly connect with farmers having no results. A first try to deliver the questionnaire in printed form proved to not be a good option, as most of subjects left unanswered questions, which lead to missing values. A printed questionnaire form would not also be an option in terms of printing and transfer cost. In the cases where questionnaires were sent back in printed form, 50% of them had missing values on nominal variables, leading to their rejection from the final results. Furthermore, the direct connection with farmers, showed that they are distrustful and hesitant people. In 90% of phone calls, they refused to answer the questionnaire. So even if a database existed it might not be useful. Finally, Attica district is mostly an urban area and Greek islands focus their economy on fishery. Even though they do have a small agricultural population, they don't focus their regional economy on agriculture. Consequently, Attica, Ionian Islands, North Aegean and South Aegean districts were excluded from the research. By this way the research could be more focused and less time-consuming. Another issue of this study is the fact that PA is indirectly implied at the questionnaire. As a result, it is not certain, if an agricultural process is improved due to PA. Maybe other innovative practices were applied, but it is regarded that nowadays most innovative activities concern PA.

In order to answer our exploratory question, automatic clustering was used and specifically the agglomerative hierarchical clustering (HAC), or else called ascending hierarchical classification (CAH). On this method a criterion based on their distance is used, to join objects in pairs. The clustering algorithm is completed, when all the original objects are merged into one.

3.2 Data Analysis Process

A data differentiation could be between categorical and continuous data. There are statistical methods, which are most suitable for continuous data and others that are suitable for categorical data. But there are datasets in which both data types are included. In terms of clustering a mixed typed dataset is an important field of research interest. This matter is tackled by applying a strategy of using a combination of sequential dimension reduction and clustering [43–45].

CAH is a clustering method based on the criterion of inertia [46–49]. The criterion of inertia is also called the generalized criterion of Ward [50] and is particularly prevalent in applications of humanities and society [51].

The result of the Ascending Hierarchical Classification is a dendrogram, i.e. a diagram in the form of a tree, inclined or inverted, depending on the program used, on which the successive compounds of the elements are depicted. The closer to the top of the tree chart, the more general the teams are, and the closer to the base of the tree chart, the more specialized the teams are.

Let I_o be the total inertia of n objects (points) with respect to their center of gravity g. We consider a grouping of n objects into s groups G_1, G_2, . . ., G_s,. Each G_i group has a g_i center of mass with mass $m_i$the sum of the masses of its objects. The inertia of the groups' center of gravity with respect to the center of gravity g is called the "order inertia" and is denoted by I_Δ.

$$I_\Delta = \sum_{i=1}^{s} mi\, d^2(g_i, g)$$

Since the center of gravity of the center of gravity is g, the order inertia measures the deviation of the groups' centers from g. The greater the extraclass inertia, the more distinct the groups are.

In each G_i group, the inner inertia is calculated, i.e. the inertia of its points with respect to the center of gravity g_i. Inner inertia measures the deviation of its points from its center of gravity, so the lower the inner inertia, the more compact the group is. The sum of the inner inertia of groups is called intraclass inertia (IE). Small intraclass inertia indicates compact groups. The Huygens theorem states that in each grouping the sum of extraclass and intraclass inertia remains constant. What is changing is the rate of extraclass and intraclass inertia. That is:

$$I_{o\lambda} = I_\Delta + I_E$$

It is worth noting that a group consisting of a single point has an inner inertia of zero as its center of gravity coincides with that point. Therefore, considering the initial state of the n objects as a trivial grouping of n groups with an element, there comes the realization, that the total inertia I_o is only intraclass inertia. Also considering the n objects as a group (final state), then its center of gravity coincides with g and therefore the extraclass inertia is equal to zero. In this case the total inertia I_o is only intraclass inertia [52].

When switching from a group s to the next s − 1 group grouping, two groups merge into one, while the s − 2 groups remain unchanged. What does not change from one grouping to another is the rates of intraclass and extraclass inertia, as their sum is constant.

Specifically, when switching from s groups to s − 1 groups, intraclass inertia increases and consequently the extraclass inertia decreases. That is, by limiting the number of groups, the groups become less compact and less distinct.

According to Ward's generalized criterion [49], "when switching from s-grouping to s − 1 grouping, we unite those groups so that the increase of intraclass inertia becomes minimal". It is proved that the increase of intraclass inertia while uniting groups G_i, G_j with mass m_i, m_j and centers g_i, g_j is:

$$\delta = \frac{mimj}{mi + mj} d^2(g_i, g)$$

where $d^2(g_i, g_j)$ is the distance of group centers G_i, G_j [53].

While the method proceeds, so as to record the most statistically significant variables, a statistical test was used based on which one is tested for group c and one answer is *j*, if the percentage of answer j is significantly higher in group *c* than the percentage of answer *j* in the whole sample (A) [54].

The process is evolving as in case of classical case studies. The null hypothesis H_0 states that the percentage of the answer *j* in group *c* is equal to the percentage of the answer *j* in the whole sample. The alternative H_0 is that the percentage of the answer *j* is greater in group c than the percentage of the answer j in the whole sample.

The statistical test used follows the standard normal distribution and at each value of this test corresponds a probability of *p*. The corresponding probability *p* is the probability of incorrectly rejecting the null hypothesis of the equality of the percentages of the answer in the group and the whole sample. The higher the percentage of response *j* in group *c*, the lower the corresponding probability and the higher the control value. The higher the control value, the higher the probability to reject the H_0 hypothesis.

It is noted that because a one-sided test is used following the normal distribution, the answers are statistically significant at a significance level of $p = 5\ \%$, if the value of the test is greater than 1.65. Consequently, for the interpretation of the groups, those answers with a probability below 5% or a control value greater than 1.65 can be implemented [55].

4 Results

In total, 1032 answers were collected. After the data collection process, the next step was to confirm that the collected answers successfully represent the population. As far as the geographical distribution of farmers concerned, the Table 1 demonstrates the results and the nationwide distribution of farmers in Greece by region. By comparing the columns named "Total Holdings (% of total population)" and "Questionnaire's answers per region (% of total answers)", the collected sample successfully represent the agricultural population of Greece.

On the arable lands use, based on ELSTAT's data 6976 hectares are used for arable crops, 256 hectares are used for horticulture and 4391 thousand hectares are used for permanent crops [41]. So, a percentage calculation can take place as it is demonstrated on Fig. 1.

As the question regarding this variable was multiple choice, from this variable three new dichotomous variables come up with the possible answers of "yes" or "no". The next pie chart (Fig. 2) shows off the distribution of agricultural land among farmers.

The results demonstrate that there is small diversification in terms of agricultural land size. The results are reasonable regarding the fact that extensive agricultural plains in Greece exist only in Thessaly and in Eastern Macedonia. The Fig. 3 demonstrates the positive answers on process innovations.

The most dominant innovation concerns fertilizer use, and the least observable innovation concerns ploughing. As permanent crops do not need ploughing and this crop type represents 54% of the answers, this is an expectable result. The Fig. 4

Table 1 Sample—population comparison

Region (DIS)	Coded name	Total holdings	Total holdings (% of total population)	Questionnaire's answers per region	Questionnaire's answers per region (% of total answers)
Eastern Macedonia and Thrace	DIS1	51,628	9	87	8
Central Macedonia	DIS2	96,482	17	163	16
Western Macedonia	DIS3	23,089	4	46	4
Epirus	DIS4	29,462	5	49	5
Thessaly	DIS5	60,323	10	126	12
Central Greece	DIS6	65,859	11	110	11
Western Greece	DIS7	80,502	14	133	13
Peloponnese	DIS8	88,410	15	158	15
Crete	DIS9	87,040	15	160	16
Summary		582,795	100	1032	100

Fig. 1 Crop type answers (% of total answers)

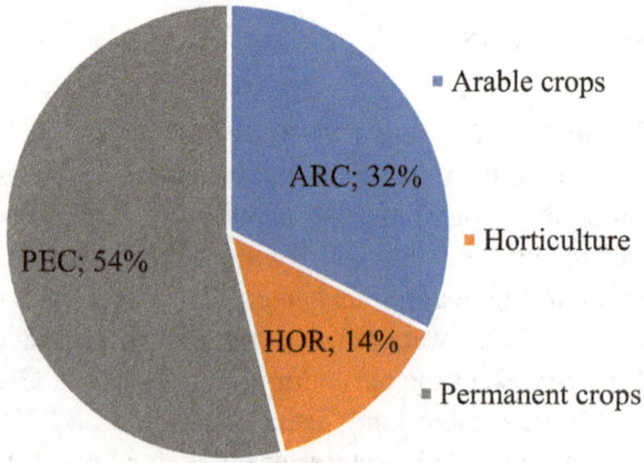

Fig. 2 Arable land answers in hectares (% of total answers)

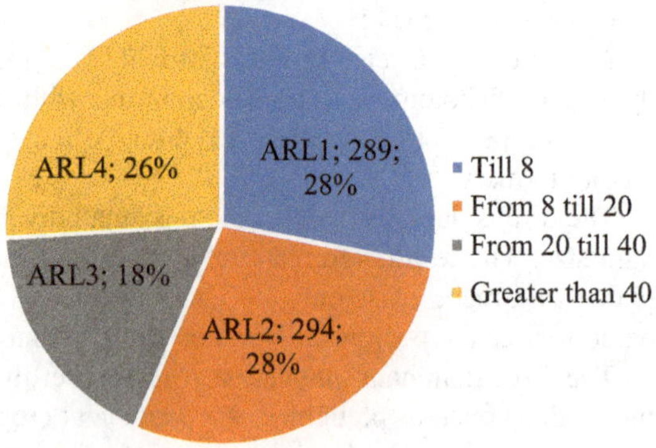

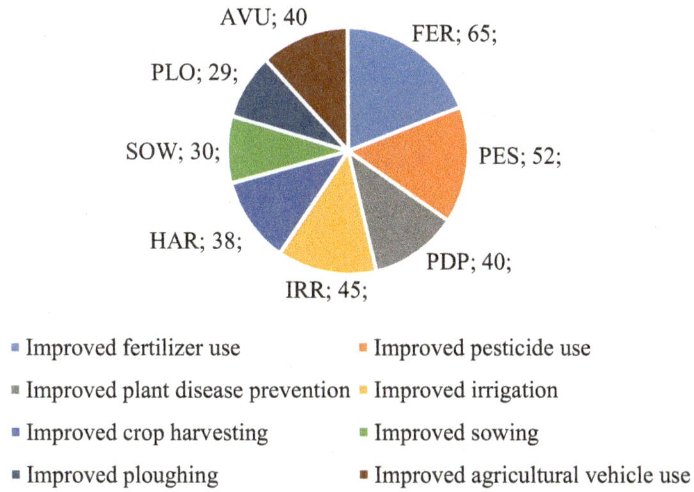

Fig. 3 Positive answers on process innovations (% of total answers)

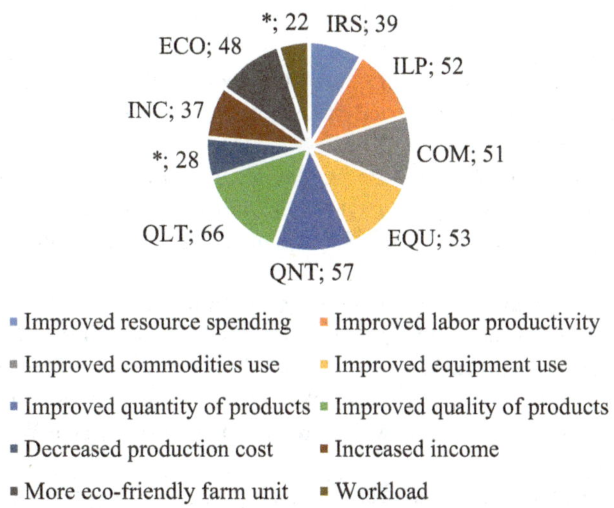

Fig. 4 Positive answers on economic benefits (% of total answers)

presents the percentage of farmers who have been benefited for each possible economic benefit.

The "decreased production cost" and the "workload" variables had one-sided answers as most farmers answered "no". So these characteristics would not be useful as variables and they were excluded from the data analysis. Most farmers regard, that they have improved their products' quality, while their workload remains at least the same.

Table 2 Nodes of CAH

N[a]	A[b]	B[c]	W[d]	δ[e]
10	1	3	0.12609	0.00066
11	4	9	0.20723	0.00087
12	6	7	0.23362	0.00122
13	2	5	0.28347	0.00216
14	11	8	0.35679	0.00272
15	10	13	0.40956	0.00405
16	14	12	0.59041	0.00646
17	15	16	1	0.0277

[a]Indicate the new nodes that are created immediately after the initial nodes—objects. Their numerical name refers to the order in which they were created
[b]The nodes in which this node is included (or in which it breaks) are displayed
[c]The nodes in which this node is included (or in which it breaks) are displayed
[d]This is the percentage of the node weight, i.e. the number of people that the node includes in percentage
[e]Records the increase in intraclass inertia in the creation of the new node (or the decrease in the order inactivity respectively)

The first statistical analysis' table shows the nodes of the Ascending Hierarchical Classification. It contains the created nodes together with some parameters, which are further analyzed (Table 2).

Therefore, as can be seen, a sharp increase of the intraclass inertia results in the level of the number 15 node, resulting in the creation of a classification with 3 groups. Next, we receive the result of CAH through the dendrogram that is presented schematically (Fig. 5).

According to the above diagram we have in the formation of three groups. The first group consists of node 15, including the districts of Eastern Macedonia and Thrace, Central Macedonia, Western Macedonia and Thessaly. The second group consists of node 14, including the districts of Epirus, Peloponnese and Crete. The third group consists of node 12, in which are included the districts of Western Greece and Central Greece.

Thus, all the characteristic answers for each group resulting from CAH emerge. These characteristic answers are recorded in Table 3. At next, the statistically significant variables are presented for each group in a decreasing order.

As regarding to the dichotomous variables encoding, the number "1" next to the encoded name means "yes" to the concerning question and the number "0" means "no" to the concerning question.

When a variable exists in more than one group then it is included in the group that is most important (e.g. HOR1 will be included in group 12 and not in group 14). In this way, groups are maintained with distinct (different) characteristics compared to the rest ones. According to the Table 3, the most important variables of group 12 (and in addition in order of importance) are HOR1, ILP1, INC1, HAR1, PLO1, AVU1, and QLT1. Respectively, the most important variables of group 14 are

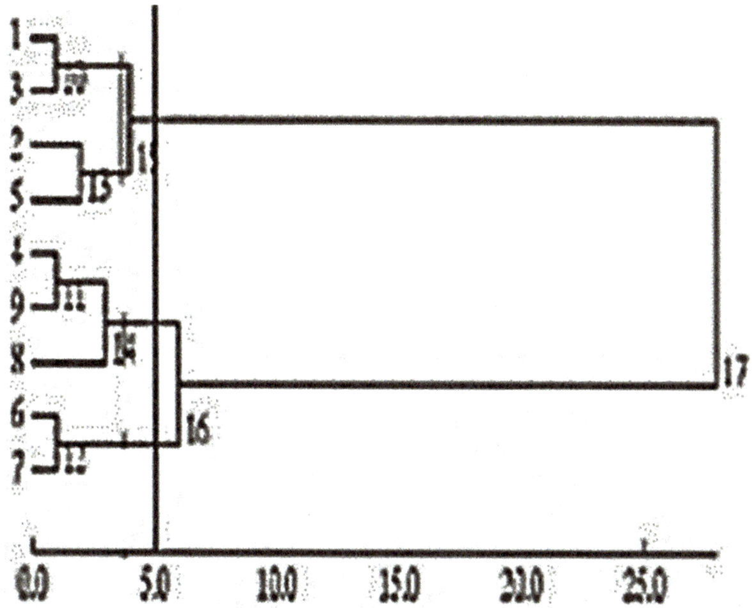

Fig. 5 Dendrogram of CAH

Table 3 Description of the most statistically significant variables of groups 12, 14, 15 in terms of significance p = 5%

Coded name	12	Coded name	14	Coded name	15
HOR1	5.3	ARC0	12.4	ARC1	13.9
ILP1	2.7	ARL1	9.5	ARL4	10.4
INC1	2.4	PEC1	7.7	PEC0	9.4
PES0	2.3	SOW0	4.3	SOW1	4.6
HAR1	2.2	ARL2	4.08	FER0	3.3
PLO1	2.2	FER1	3.5	WAT0	3.3
AVU1	2.1	IRR1	3.0	ARL3	2.6
IRR1	1.7	HOR1	2.7	AVU0	2.5
QLT1	1.7	AVU0	2.2	HAR0	2.5
ARC0		HAR1	2.1	HOR0	2.4
ARC1		ARC1		PDP1	2.1
Coded name	12	Coded name	14	Coded name	15
HOR0		HOR0		ILP0	2.0
PEC0		PEC0		QLT0	2.0

ARC0, ARL1, PEC1, SOW0, ARL2, FER1, and IRR1. The most important variables of group 15 are ARC1, ARL4, PEC0, SOW1, FER0, IRR0, ARL3, HAR0, HOR0, PDP1, ILP0, and QLT0.

5 Discussion

There are quite interesting findings regarding the results. With respect to the crop type most of the respondents cultivate permanent crops. This result can be explained by the fact that 44% of the sample operate in Western Greece, Peloponnese and Crete. These regions' climate favor permanent crops, so this fact justifies the results.

In terms of arable land, there are no recent data about arable land distribution. In addition, big agricultural plains exist only in Eastern Thessaly and Eastern Macedonia. This fact means that most farmers have not the possibility to own vast agricultural land. There is also the fact that many farmers' land is scattered through an area. In each case the results demonstrated that there are small differences between each category on this variable.

The majority of farmers are innovative in fertilizer and pesticide use. Harvesting, sowing and ploughing are processes that take place on arable crops. As arable crops represent 32% of the total sample, this justifies that the innovation percentage on these processes is low. Furthermore, an agricultural vehicle is most useful for a minimum size of arable land. Hence, in relative comparison the 40% of innovative farmers is a good innovation percentage. The rest process innovation percentages indicate that farmers in Greece keep improving their work, either the innovation refers to an improved existing method, or the implementation of a new one.

On economic benefits the quantity of produced agricultural products is regarded as improved by 66% of the sample. Indeed, Greece's climate favors agriculture. In addition, as Greece has not such vast agricultural land like United States of America, Ukraine or Australia, Greece cannot produce the amount of agricultural products like these countries. As a result, a focus on improved quality of products is suggested as the most suitable strategy for farmers in Greece. In general farmers have answered that they had half of the suggested economic benefits during the last 3 years. Nevertheless only 37% of the sample has increased their income. This fact may be a matter of further research. An interesting finding is the fact that almost half of farmers are ecologically aware, as 48% of them declared that their agricultural activities are more eco-friendly than 3 years ago. Finally, production cost and workload are two matters that most farmers are not benefited. The percentage of positive answers was so low, that these characteristics had low quality as variables, so they were excluded from the data analysis.

The cluster analysis was conducted on a Burt table in which the nine regions of Greece were the table's row and the rest of the variables as the columns. The analysis highlighted three distinct clusters. The differences between these clusters included differences on climate, crop type, size of arable land, but also differences on innovation and economic benefits. The differences on innovation and economic benefits differenced not only on kind but also in numbers.

6 Conclusions

This research aimed to seek agricultural innovation across Greece, along with the resulting economic benefits. In particular, the ultimate goal was to cluster Greek regions on crop type, size of arable land, process innovations, and economic benefits. To fulfill this purpose an e-questionnaire was created, directed to farmers on mainland Greece and Crete. Based on data provided by Hellenic Statistical Authority, there were specific region answer goals, in order to represent the agricultural population of Greece. Because it was not possible to directly communicate with farmers, people who are professionally linked with farmers were asked to forward the questionnaire. At the end of the research 1032 answers were collected and ascending hierarchical classification was applied to process the data.

The first important extracted information is that the clusters are composed of districts which each one is next to each other. This fact indicates climate similarities. By this way, a map of agricultural innovation is created, and the Greek territory is divided in three groups.

The first group consists of farmers in Eastern Macedonia and Thrace, Central Macedonia, Western Macedonia and Thessaly. This group is primarily characterized by arable crops, arable land greater than 40 hectares and not permanent crops. The category "arable land greater than 40" is a legit outcome in this group due to the fact that in Eastern Macedonia and Thessaly, the largest agricultural plains in Greece exist. This group is also characterized by no horticulture crops, improved sowing, improved plant disease prevention, but, on the other hand, not improved fertilizer use, irrigation, harvesting and not improved labor productivity and quality of final products. Although the second series of characteristics is less important than the first ones, it could be concluded that despite the fact that innovations, such as VRP/VRS and UAV, are implied, there are no positive economic outcomes, despite the advantage of having the largest arable land in Greece.

The second group consists of farmers in Epirus, Peloponnese, and Crete. This group is heavily characterized by no arable crops, arable lands below 20 hectares and permanent crops. A few less intense characteristics include improved fertilizer use and irrigation, but not improved sowing. Permanent crops and small arable land prevail in this group. Few innovations (e.g. VRNA, and VRI) have been implemented during the last 3 years, however, with no positive economic benefits.

The third group is consisted of farmers in Western and Central Greece. This group is strongly labeled by horticulture. A series of less important variables that define this group are improved harvesting, ploughing, and agricultural machinery use. Thus, improved labor productivity, increased income and quality of final products are also variables that are included in this group. In terms of PA, MG, and UAV implementation, this group is the only one that has economic benefits in comparison to the others.

In the Community Innovation Survey questionnaire, a business is regarded as innovative if it uses at least one innovation during the last 3 years. Based on this definition, Greek agriculture is innovative, but there are still no economic outcomes. In the previous clustering although every group included innovations on agricultural processes, only the last one included also economic benefits. The main difference between groups is the crop type and the number of innovative processes. The first two groups are characterized by arable and permanent crops accordingly and two innovation types on each one of them. The prevailing crops in the last group are horticulture crops and three innovation types take place. Therefore, farmers in Western and Central Greece are the most benefited.

In conclusion, horticulture crops are regarded more productive than the others and it is indicated that a minimum number of innovations is required for economic benefits to come up. Horticulture requires less arable land than other crop types and plants in greenhouses are safe from sudden weather effects which is the most serious farmers' concern. Moreover, it is interesting to note that arable land used in agriculture consists only of the 2% of total utilized arable land use in Greece [41]. Further research could study farmers' profiling, searching for characteristics that define innovative behavior like demographics or entrepreneurial mindset. Horticulture crops might also attract academic community's interest.

References

1. Eise J, Foster K (2018) How to feed the world
2. Zarco-Tejada PJ, Bubbard N, Loudjani P (2014) Precision Agriculture: An Opportunity for EU Farmers - Potential Support with the CAP 2014-2020
3. Fountas S, Aggelopoulou K, Gemtos TA (2016) Precision agriculture: Crop management for improved productivity and reduced environmental impact or improved sustainability. In: Eleftherios I, Bochtis D, Vlachos D, Dimitrios A (eds) Supply Chain Management for Sustainable Food Networks. John Wiley & Sons, Ltd., pp 41–65
4. Hunt ER, Daughtry CST (2018) What good are unmanned aircraft systems for agricultural remote sensing and precision agriculture? Int J Remote Sens 39:5345–5376. https://doi.org/10.1080/01431161.2017.1410300
5. Zude-Sasse M, Fountas S, Gemtos TA, Abu-Khalaf N (2016) Applications of precision agriculture in horticultural crops. Eur J Hortic Sci 81:78–90. https://doi.org/10.17660/eJHS.2016/81.2.2
6. Guardo E, Di Stefano A, La Corte A, et al (2018) A fog computing-based IoT framework for precision agriculture. J Internet Technol 19:1401–1411. https://doi.org/10.3966/160792642018091905012
7. Basso B, Dumont B, Cammarano D, et al (2016) Environmental and economic benefits of variable rate nitrogen fertilization in a nitrate vulnerable zone. Sci Total Environ 545–546:227–235. https://doi.org/10.1016/j.scitotenv.2015.12.104
8. Moral FJ, Terron JM, Marques da Silva JR (2010) Delineation of management zones using mobile measurements of soil apparent electrical conductivity and multivariate geostatistical techniques. Soil Tillage Res 106:335–343

9. Yao R-J, Yang J-S, Zhang T-J, et al (2014) Determination of site-specific management zones using soil physico-chemical properties and crop yields in coastal reclaimed farmland. Geoderma 232:381–393
10. Córdoba MA, Bruno CI, Costa JL, et al (2016) Protocol for multivariate homogeneous zone delineation in precision agriculture. Biosyst Eng 143:95–107. https://doi.org/10.1016/j.biosystemseng.2015.12.008
11. Pedersen SM, Lind KM (2017) Precision agriculture: Technology and economic perspectives. Springer International Publishing
12. van Evert FK, Gaitán-Cremaschi D, Fountas S, Kempenaar C (2017) Can precision agriculture increase the profitability and sustainability of the production of potatoes and olives? Sustain 9. https://doi.org/10.3390/su9101863
13. Balafoutis A, Beck B, Fountas S, et al (2017) Precision agriculture technologies positively contributing to ghg emissions mitigation, farm productivity and economics. Sustain 9:1–28. https://doi.org/10.3390/su9081339
14. Liakos V, Tagarakis A, Aggelopoulou K, et al (2017) In-season prediction of yield variability in an apple orchard. Eur J Hortic Sci 82:251–259. https://doi.org/10.17660/eJHS.2017/82.5.5
15. Yost MA, Kitchen NR, Sudduth KA, et al (2019) A long-term precision agriculture system sustains grain profitability. Precis Agric 20:1177–1198. https://doi.org/10.1007/s11119-019-09649-7
16. Lambert D, Lowenberg-DeBoer J (2000) Precision Agriculture Profitability Review
17. Hedley CB, Yule IJ (2009) Soil water status mapping and two variable-rate irrigation scenarios. Precis Agric 10:342–355
18. Timmermann C, Gerhards R, Kuhbauch W (2003) The Economic Impact of Site-Specific Weed Control. Precis Agric 4:249–260
19. Vasileiadis VP, Sattin M, Otto S, et al (2011) Crop protection in European maize-based cropping systems: Current practices and recommendations for innovative Integrated Pest Management. Agric Syst 104:533–540
20. Joao RSM, Conrado DR, da Romeu RCC, et al (2016) Study of an electromechanical system for solid fertilizer variable rate planting. African J Agric Res 11:159–165. https://doi.org/10.5897/ajar2014.9349
21. He X, Ding Y, Zhang D, et al (2019) Development of a variable-rate seeding control system for corn planters Part II: Field performance. Comput Electron Agric 162:309–317. https://doi.org/10.1016/j.compag.2019.04.010
22. Guo J, Li X, Li Z, et al (2018) Multi-GNSS precise point positioning for precision agriculture. Precis Agric 19:895–911. https://doi.org/10.1007/s11119-018-9563-8
23. dos Santos AF, da Silva RP, Zerbato C, et al (2019) Use of real-time extend GNSS for planting and inverting peanuts. Precis Agric 20:840–856. https://doi.org/10.1007/s11119-018-9616-z
24. Shockley JM, Dillon CR, Stombaugh TS (2011) A Whole Farm Analysis of the Influence of Auto-Steer Navigation on Net Returns, Risk, and Production Practices. J Agric Appl Econ 43:57–75. https://doi.org/10.1017/s1074070800004053
25. Bora GC, Nowatzki JF, Roberts DC (2012) Energy savings by adopting precision agriculture in rural USA. Energy Sustain Soc 2:1–5. https://doi.org/10.1186/2192-0567-2-22
26. Ortiz B V., Balkcom KB, Duzy L, et al (2013) Evaluation of agronomic and economic benefits of using RTK-GPS-based auto-steer guidance systems for peanut digging operations. Precis Agric 14:357–375. https://doi.org/10.1007/s11119-012-9297-y
27. Jensen HG, Jacobsen L-B, Pedersen SM, Tavella E (2012) Socioeconomic impact of widespread adoption of precision farming and controlled traffic systems in Denmark. Precis Agric 13:661–677
28. Barnes A, De Soto I, Eory V, et al (2019) Influencing factors and incentives on the intention to adopt precision agricultural technologies within arable farming systems. Environ Sci Policy 93:66–74. https://doi.org/10.1016/j.envsci.2018.12.014

29. Radoglou-Grammatikis P, Sarigiannidis P, Lagkas T, Moscholios I (2020) A compilation of UAV applications for precision agriculture. Comput Networks 172:107148. https://doi.org/10.1016/j.comnet.2020.107148
30. Maes WH, Steppe K (2019) Perspectives for Remote Sensing with Unmanned Aerial Vehicles in Precision Agriculture. Trends Plant Sci 24:152–164. https://doi.org/10.1016/j.tplants.2018.11.007
31. Bac CW, Van Henten EJ, Hemming J, Edan Y (2014) Robots for High Value Crops: State of the Art Review and Challenges Ahead. J F Robot 31:888–911
32. Ball D, Ross P, English A, et al (2015) Robotics for Sustainable Broad-Acre Agriculture. In: Mejias L, Corke P, Roberts J (eds) Springer Tracts in Advanced Robotics. Springer, Cham
33. Nuske S, Wilshusen K, Achar S, et al (2014) Automated Visual Yield Estimation in Vineyards. J Fields Robot 31:837–860
34. Wachowiak MP, Walters DF, Kovacs JM, et al (2017) Visual analytics and remote sensing imagery to support community-based research for precision agriculture in emerging areas. Comput Electron Agric 143:149–164. https://doi.org/10.1016/j.compag.2017.09.035
35. Vuran MC, Salam A, Wong R, Irmak S (2018) Internet of underground things in precision agriculture: Architecture and technology aspects. Ad Hoc Networks 81:160–173. https://doi.org/10.1016/j.adhoc.2018.07.017
36. Schumpeter JA (1935) The Analysis of Economic Change Author (s): Joseph A. Schumpeter Source: The Review of Economics and Statistics, Vol. 17, No. 4 (May, 1935), pp. 2–10 Published by: The MIT Press Stable URL: http://www.jstor.org/stable/1927845. Anal Econ Chang 17:2–10
37. OECD (2005) Oslo Manual
38. Association of Greek Regions (2011) Regions of Greece. In: Assoc. Greek Reg.
39. Hellenic Statistical Authority (2020) Greece in Figures
40. Kountios G, Ragkos A, Bournaris T, et al (2018) Educational needs and perceptions of the sustainability of precision agriculture: survey evidence from Greece. Precis Agric 19:537–554. https://doi.org/10.1007/s11119-017-9537-2
41. ELSTAT (2019) Yearly Agricultural Research
42. Milone P, Ventura F (2019) New generation farmers: Rediscovering the peasantry. J Rural Stud 65:43–52. https://doi.org/10.1016/j.jrurstud.2018.12.009
43. Moschidis O (2015) Unified coding of qualitative and quantitative variables and their analysis with ascendant hierarchical classification. Int J Data Anal Tech Strateg 7:114–128. https://doi.org/10.1504/IJDATS.2015.068745
44. Moschidis O, Chadjipadelis T (2017) A method for transforming ordinal variables. Stud Classif Data Anal Knowl Organ 285–294. https://doi.org/10.1007/978-3-319-55723-6_22
45. Markos A, Moschidis O, Chadjipadelis T (2020) Sequential dimension reduction and clustering of mixed-type data. Int J Data Anal Tech Strateg 12:28–30
46. Benzecri J-P (1973) Analyse des Données (T.1: La Taxinomie, T.2: Correspondances). Dunod, Paris
47. Benzecri J-P (1981) Practique De l'Analyse des donnees. Vol 3: Linguistique et lexicologie. Dunod, Paris
48. Benzecri J-P (1982) Construction d' une Classification Ascendante Hiérarchique par la recherché en chaîne des Voisins Réchiproques. Les Cah l' Anal des Données VII:209–218
49. Benzecri J-P, Benzecri F, Maiti GD (1992) Pratique de l' Analyse des Données en Médecine. Vol4: Medecine, Pharmacologie, Physiologie clinique. Statmatic, Paris
50. Ward JH (1963) Hierarchical grouping to optimize an objective function. J Am Stat Assoc 58:236–244
51. Aldenderfer MS, Blashfield RK (1984) Cluster Analysis. Sage Univ Pap 44:88

52. Moschidis O (2003) Contribution to comparative survey of multidimensional scales with the methods of multivariate analysis
53. Lebart L, Morineau A, Piron M (2000) Statistique Exploratoire Multidimensionnelle. Dunod, Paris
54. Morineau A (1984) Note sur la caracterisation statistique d' une classe et les valeurs-tests. Bull Tech du Cent Stat Inform Appliquées 2:20–27
55. Bechrakis TE (1999) Multidimensional Data Analysis. Nea Sinora - Livani, Athens

Part III
Diffusion of Agriculture Digital Transformation

AI-Based Chatbot System Integration to a Social Media Platform for Controlling IoT Devices in Smart Agriculture Facilities

Eleni Symeonaki, Konstantinos Arvanitis, Panagiotis Papageorgas, and Dimitrios Piromalis

1 Introduction

Sustainability is considered nowadays as one of the most imperative targets to be achieved globally in order to cope with the imminent climate change related challenges. This target leads to the stronger involvement of innovative technologies in agricultural facilities in the context of Smart Agriculture [1], enabling accordingly the active participation of the stakeholders so as to improve productivity by maximizing the efficiency inputs and minimizing their environmental impacts.

The technology of the Internet of Things (IoT), which is continuously evolving and maturing, is considered to be a valuable asset in the development of Smart

E. Symeonaki (✉)
Department of Natural Resources Management and Agricultural Engineering, Agricultural University of Athens, Athens, Greece

Department of Industrial Design and Production Engineering, University of West Attica, Athens, Greece
e-mail: esimeon@uniwa.gr

K. Arvanitis
Department of Natural Resources Management and Agricultural Engineering, Agricultural University of Athens, Athens, Greece
e-mail: karvan@aua.gr

P. Papageorgas
Department of Electrical and Electronics Engineering, University of West Attica, Athens, Greece
e-mail: ppapag@uniwa.gr

D. Piromalis
Department of Industrial Design and Production Engineering, University of West Attica, Athens, Greece
e-mail: piromali@uniwa.gr

D. D. Bochtis et al. (eds.), *Information and Communication Technologies for Agriculture—Theme IV: Actions*, Springer Optimization and Its Applications 185,
https://doi.org/10.1007/978-3-030-84156-0_10

Agriculture through the extensive use of intelligent remote-controlled production equipment such as Wireless Sensor Networks (WSNs) and mobile embedded systems [2]. The IoT concept, as it results by the research in literature, could be best defined as "a world-wide network of interconnected objects uniquely addressable, based on standard communication protocols" [3, 4]. By simple means, it can be said that the IoT enables anything to be connected anytime, at anyplace, with anything and anyone, ideally using any networks and any services [5, 6]. Along these lines, the IoT technology benefits field equipment with sensory and computational support for inter-communicating and interacting via a highly distributed public network such as the Internet [7–9].

The most novel solutions regarding IoT agricultural applications tend to adopt ubiquitous interconnectivity methods along with cost-effective cloud services granted by smart mobile devices [10]. In this context, several applications specially designed to run on smartphones, tablets and other mobile devices have been introduced up to present with the purpose to establish interaction methods among the agricultural IoT objects (physical and artificial) accessing and transacting information via a highly distributed public network such as the Internet. Mobile applications specially addressing to the IoT for Smart Agriculture have been lately presented [11], offering the opportunity to increase yields through modernized production methods with respect to the environment, contributing in this way to the global sustainable growth.

However, the adoption and usage of such innovative technology practices in agriculture facilities is still rather limited and fragmentary since, as several studies indicate, only a rather small proportion of agricultural stakeholders takes advantage of the opportunities offered by the consolidation of the IoT with smart mobile devices. This seems to be due to the fact that resistance to change remains an obstacle in agriculture and familiarity to the Information and Communication Technologies (ICT) features continues to be a challenge in rural areas.

Since social networking has been recorded as the second largest traffic volume contributor worldwide, with an average share of over 15% of total mobile data traffic, the integration of agricultural mobile applications to social media messaging platforms could be the key for overcoming the barriers of the IoT technologies penetration in agriculture facilities [12].

This work attempts to introduce a user-friendly, efficient and secure framework for controlling IoT agricultural devices in natural language dialogues, through the deployment of an Intelligent Conversational Agent (chatbot) based on Artificial Intelligence (AI) and its integration to an instant messaging application of a popular social media platform (as communication user interface), for providing end-users with context-aware services related to the monitoring, as well as the control of an agriculture facility in question–answer sessions conducted in natural language. For this purpose, the messenger application of 'Facebook'[1] has been chosen since, in comparison to other social media platforms, it encompasses the highest penetration

[1] Available at https://www.facebook.com

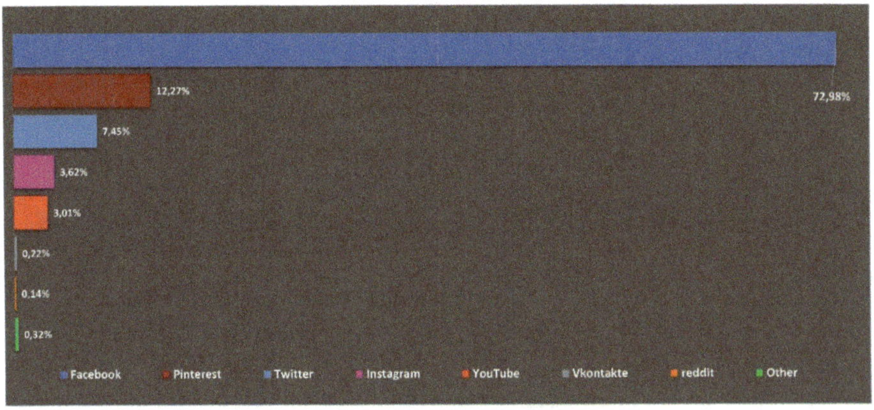

Fig. 1 Mobile social media stats worldwide from December 2018 to December 2019

to mobile users, as shown in Fig. 1. What is more, it offers several features among which, open official Application Programming Interfaces (APIs) for chatbot development, WebView objects for increased control over the user interface and advanced configuration options, unimpeded file sharing (text, audio, image, video, etc.) as well as analytics and feedback. Last but not least, the 'Facebook Messenger'[2] application employs Natural Language Processing (NLP) and Machine Learning (ML) algorithms which allow the understanding and extraction of information (entities) out of instant messaging dialogues carried out as natural language conversations.

Although Facebook comes up with 'Wit.ai' as its own chatbot development platform, in order to grant flexible further integration of the chatbot to other popular messenger applications which provide official APIs for this purpose (i.e. Twitter, Viber, Slack, Line, Telegram, etc.), the 'Dialogflow'[3] platform offered by Google was selected for its development. 'Dialogflow' (formerly known as Api.ai) is an open-source platform which enables the deployment of chatbots through several features similarly to 'Wit.ai'. In particular 'Dialogflow' integrates NLP and incorporates ML offering the end-user various ways of interaction which may engage text and voice based conversational interfaces powered by AI. Furthermore 'Dialogflow' offers recognition of more than 15 languages while it involves an in-line code editor allowing the performance of various tasks straight from the console. On top of its usage is totally free of charge and it provides developers with excellent documentation.

On this ground, the conceptual framework focusing on the features of intelligent conversational agents and the benefits of employing chatbot systems as interfaces for IoT agricultural applications are reviewed in brief. Thereafter the architecture, the

[2] Available at https://www.messenger.com

[3] Available at https://dialogflow.cloud.google.com

operating features as well as an overview of a chatbot system capable of controlling a group of IoT devices through the interaction via instant messaging in a popular social media platform is described. Finally, the conclusions deriving from this attempt are presented and some reference on the ongoing research work is made.

2 Conceptual Framework Overview

Spoken dialogue technology refers to the turn-by-turn interaction between humans and intelligent systems in terms of natural language communication ranging from only a small set of words (such as the digits 0–9 and the words yes/no) to large vocabulary dialogues [13]. Although spoken dialogue technology was not widely deployed until quite recently, the concept of imitating real human-to-human interaction between end-users and devices was first presented by Alan Turing in 1950, when he developed a test, which is widely known as the "Turing Test", in order to evaluate if and to which extent a "machine" is capable of exhibiting human-like intelligent behavior [14]. The Turing Test is so pioneering that it is still used up to present as it consists a reliable evaluation test for the performance of "machines" to behave human-likely in the terms of Artificial Intelligence (AI).

One of the earliest successful attempts to imitate human talk conversation by machines, based on the criteria of the Turing Test, is dated back in the 1960s, when "ELIZA" was developed as a natural language processing computer program capable of processing end-user inputs and engage into conversation according to a predefined script [15]. Some decades later, a project called ALICE (Artificial Linguistic Internet Computer Entity), also referred to as "Alicebot", has begun to be developed based on the "ELIZA" findings, relying on recursive programming techniques and using an XML schema named Artificial Intelligence Markup Language (AIML) in order to specify heuristic conversation rules which relied on [16]. More recent approaches in the deployment of this technology were "Watson", a project which was developed by IBM researchers through employing DeepQA as "a software architecture for deep content analysis and evidence-based reasoning" [17], as well as a neural conversational model employing a sequence-to-sequence (seq2seq) framework for modelling conversations developed by Google researchers [18]. At present, due to the progress in language processing and dialogue modelling, there is a broad variety of systems that deploy spoken dialogue technology methods ranging from simple question-answering models which can answer a single question at a time, to sophisticated dialogue systems, which allow extended conversational interaction between end-users and devices [19].

Conversational User Interfaces (CUI) are software dialogue systems which facilitate any average user to interact with any device, anywhere and at any time, without the need of special skills or training, by involving a variety of written or oral natural communication forms in order to simulate actual conversations in the end-users' native languages rather than in specific command-line syntax [20]. Advanced CUIs support the situated language understanding of any probably ambiguous, insufficient or partial multimodal inputs and the deriving of completely correlated outputs, through the implementation of Artificial Intelligence (AI) techniques, using Natural Language Processing (NLP) and Natural Language Understanding (NLU) programs [21]. CUIs differentiate, depending on the degree in which the system or the end-user are able to control the communication, into:

- Menu-based, on which the system maintains the absolute control of the entire interaction by following a predefined dialogue flow in order to generate specified prompts, accept or reject specified user responses (words or phrases) and guide actions based upon a variety of alternative paths relating to the accepted responses. This kind of applications has been effectively incorporated on a wide scale since they result to a high rate of successful interactions due to the limited options offered to the end-users [22, 23].
- Template-based, which permit some kind of adjustable communication control as the dialogue flow is not predefined but relies on a template according to the user's input content and the output that the system is going to generate. In this case, end-users are allowed to respond in a more flexible way to the prompts as well as to correct any errors of recognition and understanding through a kind of natural language input [24].
- Agent-based, which enable bilateral communication control since both the system and the end-users are regarded as agents and are allowed to verify their actions as well as time the actions of each other. The dialogue flow is dynamically progressing as a sequence of associated steps while there are mechanisms for the prediction of context as well as for the detection and correction of errors. End-users are also able to control the dialogue at any time, introduce new topics or for what is more, add contributions which are not constrained by the preceding prompts [25, 26].

A major reason for the recent impetus of CUIs in the form of conversational agents [27] is the recently growing need for extensive access to web services and online information through intelligent, effective, dynamic, flexible, multimodal and user friendly means of Human-Machine Interaction (HMI). This can be achieved through the integration of machine learning techniques [20, 28] and NLU functions in various services, such as localized search, dialogue management, remote control, and the Internet of Things.

Chatbots (also known as Chatterbots or simply as bots) are the most prominent conversational agents of current conversational user interfaces. The term "Chatbot" refers to an interactive software dialogue system, which enables real-time communication with the end-users by simulating and reproducing turn-by-turn intelligent conversations in natural language via textual (textbots) or even auditory methods (voicebots). While various chatbot architectures have been introduced for specific use cases, these do not conclude to a standardized architectural framework. A suggestion for a general architectural framework architecture for chatbot systems was proposed by Braun et al. in 2017 as shown in Fig. 2. According to this architecture a chatbox system consists of three main modules [19]: Request Interpretation, Response Retrieval and Message Generation. In the context of Request

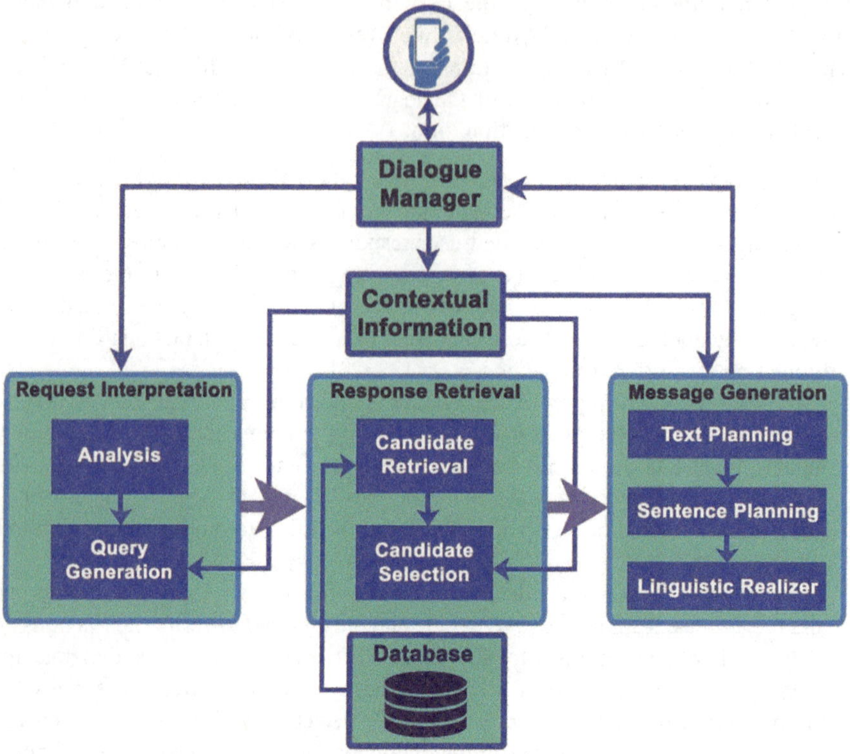

Fig. 2 General architectural framework for a chatbot system

Interpretation, a "request" is not necessarily a question, but can also be any user input, while equally a "response" to this input could be any output statement. The Message Generation follows the classical Natural Language Generation (NLG) pipeline [29].

Chatbot systems constitute a technological trend which strongly coordinates with the current IoT concept providing a highly effective and user-friendly interface solution [30]. From this prospect, there are some significant advantages in employing chatbots as an IoT interface rather than conventional applications for different platforms and versions. Some of the reasons why chatbots are such an appropriate interface solution in the field of IoT are as follows:

- Interaction in natural language as they are capable of creating triggered rules for IoT smart devices so as to activate any action requested by the end-user. Natural language processing algorithms using artificial intelligence features are responsible for unpacking the intent and pass any required instructions to the IoT gateway for processing. Moreover, the chatbot system is empowered through artificial intelligence learning techniques.

- De-parameterized environment of interaction as chatbots do not require the systematic input of all parameters in order to complete a request.
- End-users are not obliged to learn the operation of different IoT applications as they can query IoT networks of devices simply by using their native language, without having to know any interface sequence or command structure. Additionally, end-users do not have to separately download applications as these may be centrally and directly accessible through existing chat clients.
- Refining of end-user requests for subsequent interactions and control reducing in this way the problems of information abstraction which present in conventional interface solutions.
- No demand for constant application updates neither for any maintenance of older versions at the back-end of different operating systems versions or mobile platforms.
- End-users of IoT applications are offered a natural, pleasant and simple interaction environment, which operates on any messenger service platform whether this is mobile, in-app or via web chat.

3 Methodology

3.1 Chatbot System Architecture and Features

The architecture of the proposed chatbot system is presented in Fig. 3. According to the flow of this architecture any query (Input) entered into the Messenger App by the end-user is forwarded to the 'Dialogflow' agent, which is a natural language understanding module, created for understanding the expressions of human language and handling the types of conversations required for the proposed system. The agent processes and transforms the end-user's voice or text to structured data that are comprehendible by the IoT applications. In the case for example a query is "What is the soil temperature?", phrases such as "is" and "the" are going to be removed by the agent by employing a preprocessing technique. Subsequently, the system retrieves the required keywords from the query and incorporating a keyword matching algorithm, it matches them with those deposited in the knowledge base (Intent). At last, the system generates the appropriate output which is returned through the Messenger App as a response to the end-user.

As deriving by the aforementioned basic flow of conversation, the chatbot involves the aspects of: (a) the input given by the end-user, (b) the given input parsed by the agent and (c) the response returned to the end-user by the agent. In this context, the chatbot incorporates five main features that is Intents, Entities, Contexts, Events and Fulfillment, as described forth below.

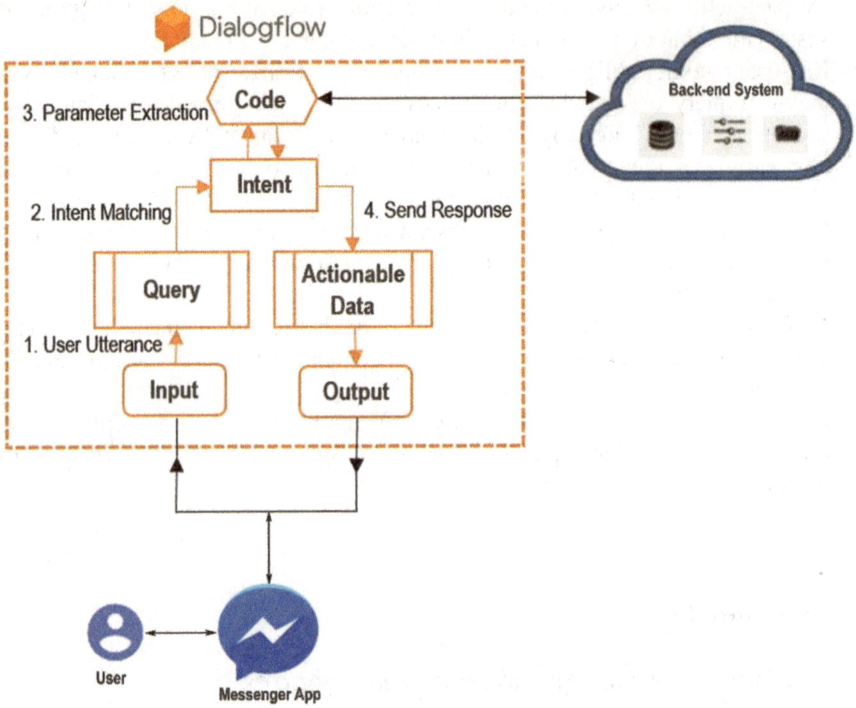

Fig. 3 Chatbot system architecture

Intents

In order to define the flow of the conversation, intents that map the end-user input to responses are created in the agent, whereas in each one of these intents, examples of user expressions triggering the intent, actions extracted from each expression as well as the ways of response, are properly defined.

For allowing the mapping of the end-user's input to responses, Intents involve the following:

(a) *Intent Name*, for identifying the matched intent.
(b) *Training Phrases*, for matching a particular intent each time an end-user expression is similar to one of these phrases. The matching of expressions to intents is performed based on the significant values, words or idioms specified within the phrases. Figure 4 presents an example of how 'Dialogflow' performs a successful matching of a user's expression to an intent.
(c) *Parameters*, which are structured data, defining the relevant information extracted from the expressions as specific values, in order to be adequately employed as input in the performance of logic operations or generate responses. For instance, parameters may include, names, dates, time or location while each

Fig. 4 Indicative conversational flow when a successful matching is performed

parameter has a characteristic dictating specifically the mode of data extraction, referred to as "entity type".

(d) *Actions* provided to the system in order to trigger certain operations in the IoT applications according to the matched intents.
(e) *Responses* which are returned to the end-users as text, speech, or visual context, providing them either with answers to their queries, requests for additional information, or termination of the conversation.

Entities

All significant data deriving from an end-user's query have a corresponding entity. While intents permit the comprehension of the motivation lying beneath specific end-users queries, entities are employed to select particular aspects of information (i.e. agricultural product names, measured units, etc.). Apart from the entities defined for the purposes of the proposed chatbot in order to match custom agricultural data, as shown in Fig. 5, predefined system entities are also provided to match several common data types. For instance, the system provides entities for matching time, date, location, and so on. Entity details, discussed in more specific terms are as following:

(a) *Entity Type*, which specifically defines the mode of data extraction from end-users queries.
(b) *Entity Entry*, which provides a set of words or phrases that are considered to be equivalent. For each entity type, there may exist several entity entries.
(c) *Entity Reference Value and Synonyms*, which are provided in the cases when entity entries come up with multiple words or phrases that are considered to be equivalent.

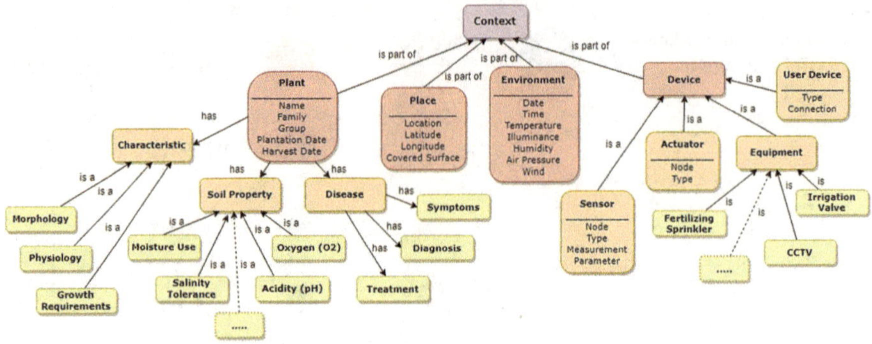

Fig. 5 Indicative custom and predefined entities in order to match agricultural data

Contexts

Contexts are used for controlling the flow of a conversation as they correspond to natural language context in order to handle disputable end-user utterances expressed by the end-users (i.e. it is orange) and match an intent in the most accurate way. In particular, contexts for an intent are configured by setting input and output contexts identified by string names. Each time an intent is matched, all corresponding output contexts, which are configured for that intent, become active. An indicative case of using context for controlling the flow of a conversation in the proposed chatbot is depicted in Fig. 6. According to this case the end-user requests information about a specific wireless sensor node deployed in the agriculture facility and the expression is matched to the *"MonitoringControl"* intent for checking the output context in order to become active. Following, the agent addresses a question to the end-user requesting a specification about the type of information needed, concerning this sensor node. When the end-user responds with "moisture", this expression is matched to the *"CheckingMoisture"* intent. As this intent has a checking input context, which is required to be active so as to be matched, a similar *"CheckingHumidity"* intent may also incur for matching the same end-user expression when the *"humidity"* context is active. Finally, after the system performs all necessary actions, the agent responds to the end-user with a message containing the moisture values measured by the specific wireless sensor node.

Events

Events are used for invoking intents based on probable occurrences, instead of actual queries expressed by the end-users. 'Dialogflow' supports events from a great number of various platforms, among which Facebook Messenger, based on the actions that end-users take on the platform. Additionally, custom events are created in order to denote ways of communication that cannot be skillfully captured by text or voice. Custom events can indicate for instance that a button has been clicked or

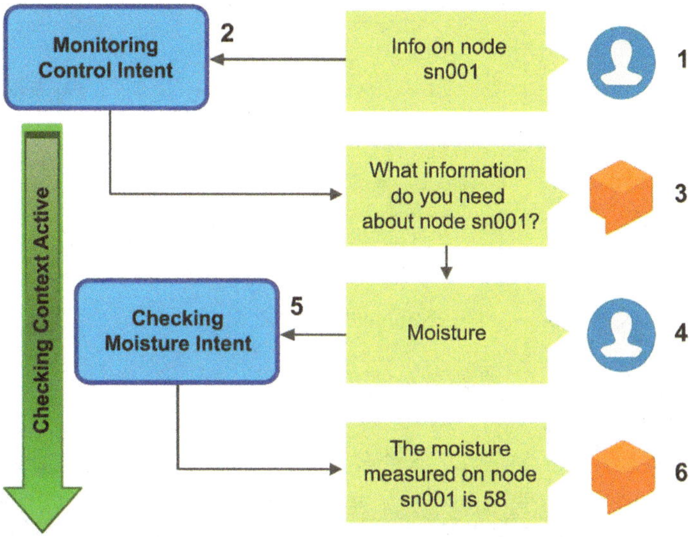

Fig. 6 Indicative case of using context for controlling the flow of a conversation

authorization has been granted by an end-user as well as that a predefined amount of time has elapsed. Finally, it has to be noted that events provide the ability of responding to the end-users actions into the agent and in fulfillment, as well.

Fulfillment

Fulfillment for an intent is a feature enabled each time this intent requires a dynamic response or some action to be undertaken by the system. In particular, when an intent with the feature of fulfillment being enabled is matched, a request containing information about the matched intent is sent to the webhook service and all required actions are performed, responding with information for how the agent should proceed. Each intent has a setting to enable the feature of fulfillment while the corresponding processing flow is shown in Fig. 7. According to this processing flow the end-user addresses a query to the agent expressed in text or voice. In the following step the end-user expression is matched to an intent and the corresponding parameters are extracted. Subsequently, a webhook request message is sent to the webhook service containing all required information about the matched intent, the parameters, the action as well as the response specified for the intent. Then the service performs all necessary actions (i.e. database queries, external API calls, etc.) and replies with a webhook response message containing the response to be sent to the end-user. Finally the end-user receives the response as a text or voice message in the application. In case an intent is matched without the feature of fulfillment being enabled then the predefined static response for this intent is incorporated.

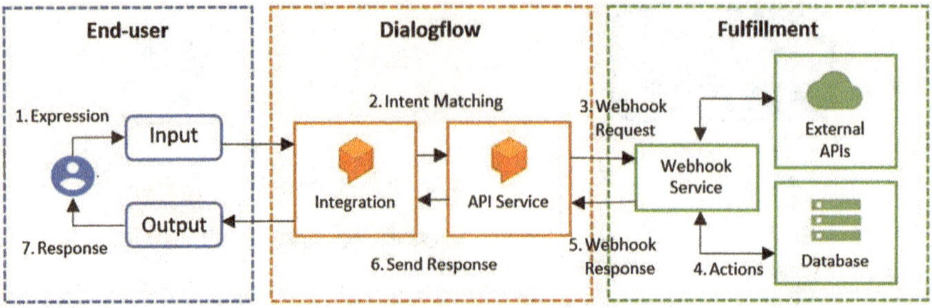

Fig. 7 Processing flow with the feature of fulfillment enabled

3.2 Chatbot System Integration

After the agent of the AI-based chatbot system has been developed and properly trained, all data are integrated to the Facebook Messenger application for establishing a user-friendly interface. As 'Dialogflow' permits integrating agents from the Api.ai natural language processing service with a web application, the 'Node.js'[4] platform has been incorporated, for providing the development of the chatbot with a server-side JavaScript runtime environment so as to create a simple webserver with two webhook (also known as HTTP push API or web callback) endpoints, the first for the initial Facebook verification and the second for being responsible for any other messages from the Facebook Messenger.

The 'Node.js' cross-platform is open-source and executes JavaScript code outside of a browser, offering an extended library of various JavaScript modules which simplify the process of web development. 'Node.js' is suitable for developing fast and lightweight web applications while it is also highly scalable because it is capable of handling a great number of simultaneous connections with high outputs. Since, Node.js applications are composed in JavaScript, the V8 engine has been incorporated for this purpose. V8, which is developed by Google, provides a high-performance open source JavaScript and WebAssembly engine, written in the C++ programming language. It runs on Windows 7 edition or later, macOS 10.12+, and Linux systems while in can compile to ×86, ARM or MIPS instruction set architectures in both their 32-bit and 64-bit editions. V8 can run as standalone, or can be embedded into any C++ application and it has been ported to PowerPC and IBM s390 for usage in servers [31].

Finally, in order to integrate this solution into the cloud and ensure its constant execution over the internet, the cloud computing platform services from Microsoft Azure platform [https://azure.microsoft.com] were used for the deployment of the chatbot.

[4] Available at https://nodejs.org

4 Implementation and Results

Based on the previously-described methodology, an AI-based chatbot system was developed and integrated to the Facebook Messenger application for monitoring as well as for controlling a Wireless Sensors and Actuators Network (WSAN) deployed in an agriculture facility. For the materialization of the project, data acquired by a WSAN [32] deployed in an agriculture facility were used. The experiment was performed during June 2018 in the 'Kopaida' farm (120 km north of Athens) which is managed by the Agricultural University of Athens. The farm covers an area of 1020 hectares, divided into 11 sections of arable land, where various crops (cereal, maize, cotton, alfalfa, etc.) are being cultivated. Moreover, it is equipped with agricultural machinery, an underground irrigation system, and a modern weather station.

The implementation process involved 300 sample sets of data (considering a sampling rate of 15 min) acquired in a maize parcel where environmental and soil wireless sensors were deployed. These raw data were transmitted via a gateway, using the LoRaWAN communication protocol, to a context-aware middleware cloud component where they were centrally processed and managed. The contextual information deriving from the middleware cloud is then made available to the end-users through the developed AI-based chatbot, which is integrated to the messenger application of Facebook, for providing them with context-aware services related to the monitoring as well as the control of the agriculture facility in question–answer sessions conducted in natural language.

The fundamental functions developed for the chatbot system, concerning its adaption to different forms of dialogue which achieve the same intents, are introduced below. It has to be noticed that although the dialogues presented in this section are in English, the chatbot system can be easily adapted to several natural languages as it is supported by appropriate NLP and ML algorithms.

In Fig. 8 some indicative results of the conversations and operations performed in the chatbot system are presented, concerning the following Intents:

- **Intent 1: Greeting**; a greeting interaction is important in the conversation in order to make the system's usage more accessible and friendly.
- **Intent 2: Menu**; the options menu is offered in order to provide the end-user with a more direct interaction interface with the chatbot and increase its usability.
- **Intent 3: Help**; a help session was developed in the conversation in order to encounter any difficulties of end-users to understand the chatbot functionalities or in case any operation details have to be confirmed. In this context, the chatbot is capable of detecting the users' probable problems or doubts and offer them the help required.
- **Intent 4: Monitoring and Control**; this intent concerns the establishment of interaction with the defined entities (IoT devices of the agriculture facility) offering monitoring information about the cultivation and control actions of the equipment.

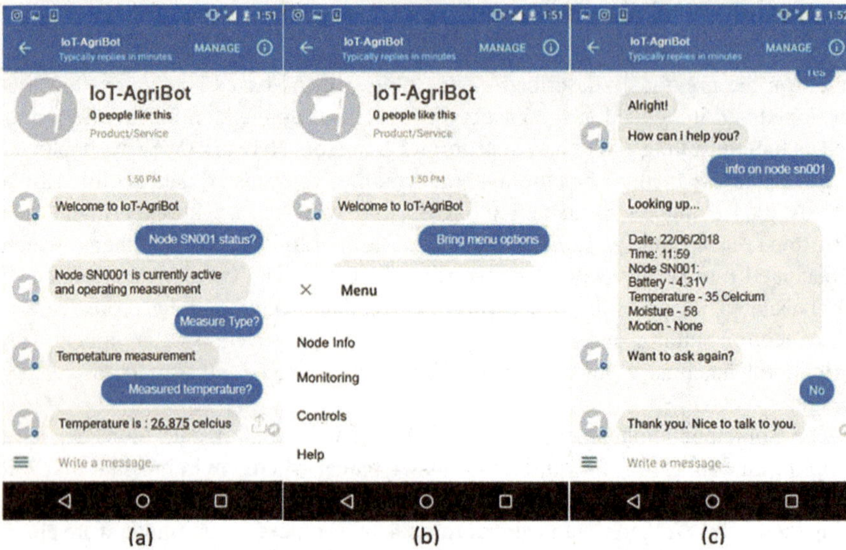

Fig. 8 Indicative results of conversations and operations in the chatbot system

With the objective to study the functionality of the AI-based chatbot system and validate its performance, a sufficient number of requests (referring to queries for actions and corresponding replies) has been sent and some metrics were provided regarding the number of the requests sent to the chatbot in total, as well as the number of successful and failed ones. It has to be noted that as successful are considered the requests that were recognized and were responded properly by the chatbot, while as failed are considered the requests that the chatbot was unable to reply or not needed to be further clarified. The results of these metrics are shown in Fig. 9.

Moreover, the Server Response Time, representing the time spent from the arrival of a request into the agent until the dissemination of the processed contextual information to the messaging applications has been evaluated, since it is a significant parameter concerning the system's performance. As derives from the metrics provided in Fig. 10, the system performed well based on the time of its response.

5 Conclusions and Further Research

The issue of establishing interaction methods among users, applications and systems involved in the agricultural sector through interfaces which are simple and user-friendly, is considered to be essential for achieving the maximum possible penetration of IoT technologies in this field for the benefit of Smart Agriculture development. The integration of an AI-based chatbot system, as an intelligent conversational agent, to the instant messaging application of a popular social

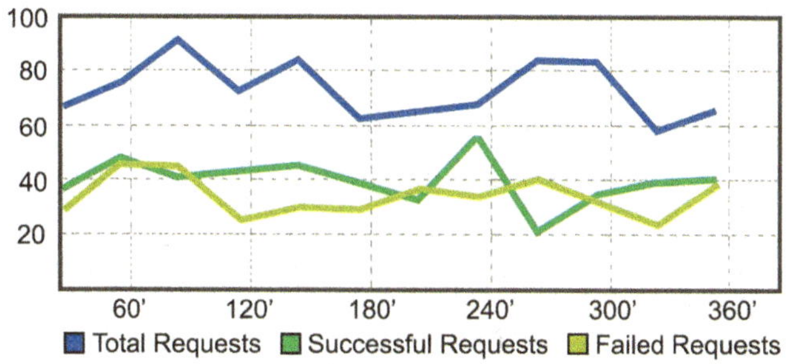

Fig. 9 Metrics results for requests performance evaluation

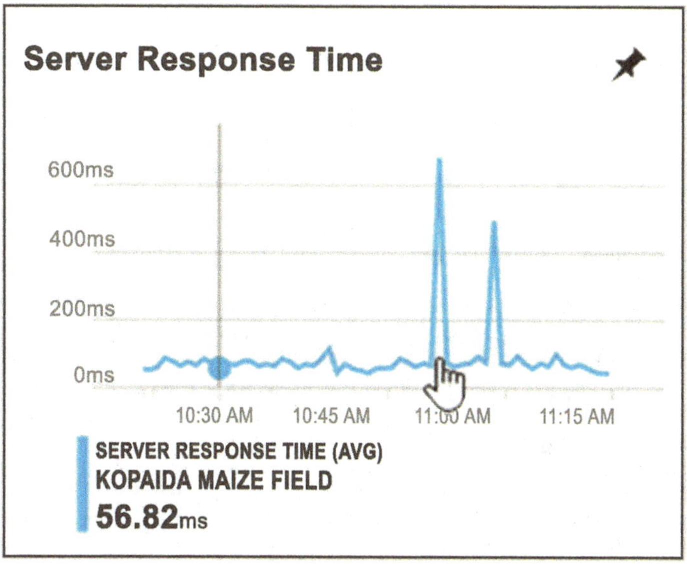

Fig. 10 Metrics results for server response time evaluation

media platform is believed to assist the development of smart agriculture in a great extent, through achieving the maximum possible penetration of the IoT technologies in this sector. Given the observed aspects in this research, it was possible to achieve this objective, presenting a promising solution which provides an efficient, effective and user-friendly mean of interaction between the end-users and the IoT devices deployed in agriculture facilities, based on the messenger application of Facebook and the cognitive services of the 'Dialoglow' platform and 'Node.js' technology. The chatbot system was evaluated based on certain parameters and it performed well. As future work, the AI-chatbot system is planned to be also integrated to other popular social media platforms and further tests will be performed in order to obtain results regarding its performance and ascertain its efficient operation.

References

1. Lipper, L. et al. (2014) 'Climate-Smart Agriculture for Food Security', Nature Climate Change Journal, 4, pp. 1068–1072, https://doi.org/10.1038/nclimate2437
2. Symeonaki E.G., Arvanitis K.G., Piromalis D.D. (2019) Cloud Computing for IoT Applications in Climate-Smart Agriculture: A Review on the Trends and Challenges toward Sustainability. In: Theodoridis A., Ragkos A., Salampasis M. (eds) Innovative Approaches and Applications for Sustainable Rural Development. HAICTA 2017. Springer Earth System Sciences. Springer, Cham; pp. 147–167, https://doi.org/10.1007/978-3-030-02312-6_9
3. Atzori L., Iera A., Morabito G. (2010) The internet of things: a survey. Comput Netw J, Elsevier, Amsterdam 54(15):2787–2805, https://doi.org/10.1016/j.comnet.2010.05.010
4. DG Infso, EPoSS (2008) Internet of things in 2020: a roadmap for the future, European Commission DG Infso & European Technology platform on smart systems integration, p 32
5. Ezechina MA, Okwara KK, Ugboaja CAU (2015) The internet of things (Iot): a scalable approach to connecting everything. Int J Eng Sci, Elsevier, Amsterdam 4(1):09–12
6. Saranya CM, Nitha KP (2015) Analysis of security methods in internet of things. Int J Recent Innov Trends Comput Commun 3(4):1970–1974
7. Lakhwani, K.; Gianey, H.; Agarwal, N.; Gupta, S. Development of IoT for smart agriculture a review. In Proceedings of the International Conference on Emerging Trends in Expert Applications & Security (ICETEAS 2018), Jaipur, India, 17–18 February 2018; Rathore, V.S., Worring, M., Mishra, D.K., Joshi, A., Maheshwari, S., Eds.; Springer: Singapore, 2019; pp. 425–432
8. Madushanki, A.; Halgamuge, M.; Wirasagoda, H.; Syed, A. Adoption of the Internet of Things (IoT) in agriculture and smart farming towards urban greening: A review. Int. J. Adv. Comput. Sci. Appl. 2019, 10, 11–28.
9. Talavera, J.M.; Tobón, L.E.; Gómez, J.A.; Culman, M.A.; Aranda, J.M.; Parra, D.T.; Quiroz, L. A.; Hoyos, A.; Garreta, L.E. Review of Iot applications in agro-industrial and environmental fields. Comput. Electron. Agric. 2017, 142, 283–297
10. Dlodlo, N. and Kalezhi, J. (2015) 'The internet of things in agriculture for sustainable rural development', in Proceedings of the International Conference on Emerging Trends in Networks and Computer Communications (ETNCC), Windhoek, Namibia, pp. 13–18
11. Costopoulou, C., Ntaliani, M. and Karetsos S. (2016) 'Studying Mobile Apps for Agriculture', IOSR Journal of Mobile Computing & Application, 3(6), pp. 44–99
12. Lathiya, A., Rathod, A., and Choudhary, K. (2015), 'Role of social media in agriculture' International Journal of Commerce and Business Management, 8, pp. 268–273, https://doi.org/10.15740/HAS/IJCBM/8.2/268-273.
13. Dybkjær, L., Bernsen, N.O. and Minker, W. (2004) 'Evaluation and usability of multimodal spoken language dialogue systems' Speech Communication, 43(1–2), pp. 33–54, https://doi.org/10.1016/j.specom.2004.02.001
14. Turing, A.M. (1950) Computing Machinery and Intelligence, Mind, 59(236), pp. 433–460
15. Weizenbaum, J. (1966) ELIZA - A Computer Program for the Study of Natural Language Communication between Man and Machine. Communications of the ACM, 9(1), pp. 36–45, https://doi.org/10.1145/365153.365168
16. Wallace, R.S. (2009) The anatomy of ALICE. In Parsing the Turing Test; Epstein, R., Roberts, G., Beber, G., Eds.; Springer, Netherlands; pp. 181–210, ISBN. 978-1-4020-9624-2
17. Ferrucci, D.A. (2012) Introduction to "This is Watson". IBM Journal of Research and Development, 2012, 56(3.4), pp. 1–15, https://doi.org/10.1147/JRD.2012.2184356
18. Oriol, V.; Quoc, L. (2015) A Neural Conversational Model. In Proceedings of the 31st International Conference on Machine Learning, Lille, France, 2015; JMLR: W&CP 37, arXiv:1506.05869 [cs.CL]
19. Braun, D., Mendez, A.H., Matthes, F. and Langen, M. (2017) 'Evaluating natural language understanding services for conversational question answering systems', in proceedings of the

18th Annual SIGdial Meeting on Discourse and Dialogue, Association for Computational Linguistics, Saarbrücken, Germany, pp. 174–185
20. Schnelle-Walka, D., Radomski, S, Milde B., Biemann, C. and Muhlhauser, M. (2016) 'NLU vs. Dialog Management: To whom am I speaking?', in Proceedings of the Joint Workshop on Smart Connected and Wearable Things (SCWT'2016), Sonoma, California, USA, https://doi.org/10.13140/RG.2.1.1928.4247
21. McTear, M., Callejas, Z. and Griol, D. (2016) 'Implementing spoken language understanding', The conversational interface: Talking to Smart Devices, Springer International Publishing, Switzerland, pp. 187–208, https://doi.org/10.1007/978-3-319-32967-3_9
22. Barnard, E. et al. (1999) A Consistent Approach to Designing Spoken-Dialog Systems. In Proceedings ASRU Workshop, Keystone, Colorado, USA
23. Billi, R. et al. (2002) Automation of Telecom Italia Directory Assistance Service: Field Trial Results. In Proceedings of Proceedings 1998 IEEE 4th Workshop Interactive Voice Technology for Telecommunications Applications (IVTTA '98), Torino, Italy, 29–30 September 1998; IEEE: 06 August 2002, pp. 11–16, https://doi.org/10.1109/IVTTA.1998.727682
24. Rosset, S. et al. (1999) Design Strategies for Spoken Language Dialog Systems. In Proceedings of the 6thEuropean Conference on Speech Communication and Technology (Eurospeech'99), Budapest, Hungary, 5–9 September 1999; pp. 1535–1538
25. Glass, J. (2003) A Probabilistic Framework for Segment-Based Speech Recognition. Computer, Speech, and Language, 17, pp. 137–152, https://doi.org/10.1016/S0885-2308(03)00006-8
26. Polifroni, J. et al. (2003) Towards the Automatic Generation of Mixed-Initiative Dialogue Systems from Web Content. In Proceedings of the 8th European Conference on Speech Communication and Technology, (Eurospeech 2003 - Interspeech 2003), Geneva, Switzerland, 1–4 September 2003; pp. 193–196
27. Ferrara, E., Varol, O., Davis, C., Menczer, F. and Flammini, A. (2016) 'The Rise of Social Bots', Communications of the ACM, 59(7), pp. 96–104, https://doi.org/10.1145/2818717
28. Stoyanchev, S., Lison, P. and Bangalore, S. (2016) 'Rapid prototyping of form driven dialogue Systems using an open-source framework', in Proceedings of the 17th Annual Meeting of the Special Interest Group on Discourse and Dialogue, Association for Computational Linguistics, Los Angeles, USA, pp. 216–219
29. Reiter, E. and Dale, R. (2000) 'Building Natural Language Generation Systems', Cambridge University Press, Cambridge, UK, ISBN:0-521-62036-8
30. Kar, R. and Haldar, R. (2016) 'Applying Chatbots to the Internet of Things: Opportunities and Architectural Elements', International Journal of Advanced Computer Science and Applications (IJACSA), 7(11), pp. 147–154, https://doi.org/10.14569/IJACSA.2016.071119
31. v8.dev. (2019), Documentation V8. [online] Available at: https://v8.dev/docs.
32. Piromalis, D. and Arvanitis, K. (2016) 'SensoTube: A scalable hardware design architecture for wireless sensors and actuators networks nodes in the agricultural domain', Sensors, 16(8), p. 1227, https://doi.org/10.3390/s16081227

IT in Education: Developing an Online Course

Fedro S. Zazueta, Patrizia Busato, and Remigio Berruto

1 Introduction

Information technology has enabled new models that create effective learning environments and permit the successful application of learning theory principles and innovative educational technology. In addition, due to external drivers such as pandemics, institutions are faced with difficulties in ensuring continuity of education [1] which has resulted in a renewed interest in online education. This is often implemented with little preparation and lacking the understanding that online education requires a different pedagogical approach to conventional modes of teaching.

For decades, online learning has been broadly recognized as strategically important to address global needs of education. As early as 1998 UNESCO articulated a vision and framework for priority action for change and development in higher education [2]. As information technology (IT) made access to information ubiquitous, its importance to support and enable strategic actions at national levels became evident. The US National Technology Plan [3] presented a model for learning powered by technology based on the premise that advances in learning sciences and understanding how people learn coupled with rapidly evolving developments in technology create new challenges and opportunities for higher education. The European Commission [4] articulated the importance of the innovation and modernization as fundamental to transform Europe into a competitive and inclusive

F. S. Zazueta (✉)
University of Florida, Gainesville, FL, USA
e-mail: fsz@ufl.edu

P. Busato
CIGR Information Technology Section Chair, DISAFA, University of Turin, Turin, Italy

R. Berruto
DISAFA, University of Turin, Turin, Italy

D. D. Bochtis et al. (eds.), *Information and Communication Technologies for Agriculture—Theme IV: Actions*, Springer Optimization and Its Applications 185,
https://doi.org/10.1007/978-3-030-84156-0_11

economy. In a similar manner, other countries such as Italy [5] and [6] incorporated IT into their education strategy as well as programs enabled by IT to improve outcomes of research and education institutions.

Although some institutions have embraced online education for their resident and distance programs, many institutions across the globe have been reluctant to adopt it as a learning platform. The slow of progress towards the desired outcomes related to adoption of online delivery is often adjudicated to budgetary constraints. However, it is more likely that this is caused by reluctance to change by those involved in education and failures caused by an insistence to replicate the face-to-face experience of the classroom in an online environment in lieu of appropriate pedagogical models.

This general state of use of IT in education is also reflected in agricultural and biological engineering programs. It is thus necessary that investments in IT in education not only improve learning outcomes, but also reduce the cost of instruction. Experience demonstrated that this is achievable given the right investments and adoption of IT in education. A review of 156 redesigned courses involving 195 institutions and—250,000 students showed that in 72% of the courses learning outcomes were improved, while in 28% there were no improvements. In addition, the cost of instruction was reduced on the average by 34% instruction [7]. Online delivery is now commonplace in strategic plans related to teaching and learning in higher education for top-ranked universities. This is often associated to improving learning outcomes, reducing the cost of instruction and innovation in teaching/learning [8].

Online teaching/learning is generally accepted as a direction for higher education institutions as an opportunity to modernize their work and create new channels that improve creative, entrepreneurial, and critical thinking skills of students. The issues that remain are related to finding the most effective and efficient ways to deliver this form of instruction [9]. For higher education in agricultural and biological engineering programs, challenges remain because of scarce budgetary resources for initial investments and the disruptive nature of the technology stemming from the cultural, historic, and economic context.

1.1 The Learning Environment in Online Education

Online education is like other learning environments in that it requires that the following components be effectively integrated and clearly articulated for student success:

1. The behavior that the student will exhibit to demonstrate he/she has acquired the knowledge, in the appropriate context.
2. The means by which the student will be assessed and evaluated.
3. A clear and logical progression of quality educational materials that ensure that the student acquires the knowledge.

4. A platform that allows communication and collaboration amongst students and instructors.
5. A platform or collection of tools that allow the student to focus on the learning experience, and allows the faculty to effectively create, maintain and manage the course.

2 Desired Student Behavior: Learning Outcomes and Learning Objectives

What the student should learn is conveyed as learning outcomes and learning objectives.

Learning outcomes articulate what the students will learn in a specific lesson or on a general scale of the entire course. Learning outcomes target knowledge, skills, or attitudes for change. Learning outcomes may be defined by someone other than the instructor, such as resulting from professional licensing requirements, the overall academic program, or departmental standards. An example learning outcome is: "an ability to identify, formulate and solve complex engineering problems by applying principles of engineering, science and mathematics" [10].

Learning objectives are more specific when compared to learning outcomes. Learning objectives focus on a collection of precise learned behaviors, particularly in the knowledge domain, that are designed in a progressive manner (scaffolded), and result in the student achieving the desired learning outcome. A learning objective states what the learner will be able to do in terms of observable behavioral change after completing an educational activity. It is important to note that the outcome of the educational exercise a is a clear and observable terminal behavior.

The sequence of activities that is related to a specific learning objective is designed in such a way that the student can exhibit the terminal behavior after completing these activities. Learning objectives as well as activities should implement instructional scaffolding [11]. This consists of sequencing learning tasks into discrete and manageable parts. As students develop their understanding, build skills, and gain confidence, towards the student being able to work independently. Activities may be very diverse and include elements such as readings, problem sets, essays, discussion, collaborative problem solving, and many others.

To develop clear and well scaffolded learning objectives it is important to understand some basic concepts related to the taxonomy of learning and theory.

2.1 Blooms Taxonomy

To promote higher forms of thinking in education a taxonomy was created [12] in three domains of educational activity:

1. Cognitive: Mental skills (knowledge).
2. Affective: Growth in feelings or emotional areas (attitude or self).
3. Psychomotor: manual or physical skills (skills) used.

Over time, Bloom's knowledge taxonomy was revised into its current form and is often represented as shown in Fig. 1 [13]:

1. ***Remembering:*** Recall or retrieve previous Learned information. (The student defines, describes, identifies, knows, labels, lists, matches, names, outlines, recalls, recognizes, reproduces, selects, states).
2. ***Understanding:*** Comprehending the meaning, translation, interpolation, and interpretation of instructions and problems. State a problem in one's own words. (The student comprehends, converts, defends, distinguishes, estimates, explains, extends, generalizes, gives an example, infers, interprets, paraphrases, predicts, rewrites, summarizes, translates).
3. ***Applying:*** Use a concept in a new situation or unprompted use of an abstraction. Applies what was learned in the classroom into novel situations in the workplace. (The student applies, changes, computes, constructs, demonstrates, discovers, manipulates, modifies, operates, predicts, prepares, produces, relates, shows, solves, uses).
4. ***Analyzing:*** Separates material or concepts into component parts so that its organizational structure may be understood. Distinguishes between facts and inferences. (The student analyses, breaks down, compares, contrasts, diagrams, deconstructs, differentiates, discriminates, distinguishes, identifies, illustrates, infers, outlines, relates, selects, separates).
5. ***Evaluating:*** Make judgments about the value of ideas or materials. (The student appraises, compares, concludes, contrasts, criticizes, critiques, defends, describes, discriminates, evaluates, explains, interprets, justifies, relates, summarizes, supports).

Fig. 1 Revised bloom's taxonomy in the cognitive domain

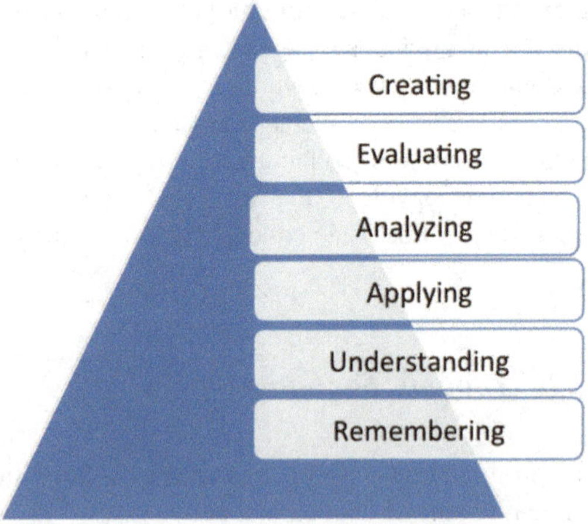

6. ***Creating:*** Builds a structure or pattern from diverse elements. Put parts together to form a whole, with emphasis on creating a new meaning or structure. (The student categorizes, combines, compiles, composes, creates, devises, designs, explains, generates, modifies, organizes, plans, rearranges, reconstructs, relates, reorganizes, revises, rewrites, summarizes, tells, writes).

2.2 The Cognitive Dimension

Bloom's revised Taxonomy also added the concept of a knowledge matrix to add a cognitive dimension as shown in Fig. 2:

Where:

1. ***Facts:*** A specific and unique data or instance.
2. ***Concepts:*** A class of items, words, or ideas that are known by a common name, includes multiple specific examples, shares common features. There are two types of concepts: concrete and abstract. It includes knowledge of terminology and of specific details and elements.
3. ***Processes:*** A flow of events or activities that describe how things work rather than how to do things. There are normally two types: business processes that

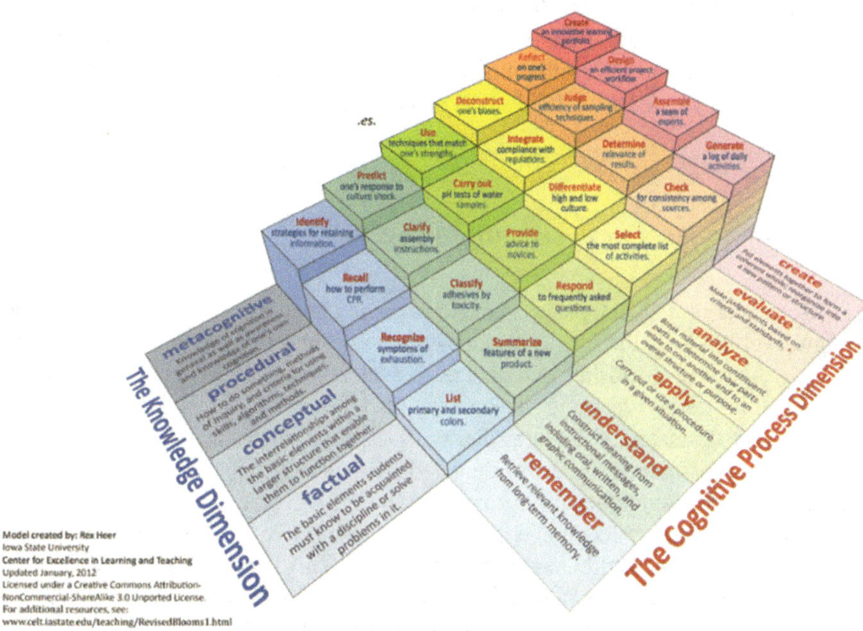

Fig. 2 Example of action verbs in the cognitive and knowledge domains. Attribution: https://slcc.instructure.com/courses/339717/pages/interactive-blooms-taxonomy

describe workflows and technical processes that describe how things work in equipment or nature. They may be thought of as the big picture, of how something works. It includes knowledge of classifications and categories, principles and generalizations, theories, models, and structures.

4. ***Procedures:*** A series of step-by-step actions and decisions that result in the achievement of a task. There are two types of actions: linear and branched. It includes knowledge of subject-specific skills and algorithms, techniques and methods, and the criteria for determining when to use appropriate procedures.
5. ***Principles:*** Guidelines, rules, and parameters that govern. It includes not only what should be done, but also what should not be done. Principles allow one to make predictions and draw implications. Given an effect, one can infer the cause of a phenomena. Principles are the basic building blocks of causal models or theoretical models (theories).
6. ***Metacognition:*** Includes strategic knowledge, knowledge about cognitive tasks including appropriate contextual and conditional knowledge, self-knowledge.

Note that knowledge of a subject can vary widely within Bloom's taxonomy. On any subject and individual may be performing at the lowest level, in which an individual retrieves relevant knowledge from memory to recall basic elements of the subject (remember, factual). This is can be compared to cognition awareness and knowledge on the same subject to create new knowledge (create, metacognitive), which is a much more complex level of behavior in the taxonomy.

Thus, it is very important that a clear articulation of learning behaviors (objectives) be made to ensure that the student can perform at the desired cognitive and knowledge level. The traditional "the material was covered in the class" provides no clear indication of the level of competence of the student. In addition, these objectives should be aligned with the desired outcomes.

2.3 Components of a Learning Objective

The use of learning objectives in a course has many benefits and uses. At the course design level it allows the faculty to focus on the desired learning outcomes, design the progression of acquiring knowledge in a clearly articulated progressive way, ensure that all desired learning takes place, provides a clear basis for the development of didactic materials, ensures clarity of expectations and performance during evaluation and assessment of the student and the course, and is helpful in identifying elements of a course that need improvement.

Most important is to recognize that a learning objective is a clear, unambiguous, statement of the expected behavior of the student. This benefits the student particularly in situations in which learning activities are conducted in isolation, as is often the case in blended and distance education models [14, 32].

A learning objective should include the following:

1. The behavior (or skill) the student will perform.
2. The conditions under which the student will perform.
3. The criteria used to measure student performance.

The behavior is articulated by an action verb which communicates the performance by the learner. Action verbs describe and a behavior that can be observed and that are measurable within the teaching time frame. Examples of verbs related to a level in the taxonomy are shown in Fig. 2. There are numerous sources in the literature where recommended action verbs associated to Bloom's taxonomy can be found.

An example of a learning objective is: "Given the geometry of an isothermal closed-pipe hydraulic system, the student will draw a diagram showing the distribution of enthalpy, kinetic and potential energies per unit weight, including friction losses."

3 Learning Theory[1]

This material is not intended to review learning theory in depth. However, it is important for the practitioner to have some basic understanding of it. Clarity on what the students should learn is essential. In addition, it is equally important to understand how students learn to create an effective learning environment that uses resources and the time of faculty and students efficiently. The three most prominent learning theories are known as behaviorism, cognitivism, and constructionism [33–35].

1. ***Behaviorism*** is a world-view that operates on a principle of stimulus-response. It assumes that a learner is essentially passive and responds to external stimuli. The learner starts as a "tabula rasa" and behavior is shaped through positive or negative reinforcement. Learning is defined as a change of behavior in the learner.
2. ***Cognitivism*** is a paradigm where the learner is viewed as an information processor. Knowledge is seen as a schema, or symbolic mental construction. Learning is a change in in a learner's schemata. Cognitivism responds to behaviorism by recognizing that people require active participation to learn and changes in behavior are an indication of what occurs within the learner's brain.

[1]For further resources the readers are referred to the following material.

- https://citt.ufl.edu/resources/
- http://web.calstatela.edu/dept/chem/chem2/Active/main.htm
- http://josotl.indiana.edu/article/view/1744
- http://www.icte.org/T01Library/T01 245.PDF
- http://www.league.org/gettingresults/web/module3/active/index.html
- https://odee.osu.edu/active-learning

3. ***Constructivism*** postulates that learning is an active and constructive process in which the learner is the information constructor. (An individual construct his/her own subjective reality linked to prior knowledge). It views learning as an active and contextualized process in which knowledge is constructed (as opposed to acquired). This construction is based on the learner's personal experience and hypothesis about the environment, bringing past experiences and cultural factors into a learning situation [31].

The takeaway from these learning theories as it relates to pedagogy in engineering is that behaviorism helps in understanding and articulating learning expectations in terms of conduct. Constructivism, on the other hand, helps understand how higher levels of learning can be achieved through social interaction.

3.1 Passive and Active Learning

To develop an effective online learning environment for the student it is important to understand the learner. Not only the knowledge that the learner is coming to the course with, but also what learning behaviors are more effective. Behaviors during learning are corelated to the retention of knowledge and can be broadly classified into passive and active.

Passive learning occurs when students are engaged solely in taking in information. Examples of this include: Reading materials, listening to a lecture, watching a video, and looking at photos, diagrams, or PowerPoints. Passive learning is primarily an individual activity in which students learn by assimilating the information presented. The traditional college classroom is primarily passive. Active learning occurs when the students are focused on doing, with the course content and activities designed to increase and enhance their understanding of a topic.

Overall, students in an active learning environment display improved conceptual learning and understanding of scientific reasoning, and greater motivation and involvement in learning activities than students taught in a, more traditional, passive style [14–16].

Some examples of activities that encourage active learning are:

Online discussions/debates, group projects, concept mapping, role playing, content related games, and problem solving. Active learning includes activities that encourage the application, deeper understanding, and discovery of new knowledge. In engineering, for example, this may take the form of providing a solution to an engineering problem or designing a system.

Social activities are particularly suited for active learning. Where students critique, collaborate and generate a deep understanding of the knowledge acquired. In this context, the role of the instructor is one of directing and supporting. This puts the responsibility of learning on the shoulders of the students, with instructor as support.

A well designed and scaffolded set of activities for a given learning object should consider carefully the combination of passive and active activities that the student must carry out.

3.2 Best Practices

The best practices can be summarized in the following principles.

- Focus on the student, make learning student-centered.
- Create an environment where students are thinking about what they are learning.
- Ask meaningful questions that focus on the deeper meaning instead of the minor details.
- Give students opportunities to collaborate and learn from each other.
- Create meaningful activities that give students the opportunity to apply new knowledge.
- Create multiple ways of interacting with students. Be available to guide and assist as students work through the coursework.
- Faculty Interaction.

4 Creating a Course

4.1 Anatomy of a Course

Creating a course for online delivery requires careful design, planning and execution. A formal course is typically composed of a series of modules that are executed every 1 or 2 weeks. The structure in Table 1 is an example and recommended for a

Table 1 Example structure of an online course

Modules (weekly or biweekly)	Content
Course introduction	Contains information about the course and instructors. Course management policy. Tutorials for online learning. Help resources. Tips for success
Module 1	Subject matter module containing learning objectives, activities the student will conduct. Student evaluation. Feedback to the student
Module 2	" "
...	...
Module n	" "
Final module	Learning objectives and activities that integrate the learned behaviors towards achievement of learning outcomes
Optional: High stakes assessment	Comprehensive evaluated activity, such as a final exam

typical quarter or semester course. Figure 3 shows the top portion of a typical introductory module to the course.

Note that a high stakes assessment at the end of the course is shown as optional. The authors believe that well designed, frequent assessments that provide immediate feedback to the student are beneficial. These ensure that the student progresses through the materials with a good grasp of what was learned and reduces the stress on the student resulting from high stakes assessments. It also appears to reduce academic dishonesty. In addition, it allows the instructor to identify students at risk early on and to take remedial action to ensure student success.

Welcome to ABE6933: Logistics of the Agri-Food Chain

Please watch the short introductory video below:

- Introduction to ABE6933 (4:20) (Notes PDF) (Transcript)

Your instructors for this course are:

Dottssa. Patrizia Busato	Dr. Remigio Berruto	Dr. Fedro S. Zazueta
Dr. Patrizia Busato is Assistant Professor at DISAFA, University of Turin, Italy. Her research program has focused primarily on logistics and production optimization, resource use in agricultural processes, bioenergy and sustainability. She has conducted research on active learning methodologies, focusing on flipped classroom instruction, related to learning effectiveness and student preferences. She is a member of ASABE and Associate Editor of the CIGR Journal.	Dr. Remigio Berruto is Associate Professor at DISAFA, University of Turin Italy. His research and teaching program is focused on agricultural mechanics, transport optimization and sustainability, traceability and logistics of the agrifood chain, and agricultural information systems. He has taught engineering courses using active learning methods. He is a member of ASABE and Associate Editor of the CIGR Journal.	Dr. Fedro Zazueta is Associate CIO and Professor at the University of Florida, USA. He is charged with the direction of the Office of Academic Technology that includes the Center for Instructional Technology and Training that supports course development and the implementation of new pedagogical models at the enterprise level. He is an ASABE Fellow and lifetime honorary President of CIGR.

Office Hours

Because this is an online course, with students and instructors in different time zones it is best to set up an appointment with the instructor. If you are a student at UF you should contact Prof. Zazueta. If you are a student at UT, you should contact Prof. Busato or Berruto.

Important:

1. Questions related to the subject matter of the course should be posted on the discussion board for everyone to see. This will not only help everyone but is also part of the participation that contributes to your final grade. If you have a question, ask. Most likely your peers will have a similar question and you will be helping everyone. Likewise if you have an answer to a question.
2. Questions not related to the subject matter and need privacy may require a conference (web or face-to-face) with the instructor. If you feel your issue is personal and not related to the subject matter of the course, set up an appointment with one of the instructors.

Telephone Office Hours: By appointment only (please email to set up).

Fig. 3 Partial student view of the introduction to an online course using a course management system

The Introductory Module

It is important that the first module of the course focus on at least the following:

1. Learning outcomes related to the course.
2. An introduction to the faculty and instructors teaching the course.
3. Course syllabus.
4. A tutorial on how to use the online tools. This may consist of using a course management system and an explanation of how the materials are organized and accessed. This is an important component of the course as it allows the student to focus on learning in later stages of the course rather than focusing on the technology. Also, it will reduce questions from students resulting from lack of understanding on how to use the online tools. It is recommended that students are not allowed to continue into the subject matter unless they are proficient in using the related online tools.
5. Office hours. Specify when the Faculty, Instructors of Teaching Assistants will be available to help the students with help related to the academic content of the course or related private issues that may require attention. This avoids the expectation that these individuals are available anytime and will respond immediately.
6. Means by which the student can request help should technical problems arise. Most technical problems can be resolved by directing students to a Help Desk.
7. Academic honesty and other university policy.

Content Modules

The components of modules related to the subject matter of the course should include at a minimum the following components:

1. An introduction to the subject matter in the module.
2. The learning objectives (Fig. 4).
3. For each learning objective the scaffolded activities that lead the student to learn the material (Fig. 5).
4. For each learning objective an evaluation of the student's ability to perform as indicated in the learning objective. It is important that the student understand clearly how the evaluation will take place.

It is highly recommended that the services of an instructional technologist be recruited to select the best means by which a learning objective can be reached. This education professional can contribute greatly to creating an engaging learning environment that results in the desired outcomes and engages the student.

Module 2

Competing through Logistics and Supply Chain Performance

The focus of this unit is to understand: 1) The need for a strategic approach to logistic activities to improve supply chain performance, and 2) challenges to achieving a strategic fit of logistics with a company or a supply chain.

By completing the activities and exercises in this module, you will have achieved the following learning objectives:

1. Describe the role and relationship of logistic-value-added with the other areas of the firm.
2. Describe the characteristics of products and their relationship to main logistic activities.
3. Identify how supply chains compete in terms of time, cost, quality and sustainability as well as supporting capabilities and soft objectives.
4. Illustrate challenges and key elements required to achieve a strategic fit.
5. Describe how risk factors influence the achievement of strategic fit.

Introduction to the Module

- View introductory video to the Module (1:50) (PDF)

Evaluation:

The activities and assessments for this module will add to 100 points. You may view a detailed explanation of the grade breakdown and assignment due dates for this course on the Course Syllabus.

Fig. 4 Example of the learning objectives in a 2-week module of a logistics course

Objective 2: Describe the characteristics of products and their relationship to main logistic activities.

Activities:

- Watch Logistics in the Firm: Product Classification (7:06) (PDF)
- Watch Logistics in the Firm: Product Characteristics (18:20) (PDF)
- Watch Logistics in the Firm: Product Pricing (18:20) (PDF)
- Read article on Iphone Journey and Transport Costs
- Watch Logistical Activities (15:48) (PDF)
- Read The Perfect Order by Blanchard (PDF) (Read Online)

Assessment:

- Complete Module 2 Diagram assignment.

Fig. 5 Example of a learning objective, scaffolded activities, and assessment for objective 2 in Fig. 4

Final Module

The final module of the course should integrate the materials learned. For this purpose it is convenient to thin in terms of the activities the student can undertake that will achieve this. For example, a group project or an extended essay that demonstrates higher level knowledge and cognition.

Evaluation

Evaluations can take several forms, such as essays, quizzes, tests, projects, participation, and others. It is key, however, that these be focused on the behavior that is expected from the student as articulated in the learning objective and that the evaluation use as much as possible an objective criterion. For this purpose, it is good practice to use rubrics that clearly outline the expectations on how the evaluation will be done (Fig. 6). Also, well-designed evaluations should provide feedback to the student about their performance.

Evaluating the student on each learning objective as the course progresses allows for the student to clearly see what his/her performance is and allows the instructor to take remedial action on a timely basis if required.

Furthermore, these micro evaluations reduce the risk for the student in not performing well in a high-stakes evaluation, such as a midterm or final exam.

Discussion Rubric (25 points)

Discussion Rubric (25 points)					
Criteria	Ratings				Pts
Quality of posts and responses	7.0 pts Comments are appropriate, thoughtful, reflective and respectful of others' postings. Posts include supporting content to justify/explain thoughts/opinions.	5.0 pts Comments are appropriate thoughtful, reflective and respectful of others' postings. Posts do not include sufficient supporting content to justify/explain thoughts/opinions.Rating Description	2.0 pts Responds with minimum of effort; states thoughts/opinions without supporting content, but is respectful of others' postings.	0.0 pts No posting, or disrespectful post(s).	7.0 pts
Relevance of posts and responses	6.0 pts Posts and replies are related to discussion topic and prompt further discussion when appropriate	4.0 pts Posts and replies related to discussion topic, but may not fully address the topic or prompt.	0.0 pts No posting, or post not related to discussion content.		6.0 pts
Quantity of responses	7.0 pts Responds thoughtfully to specified number of posts (varies each week);	4.0 pts Responds thoughtfully, but responds to one fewer post than required	3.0 pts Responds to specified number of posts with minimal effort OR responds to two fewer posts than required	0.0 pts Does not respond to other posts, or responds after the discussion due date.	7.0 pts
Grammar/Mechanics/ Word Count	5.0 pts Writing is coherent and free of noticeable grammar, spelling, or punctuation errors	4.0 pts Writing is coherent, but there are noticeable grammar, spelling, or punctuation errors.	3.0 pts Writing is incoherent, or post is difficult to read due to multiple grammar, spelling, or punctuation errors.	0.0 pts No Post	5.0 pts
					Total Points: 25.0

Fig. 6 Example of a rubric that shows the criteria that will be used by the instructor in the evaluation of an on-line discussion

5 Instructional Design in a Nutshell

Instructional design is a methodology used to produce learning environments. In contrast to curriculum design which focuses on what the student will learn, Instructional design focuses on how the student will learn. Instructional design is systematic and uses learning theory and best pedagogical practice to ensure the quality of learning.

5.1 The Development Team

In applying this methodology for an engineering course, it is important for the instructor to understand what his/her role is in the process. That formal development of a course is done by a team of individuals with different competencies that contribute to a successful product. There are three primary roles:

Instructor The instructor is a subject matter expert knowledgeable of the curriculum, responsible for articulating the learning objectives, assessment items and learning activities that compose the course. The role of the instructor is to define what is to be learned and work with the instructional designer on the best way on how this can be done. In addition, to engage in a process of continuous improvement of the course.

Instructional designer The instructional designer is an expert in education, skilled in educational technology, pedagogy, and project management. Responsible for management of the project, ensuring the quality of the content and assisting the instructor in developing high quality learning objectives, suitable assessments, learning assets, and pedagogically sound delivery of the course. A competent and experienced instructional designer is key to the success of a course in producing the desired outcomes.

Support staff Depending on the specifics of the pedagogy selected and the type of learning assets used in the course, the team may require web developers, programmers, graphic artists, videographers, transcribers, etc.

5.2 The ADDIE Methodology

ADDIE is a common methodology for instructional design. This methodology is well tested and is composed of the five phases shown in Fig. 7. This methodology is standard in instructional design. It is efficient and with the help of an instructional designer invariable results in quality instructional materials.

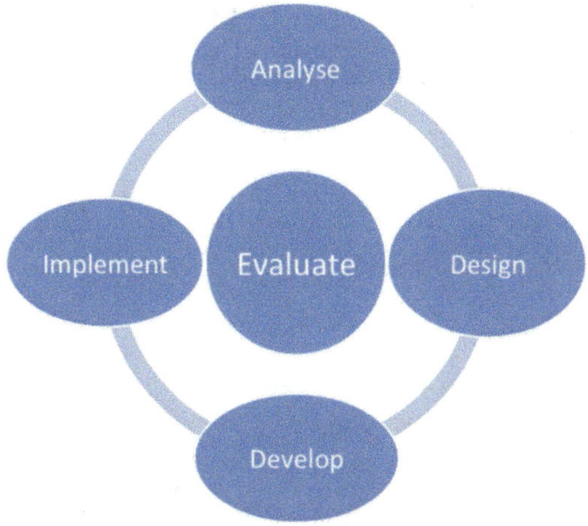

Fig. 7 ADDIE methodology

The major activities that take place during each of the phases of ADDIE are:

1. **Analysis.** The primary purpose is to articulate clearly what the instructional problem is. Instructional goals and objectives are established at this level, as well as identifying learner knowledge and skills.
2. **Design.** This stage focuses on developing learning objectives, assessment items, and learning assets. It follows a logical and orderly method for identifying, developing, and evaluating strategies to attain the course's goals. This stage requires great attention to detail.
3. **Development.** In this stage the instructional designer works with the staff to create and assemble the learning assets that were designed in the previous phase. Includes testing and debugging. In general, this process will move forwards quickly if the design phase is executed carefully.
4. **Implementation.** During the implementation phase all functional components of the course are assembled. Also, training for the instructor is provided. It is a good practice to develop a manual that covers course curriculum, learning outcomes, methods of delivery and student assessment procedures. It may also be required to conduct training for the learners.
5. **Evaluation.** Performance methods are used to measure how well the objectives were achieved. That is, the level of success the learner reaches in retaining and demonstrating acquired skills and understanding. As a general guideline, the evaluation focus can be on understanding of the material, long term retention, and critical thinking skills. Also important at this stage, is to measure how well the course materials facilitate effective learning by the student.

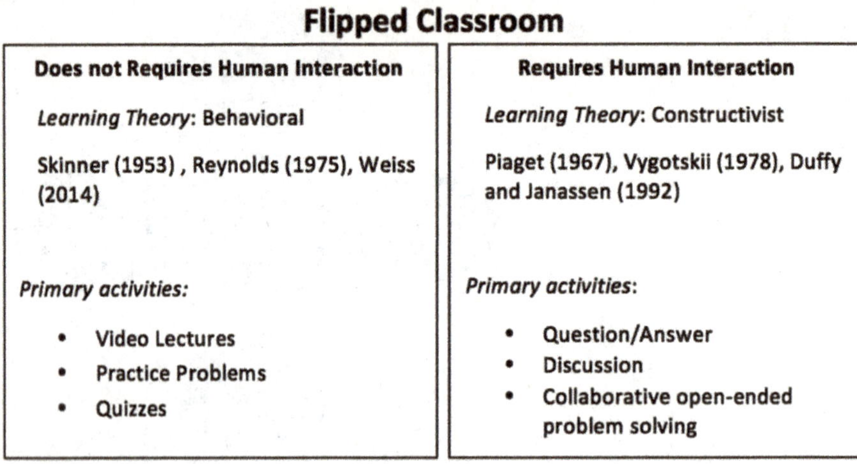

Fig. 8 A framework for activities in the Flipped Classroom

6 Active Learning: The Flipped Classroom[2]

Advances in technology and learning theory and practice have created new directions and opportunities for pedagogy in engineering education. A pedagogy currently receiving much attention is the flipped classroom. The flipped classroom is unique in its combination of active, problem-based learning constructivist ideas and direct instruction methods based on behaviorist principles [17]. This pedagogical approach is enabled by technological advances that permit the transmission and duplication of information at very low cost and various means, and the trend in education to make learning student-centered.

Consensus on a flipped classroom definition is lacking [18]. A simple definition of inverted classroom is given by Lage et al. [19]. By this definition, activities that traditionally take place in the classroom, take place outside the classroom in a flipped classroom, and vice versa. In this this workshop, a definition of flipped classroom will be used that accommodates theoretical frameworks by defining the flipped classroom not in terms of what is done in the traditional classroom, but in terms of human interaction. Thus, a flipped classroom is one in which learning activities not requiring human interaction take place outside the classroom (enabled by technology) and learning activities requiring human interaction take place in the classroom (virtual or physical). Figure 8 illustrates this definition of the flipped classroom. Note

[2]For further resources the readers are referred to the following material.

- https://www.educationnext.org/the-flipped-classroom/
- http://www.ncbLnlm.nih.gov/pmclartides/PMC2665262/
- http://ieeexplore.ieee.org/stamp/stampjsp?arnumber=6481483

that by this definition of a flipped classroom activities requiring human interaction may occur face-to-face or virtually and in synchronous and asynchronous manners.

The focus of activities not requiring human interaction is for the student to understand and apply basic concepts related to the subject matter of the course in preparation for activities requiring human interaction that focus on higher levels of learning in Bloom's taxonomy [20].

Some activities that do not require human interaction are readings, video, closed problem solving and quizzes. Early studies show that quality video lectures outperform traditional lectures [21]. Also, online homework is equally effective as paper and pencil [22, 23]. These, coupled with quizzes for self-evaluation [24] provide a solid basis for the student to engage in activities requiring human interaction focused on higher level skills such as communicate effectively; identify, formulate and solve engineering problems; and, work in teams.

Specific activities requiring human interaction include the use of face-to-face and online discussion boards used to post and answer questions (students and faculty alike) and carefully crafted open-ended problems. This approach provides an opportunity to develop activities for active learning [25], cooperative learning [26], peer-assisted-learning [27], and problem-based learning [28].

It is important to note that activities are not limited to those shown in Fig. 7. The number and type of activities can be diverse provided they focus on efficiently achieving a learning outcome and the learning style of the students [29].

6.1 Research Findings on the Performance of the Flipped Classroom

There is abundant literature on use of the flipped classroom mode of instruction. However, little work has been done in using this approach in agricultural engineering education. Research carried out by the workshop conductors focused on evaluating preference and performance of students in a Flipped-Classroom mode of instruction when compared to traditional face-to-face. Students in an Agri-food Chain Logistics course where subjected to both forms of instruction. After completion, students were asked to fill a survey on questions related to their perception and preference about the modality of the course. The survey used a Likert scale and the results are shown in Fig. 9. In addition, students in both modes of instruction were subjects of the same high-stakes assessment. Their performance is shown in Table 2.

The relevant conclusions of the study were:

- Students show a strong preference for the flipped-classroom over the face-to-face delivery of the course.
- Students performed better in a high stakes assessment when learning the course materials in the flipped-classroom mode of instruction.

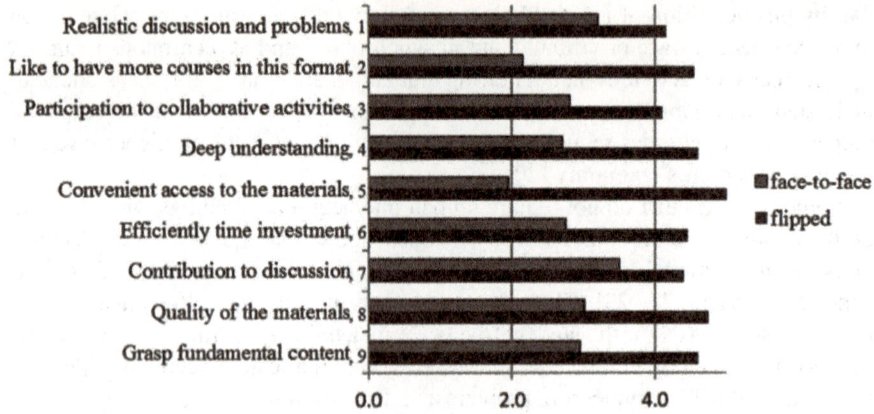

Fig. 9 Comparison of answers to student preference between flipped classroom and face-to-face instruction

Table 2 Student performance (grades) in high-stakes tests[a]

	Module 1			Module 2			Wilcoxon Signed-Rank. Test		
	Mean median		Standard deviation	Mean median		Standard deviation	N	Standardized Test statistic	P-vakte
Group A	100	100	0	78.47	9175	25.9	6	-2.226	0.026
Group B	62.38	66.67	20.2	97.62	95.83	2.2	7	2.205	0.027

[a]Highlighted cells correspond to student's test scores for the Flipped Classroom modality of instruction

It is also notable, that the students learning under the flipped classroom format performed, not only at a higher level, but also more uniformly as is shown by the standard deviation of the grade scores.

6.2 Best Practice

- Create quality pre-recorded lectures that relay the course content effectively (substantial pre-planning and prep work required before pilot semester).
- Reduce lectures to manageable segments (about 15 min).
- Develop classroom activities that promote Active Learning. Students should be applying the knowledge gained from lectures and readings (i.e., case studies, debates, discussions, group projects, problem solving, presentations, individual assignments, educational games).
- Avoid "busy work" to simply fill the time.

- Be available during class time to assist and facilitate. Circulate, be prepared to guide and entourage active learning in a student-centered environment. Interact with the class.

6.3 Creating Engaging Activities

Implementation of the Flipped classroom, online or face-to-face, requires carefully designed activities that are associated to the human interactive component. This task is more difficult that it appears at first hand. It is essential to keep the following in mind when designing an activity:

1. The activity must be closely associated to one or more learning objectives.
2. Student engagement is important to ensure that students remain focused and interested in the learning tasks. Activities should be entertaining, interactive, and meaningful for the learner.
3. The activity must be presented to the learners in such a way that they see its value (metacognition).

Defining an activity that meets the three criteria above is a task that requires careful thought and attention to detail. It is important to ensure that the learner is given the opportunity to reflect about the problem (with self and with others) and to establish an open dialogue with other students and instructors.

For this exercise follow the steps below to create an activity associated to the learning object you previously created:

1. Make a list of significant problems that drive your discipline.
2. Identify some open-ended problems that are central to the course you are teaching. Problems where the instructor has served as a consultant can be of high interest to students.
3. Make a list of ideas that are engaging problems that will drive the students into the content related to the learning objective.
4. Select the idea that best meets the three criteria listed above.
5. Define the type of activity that would best achieve the intended results (group discussion, design problem, etc.).

Draft a statement of the problem as it would be presented to the students. A context should be included so the student understands the value of resolution of the problem.

7 Tools for Online Education

Authoring and managing an online course requires a comprehensive set of tools. At the time of writing this document there were at least 537 different learning management system related software available in the market [30]. Including course authoring software that is used develop courses or customized training and authoring tools used to create interactive educational content for use in digital media and electronic devices.

There are numerous reviews and comparisons of widely used learning management systems in higher education. These tools, because of their cost and complexity as well as the characteristics of the environment where they will be used are best funded, implemented and maintained at the institutional level. Ease of use for students and instructors is the most important characteristic. Also, features vs. cost should be carefully considered when selecting an open source or commercially available system.

8 Final Comments

Online education is a commonplace in the twenty-first Century. The new understanding of how people learn, and technological developments have created an opportunity for improving educational outcomes that many major universities have embraced. Increasing understanding, retention, recall and critical thinking has been demonstrated when active pedagogies and on-line learning are implemented, from blended courses to asynchronous distance education.

References

1. Shehan, Tony (2020). COVID-19 in Higher Education: Responding to the Short-Term Challenge. Gartner Inc. Retrieved Sept. 17, 2020, from https://www.gartner.com/document/3982312?ref=algobottomrec&refval=3982205
2. UNESCO. (1998). World Declaration on Higher Education for the Twenty First Century: Vision and action and Framework for Priority action for Change and Development in Higher Education. Retrieved Dec 20, 2014, from http://www.unesco.org/education/Pducprog/wche/dedarationentihtmliwnr,eclaration
3. US Department of Education. (2010). Transforming American Education: Learning Powered by Technology. National Education Technology Plan 2010. Retrieved January 2nd, 2015, from http://tech.ed.gov/wp-content/uploads/2013/10/netp2010.pdf
4. European Commission. (2010). Europe 2020: A Strategy for Smart, Sustainable and Inclusive Growth. Available at: http://www.emeuropaeu/policy-documents/com-2010-2020-europe-2026 [Accessed Dec. 20, 2014]
5. MIUR (Producer). (2013). Piano nazionale scuola digitale (PNSD). Retrieved February 15 from http://www.istruzione.it/allegati/Avvison.2945 25-11-2013 DGSSSI.pdf

6. World Bank. (2007). China and the World Bank : a partnership for innovation. Retrieved January 2nd, 2015, from http://documents.worldbank.org/curated/en/2007/01/9739957/china-world-bank-partnership-innovation
7. NCAT. (2014). A summary of NCAT Program Outcomes. 2014). Retrieved October 10, 2014, from http://thencatorg/ProgramOutcomesSummary.html
8. Williams, R. J., Hartman, J. C., Zazueta, F. S., & Smith, J. (2012). Development for Student Success: An undergraduate Online Course in Engineering. In G. C. Clark & S. A. Clouser (Eds.), Teaching with Technology, Volume 2:The Stories Continue (Vol. 2, pp. 101–106): Learning Technology Consortium. Retrieved from http://Itcessays.files.wordpress.com/2012/08/teachingwithtechnologyv2a.pdf.
9. Bateman, J., and Davies, D. (2014). The challenge of disruptive innovation in learning technology. Medical Education, 48(3): 227–228.
10. Sweet Briar College (n.d.). Engineering Educational Objectives and Student Outcomes. Downloaded Sept 18, 2020 from https://sbc.edu/stem/engineering-educational-objectives-and-student-outcomes/
11. Frey, N., Fisher, D. (2010). ASCD. *Guiding Instruction: How to Develop Confident and Successful Learners.*
12. Bloom, B.S., Engelhart, M.D., Furst, E.J., Hill, W.H., Krathwohl, D.R. (Ed.) (1956). Taxonomy of Educational Objectives, Handbook I: The Cognitive Domain. New York: David McKay Co Inc.
13. Anderson, L.W., Krathwohl, D.R., Airasian, RW., Cruikshank, K.A., Mayer, R.E., Pintrich, RR., Raths, J., Wittrock, M.C. (2001). A Taxonomy for Learning, Teaching, and Assessing: A revision of Bloom's Taxonomy of Educational Objectives. New York: Pearson, Allyn & Bacon.
14. Deslauriers, L., L. S. McCarty, K. Miller, K. Callaghan, and G. Kestin (2019) Measuring actual learning versus feeling of learning in response to being actively engaged in the classroom PNAS, 116(39): 19251–19257. Downloaded September 9, 2020 from https://doi.org/10.1073/pnas.1821936116
15. Theobald, Elli J., Mariah J. Hill, Elisa Tran, Sweta Agrawal, E. Nicole Arroyo, Shawn Behling, Nyasha Chambwe, Dianne Laboy Cintrón, Jacob D. Cooper, Gideon Dunster, Jared A. Grummer, Kelly Hennessey, Jennifer Hsiao, Nicole Iranon, Leonard Jones, Hannah Jordt, Marlowe Keller, Melissa E. Lacey, Caitlin E. Littlefield, Alexander Lowe, Shannon Newman, Vera Okolo, Savannah Olroyd, Brandon R. Peecook, Sarah B. Pickett, David L. Slager, Itzue W. Caviedes-Solis, Kathryn E. Stanchak, Vasudha Sundaravardan, Camila Valdebenito, Claire R. Williams, Kaitlin Zinsli, Scott Freeman. (2020) Proceedings of the National Academy of Sciences Mar 2020, 117 (12) 6476–6483; https://doi.org/10.1073/pnas.1916903117
16. Udovic, D., D. Morris, A. Dickman, J. Postlethwait, P. Wetherwax (2002), Workshop Biology: Demonstrating the Effectiveness of Active Learning in an Introductory Biology Course, *BioScience*, Volume 52(3), Pp. 272–281, https://doi.org/10.1641/0006-3568(2002)052[0272:WBDTEO]2.0.CO;2
17. Bishop, J. and Verleger, M. (2013) The Flipped Classroom: A survey of the research. 120th ASEE Annual Conference & Exposition. Atlanta, GA. Paper 6219. Retrieved Dec 18, 2014 from http://www.studiesuccesho.nl/wp-content/uploads/2014/04/tlipped-dassroom-artikel.pd1
18. Chen, Y., Wang, Y., Kinshuk, and Chen, N.-S. (2014). Is FLIP enough? Or should we use the FLIPPED model instead? Computers & Education, 79(0): 16–27.
19. Lage, M. J., Platt, G. J., and Treglia, M. (2000). Inverting the Classroom: A Gateway to Creating an Inclusive Learning Environment. The Journal of Economic Education, 31(1), 30–43.
20. Krathwohl, D. R. (2002). A revision of bloom's taxonomy: An overview. Theory into Practice 41(4), 212–218.
21. Cohen, R A., Ebeling, B. J., and Kulik, J. A. (1981). A meta-analysis of outcome studies of visual-based instruction. Educational Communication and Technology Journal, 29(1): 26–36.

22. Bonham, S. W., Deardorff, D. L., and Beichner, R. J. (2003). Comparison of Student Performance Using Web and Paper-Based Homework in College-Level Physics. JOURNAL OF RESEARCH IN SCIENCE TEACHING, 40(10): 1050–1071.
23. Fynewever, H. (2008). A comparison of the effectiveness of web-based and paper-based homework for general chemistry. The Chemical Educator, 13: 264–269.
24. Stallings, V., and Tascoine, C. (1996). Student Self-Assessment and Self-Evaluation. The Mathematics Teacher 89(7), 548–554.
25. Michael, J. (2006). Where is the evidence that active learning works? Advances in Physiology Education 30(4), 159–167.
26. Foot, H., and Howe, C. (1998). The psychoeducational basis of peer-assisted learning. S. E. Keith Topping (Ed.), Peer-assisted Learning: Lawrence Erlbaum Associates.
27. Topping, K. J., and Ehly, S. W. (1998). Peer-assisted learning. Mahwah, N.J., L. Erlbaum Associates.
28. Barrows, H. S. (1996). Problem-based learning in medicine and beyond: A brief overview. p. 3–12 In New Directions for Teaching and Learning.
29. Zimmerman, A. P., Johnson, R. G., Hoover, T. S., Hilton, J. W., Heinemann, P. H., & Buckmaster, D. R. (2006). Comparison of personality types and learning styles of engineering students, agricultural I systems management students, and faculty in an Agricultural and Biological Engineering Department. Transactions of the ASABE, 49(1), 311–317.
30. Capterra (n.d.), LMS Software. Retrieved September 20, 2020 from https://www.capterra.com/learning-management-system-software/
31. Duffy, T. M. and David H. J., eds. (1992). Constructivism and the Technology of Instruction: A Conversation. Hillsdale, NJ: Lawrence Erlbaum Associates Publications.
32. Omelicheva, M. Y. and O. Avdeyeva. Teaching with Lecture or Debate? Testing the Effectiveness of Traditional versus Active Learning Methods of Instruction, PS: Political Science and Politics, (2008): 603–7. Retrieved September 9, 2020 from https://kuscholarworks.ku.edu/bitstream/handle/1808/10504/teaching%20with%20lecture%20or%20debate.pdf?sequence=1
33. Piaget, J. (1967). Six psychological studies. New York, Random House.
34. Skinner, B. F. 1953. Science and human behavior. New York, Macmillan.
35. Vygotskii, L. S., and Cole, M. (1978). Mind in Society: The development of higher psychological processes. Cambridge, Harvard University Press.

Assisting DIY Agricultural Robots Towards Their First Real-World Missions

Dimitrios Loukatos and Konstantinos G. Arvanitis

1 Introduction

Agriculture is one of the most important sectors of primary industry, characterized as sensitive, unstable, complex, dynamic, and highly competitive. In the twenty-first century, according to Food and Agriculture Organization (FAO), agricultural productivity should be increased by 60% in order to ensure a safe food supply which would adequately satisfy the nutritional needs of the constantly growing world population [1]. This goal has to be achieved despite the fact that the required resources are already stretched, as the amount of available agricultural land is decreasing due to increasing urbanization, soil erosion, and high salinity levels, while 70% of the world's freshwater supplies are consumed for agricultural purposes. In addition, it is expected that in the agricultural sector, greenhouse gas emissions will be reduced while the extreme weather conditions that impact the quantity and quality of the crops will be counterbalanced [1]. To successfully tackle these issues, the sector of agriculture has to become more productive and "climate-smart", by successfully exploiting a variety of existing and emerging technologies [2]. At this point, in terms of robotics, while the effectiveness of the modern, yet conventional, agricultural machinery is difficult to be further expanded [3], the advances in Information and Communication Technologies (ICT) allow for many significant improvements in agri-production [4], and thus Robotics and Autonomous Systems (RAS) technologies could positively contribute to the transformation of the agri-food sector [5–7]; UK-RAS Network [8].

D. Loukatos (✉) · K. G. Arvanitis
Department of Natural Resources Management and Agricultural Engineering, Agricultural University of Athens, Athens, Greece
e-mail: dlouka@aua.gr; karvan@aua.gr

D. D. Bochtis et al. (eds.), *Information and Communication Technologies for Agriculture—Theme IV: Actions*, Springer Optimization and Its Applications 185,
https://doi.org/10.1007/978-3-030-84156-0_12

In the context of Precision Agriculture (PA), robotic platforms equipped with a variety of sensors and making use of low cost Internet of Things (IoT) technologies, advanced analytics, computational intelligence tools, machine learning techniques, advanced automatic control schemes, future telecommunications, and Cloud computing technologies could provide information about soil, seeds, livestock, crops, costs, farm equipment, and the use of water and fertilizers. Furthermore, real-time stream processing, analysis, and reasoning are very important factors towards automation in the agricultural field [9], resulting in robots able to adapt to drastically varying conditions (in terms of space and time) with minimal delay. The farming robots increase the profit of farmers as they can reduce 20% of the scouting costs for cereals, and 12% for sugar beet weeding, and 24% for inter-row weeding [10, 11].

Robotic vehicle systems are using Global Navigation Satellite System (GNSS) techniques for precise positioning and, in commercial systems, the GNSS receivers are frequently combined with fixed Real-Time Kinematic (RTK) stations [12], for providing remarkable accuracy. Furthermore, these vehicles are exploiting machine-vision methods to improve their navigation in the environment, like hyper-spectral imagery. The robotic platforms equipped with quite expensive devices are able to capture both visible and Near Infrared Imagery (NIR) bands and stereovision systems. Robots are now able to perform quite precise operations [13]. A considerable tradeoff is the increased cost that the commercial versions of these systems have. This cost can be significantly reduced, if smaller autonomous robotic machines work in swarms to accomplish the farm operations [14, 15].

In general, robots destined for agricultural use should counterbalance many diverse factors ranging from more generic ones, like path planning [16], terrain assessment [17], secure operation [18] and fleet management [13, 19], to more specific ones, according to their purpose of operation. Indeed, the robotic systems in agricultural field operations have been classified according to major field operations they are involved in [20]. In this concept, considerable efforts are made, from many teams both commercial and scientific, to provide robots for weeding [21], seeding [22], disease and insect detection [13, 23], crop scouting [24], spraying [25], harvesting [26], plant management robots [27], and multi-purpose robotic systems [28]. Further details can be found in [29], where the authors provide a state-of-the art review of the current status in agricultural robotics, indicating the that the production of similar robots is coming from both companies and scientific research teams (academia). Furthermore, this research reveals that weeding, harvesting, crop scouting and multipurpose robots are drawing most of the attention, due to their challenging nature and to the quick return of investment prospects.

The robotic automation in the field of agriculture would also help attract skilled workers and graduates to the sector. Therefore, it is very important to educate future agricultural engineers in the disciplines and technologies that are involved in modern robotics but is not always easy for students to catch up with this fast-evolving process. A good way to tackle such difficulties is to encourage the students to participate in the design and implementation stages of any similar robotic vehicle, assisted by the innovative systems that have recently made the scene and are becoming very popular among the student communities, without the barriers that strictly commercial educational robotic solutions are posing [30].

There is a considerable gap, mainly in terms of size, cost, and complexity between educational robotic platforms used in interdisciplinary approaches like the ones described in ([31–33, 64]) and commercial agricultural robotic solutions like the ones described in [15, 34] and their successor models. The existence of this gap and the necessity to be bridged, become apparent by inspecting the wide acceptance that any effort to provide products in the middle has, like the Thorvald robotic platform [28], that started as a master thesis project and became a successful start-up company story. In general, this gap is partially attributed to the gap in robotic skill sets continuity experienced between high school and university level education practices [35], situation that should be improved toward fostering students' future engineering careers [36, 37]. For these reasons, it is quite challenging to investigate the extent to which university laboratory level, DIY solutions of non-negligible size, based on simple materials and wide-spread electronic components may result in actually products able to undertake real-world missions. Apparently, the robotic vehicles being created should have enough robustness and accuracy, in order to be capable of performing meaningful agricultural tasks. A very promising fact is that the boost in the electronics industry resulted in a plethora of devices of amazing characteristics that are offered at very affordable prices. The results of such a bridging approach could be encouraging both for the students, who will realize that their efforts are not far away from their forthcoming careers and for individuals working in industry sector, that can save research efforts and manufacture more tailored and price-friendly products.

This chapter describes the recent improvements based on previous work [36, 38] towards creating/upgrading electric vehicles capable of performing light-duty "real-world" agricultural operations. The whole approach is following the project-based learning model (PBL) [39]. Priorities during their construction were both the cost-effectiveness and the simplicity, using metal, wood, recyclable materials, gears, motors and wide-spread electronic components (i.e., motor drivers and sensors). Two robotic variants, of diverse nature, are presented, one for performing all-terrain soil-specific measurements and another for spraying over the crops with fertilizer, pesticides or herbicides. Initially, the "core" of these platforms was an arduino uno board [40], while, later on, composite scenarios involving raspberry pi [41] or WeMos units [42], combined with navigation components like navio2 units and pixy2 cameras were also implemented. Efficient methods for controlling them, for instance via smart phones, have been provided. Fundamental automatic control functions (in terms of speed and direction stabilization) and remote operation issues were implemented. The tests also examined simple artificial intelligence (AI) features that are now offered in educationally comprehensive forms. The "logic" of the robots was created using both visual and textual programming environments. Most of the remote interaction scenarios have been carried out through Wi-Fi [43] interfaces, while some of them involved LoRa [44] interfaces to achieve longer controlling distance. For better efficiency and autonomy, provision for small solar panel units has been taken and a custom mechanism for evaluating the energy consumption, in different configurations, has been implemented as well.

Apart from this introductory section, the rest of this chapter is organized as follows. Section 2 aims to understand the requirements that should be met by the robotic platforms and also provides the design guidelines to be followed. Section 3 highlights the most characteristic implementation details and also important enhancements that have been derived from the basic robotic layout. Section 4 reports on the lessons learned during this study and provides characteristic results, referring to the robotic vehicles being created and tested. Finally, Sect. 5 contains concluding remarks, exposes issues still open and presents plans for future work.

2 Requirements and Design Overview

The trials being highlighted in this research work are willing to bridge the gap in the performance and effectiveness between educational approaches and commercial solutions and to further lower the cost barrier. This priority becomes even more important, as swarms of smaller and cheaper vehicles tend to outperform larger single agricultural machines [14, 15]. According to the relevant scientific research findings, the performance of the robot can be better assessed, if analyzed in four categories according to the robot's components. Indeed, robots usually consist of four components: (a) sensing and localization (sensing), (b) path planning (planning), (c) mobility and steering (mobility), and (d) end effector manipulation (manipulation) [5]. The effectiveness of the agricultural robots relies heavily on those four components [5]. Efforts being made aim to improve this situation, by using cost-effective innovative components, while keeping the whole process comprehensive for the quite inexperienced students of agricultural engineering.

In this regard, the initial goal being set was to design, implement, and test a DIY (Do-It-Yourself) robotic vehicle that could host various sensors for environmental measurements or to perform light farming activities, like seeding, spraying, or carrying light crop cargos [38], in a realistic manner. The manufacturing of this robotic vehicle had to be simple, durable, and cost-effective. Furthermore, it should have a moderate size and torque, at least 1-h autonomy and the speed of a walking man. Furthermore, the robot should not be too heavy or too greedy, in terms of energy consumption. The vehicle should be able to roll on slightly anomalous or inclined terrains and surpass small obstacles.

The low-level controlling tasks had to be addressed locally by an arduino uno unit, installed on the robot and accompanied by simple electronic components, like potentiometers and suitable motor-driving equipment. For the high-level tasks, like providing remote human-robot interaction or advanced navigation, the robot had to be equipped with a raspberry pi unit able to run python (or C) code and to act as a bridge between the fixed on the robot arduino and the equipment of the user.

The raspberry pi is a quite powerful credit card–sized computer that drastically facilitates the design and implementation process. Indeed, it supports Linux shell commands and scripts, an abundant set of python and C programming capabilities, useful packages for network and data base services and image processing while there is a very active community to provide abundant material and assistance in any possible configuration problem, which is valuable and time-saving especially for the non-expert developers. The raspberry pi also provides easy storage of data and many connectivity options, from conventional ones via USB (Universal Serial Bus), Wi-Fi, Bluetooth to more microcontroller-oriented ones like SPI (Serial Peripheral Interface) [45], I^2C (Inter-Integrated Circuit) [46], serial TTL (Transistor-Transistor Logic) [47], etc. Among its trade-offs is its increased (average) power consumption that reaches the 500 mA border and the absence of sleep mode that sometimes is valuable for energy saving purposes and its size, which is not the smaller possible.

The considerable gap in roles and capabilities between the arduino uno and the raspberry pi units leaves plenty of space for intermediate-level systems like ESP-8266 (ESP8266, 2019) or ESP-32 based modules. Indeed, modules like the WeMos [42] board (an ESP8266 variant) are much faster and have more memory that the arduino uno board. Furthermore, they have similar to the arduino uno layout, they can be programmed using the same well-known environment and provide built-in Wi-Fi connectivity. Their price is also very attractive, varying around 10€. For these reasons, the adoption of WeMos units instead of raspberry pi or arduino uno units was an attractive choice for the less computationally demanding robotic implementations under testing. However, the WeMos is a 3.3 V system and its configuration and programming are slightly trickier than the arduino's and thus, it is not the ideal board for the less experienced users to start working with. In general, the cooperation among more than one comparatively simple controlling units, instead of hiring a single large one, has been favored. This arrangement provides better modularity and allows for fast configuration changes during the diverse experiments. A variety of sensor components, from potentiometers, proximity and IMU (Inertial Measurements Unit) elements to GPS (Global Positioning System), cameras (optical, NIR (Near InfraRed), thermal, etc.) and specific-purpose sensors can be easily supported. Similarly, various actuator components can be attached and controlled, like geared motors, motor drivers, servos, fluid pumps, etc.

A mechanism for supervising the robot's operation should be incorporated as well. More specifically, an android-based smart phone (or tablet) was the most convenient device to monitor and control the operation of the robotic vehicle, based on the methods described in [36, 48, 49]. This device may provide commands via its touch screen, its accelerometer sensor or via cloud-based voice recognition services. Furthermore, special software components provide valuable feedback data from the robot to the user. Both visual (e.g. Ardublock [50] and MIT App Inventor [51]) and textual (e.g. Arduino IDE, C, python) programming environments are used to guarantee a satisfactory interaction with the robotic vehicle.

Finally, as the robotic vehicle is intended to serve as a "vanilla" platform, able to host various multipurpose attachments, for testing environmental sensing equipment and for performing light-duty agricultural tasks, its construction should consist of modular software and hardware parts that are easy to combine in diverse configuration arrangements. For this reason, the overall selection of both the hardware and the software components tries to exhibit simplicity, to minimize the size and the cost and to maximize their reusability. These initial robotic vehicle design principles are reflected, in a simplified manner, in the left part of Fig. 1.

3 Implementation Details and Derived Enhancements

Taking under consideration the requirements discussed in Sect. 2, the result was a layout involving two independent electric motors giving motion to the wheels of each side, through a chain drive system, providing differential steering (actually skid-steering). This setup eliminates the need for extra mechanisms dedicated in steering tasks, provides simplicity, robustness and better maneuverability, at the cost of increased energy consumption and tire wear during the turning process, thus requiring stronger wheels and bearings. These trade-offs that the skid-steering technique has are not significant, as long as the vehicle is kept lightweight and wide enough (i.e., having its width comparable to its length dimension). The proposed robotic layout mainly consists of two separate parts (i.e., left and right) that are connected via threaded rods, of 3 mm in diameter, thus providing suspension flexibility in a cost-effective manner. One of the initial robotic vehicle implementations following the above specifications, using one stepper DC motor per side, is depicted in the right part of Fig. 1.

As described in [36, 38], the control of the motors was initially done by a microcontroller (arduino uno), via suitable power electronics (motor drivers mainly of L298 type). The necessary power was provided by deep discharge sealed acid lead batteries. The size specifications tend to a 40 cm by 40 cm layout. The motors were delivering a torque of around 5 kg cm, at 100 rpm, in order to achieve a combined

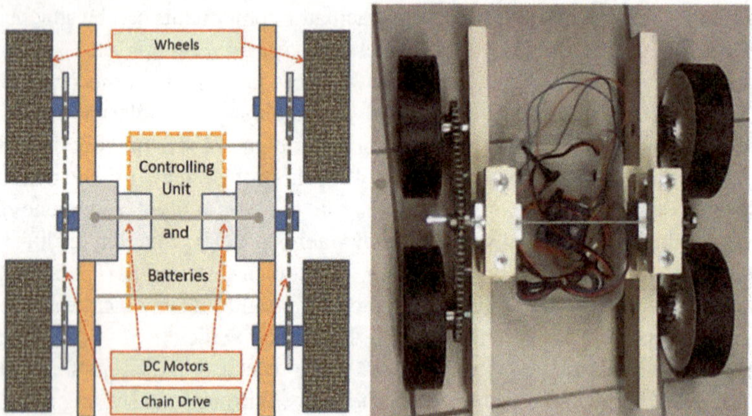

Fig. 1 Initial robotic vehicle design providing one independent motor per side and all-wheel drive (left). An early implementation using stepper DC motors (right)

dragging force of around 4 kg, taking into account the chain drive revolution reducer (:3) and the given wheel radius (approximately 7 cm).

The modular nature of this experimental robotic layout allowed for detailed components testing highlighting the strengths and the weaknesses of each candidate variant. In this regard, the accurate, in their rotation, stepper motors (operating at 8 V) that were initially fitted on the robot, were replaced by simple brushed DC motors (operating at 12 V) that produced less heat and consumed considerably less energy. The trade-off for adopting the latter DC motors was the need for a feedback control system, in order to provide accurate operation. Indeed, such a mechanism was added, using photo interrupters that were initially fixed around the driving sprockets. Furthermore, the robotic vehicle was enhanced by incorporating a direction stabilization mechanism, based on data fusion of signals provided by a nine-degree of freedom IMU/compass unit [52] fixed on the top of the robot and connected with the arduino. The system consumed without load, at low speed 0.14 A, at average speed 0.16 A, while at high speed it consumed 0.18 A. In the case of operation under load at high speed, it consumed 0.32 A with low load, 0.48 A with medium load and 0.80 A with high load. The vehicle's electronic system (microcontroller and motor driver circuits) consumed about 50–60 mA to operate, a comparatively small additional quantity.

More ambitious enhancements involved motors of higher torque (i.e., more than the double), larger batteries and required the replacement of the initial motor driver (L298 chip) with an improved circuit, based on MOSFET (Metal Oxide Semiconductor Field Effect Transistor) type transistors, thus being able to handle higher currents with lower voltage drops and heat. This layout allowed for greater wheels (of about 26 cm in diameter) with pneumatic tires to be fixed on the chassis. In addition, a small solar panel unit has been adapted on the top of the robot. This 15 W solar panel was able to deliver at about 0.8 A at 18 V under good sunlight conditions, amount that could assist and charge the batteries. Figure 2 summarizes the discussed enhancements involving DC motors and photo interrupters attached to the driving sprockets (left), and layouts with IMU/compass sensors, solar panel assistance, bigger motors and wheels (right).

The overall implementation provided satisfactory results, but the need for delivering even higher torque values and for overcoming vertically opposed obstacles (requiring the wheels of the same side to be able to turn at different speeds) led to an updated version having two independent motors per each side (i.e., one per each wheel) assisted by a dedicated chain-drive system. The new robot is slightly wider than the initial implementation in order to turn more flawlessly, without experiencing very high peak torques that lead to high peaks in power consumption. The overall construction has a 50 cm by 50 cm layout. The updated design specifications are reflected in Fig. 3.

The gearbox equipped motors are rated of 100 rpm, they have a stalling torque of 40 kg cm and a stalling current of 2 A. The chain drive system reduces the rotation speed by a factor of 3, increasing in parallel the torque by the same amount. The back side motors are equipped with an improved rotational speed feedback mechanism based on photo interrupters to provide speed stabilization to the selected target value.

Fig. 2 Intermediate layouts involving speed feedback mechanism (left). Further enhancements involving IMU/Compass, solar panel assistance, bigger motors and wheels (right)

As with the initial vehicle version, an IMU and compass unit plus a GPS can be used to provide the necessary direction stabilization and location information. A battery unit of 12 V, 7.2 Ah is the main power source for the updated robotic vehicle. In the previous cases, a combination of an arduino uno unit plus a raspberry pi (model 3B) is used to efficiently control and remotely monitor the robot. An application using a WeMos D1 R2 unit has also been implemented in order to reduce the power consumption and increase the efficiency of the robot. The total weight of the vehicle is around 10 kg. If a solar panel is hosted for better autonomy, 1.25 kg is added to this weight. A system of thicker (i.e., of 4 mm in diameter) threaded rods are used to keep the two sides of the robot together and provide the necessary flexibility to make the vehicle able to roll on anomalous terrains. The external wheel diameter is 26 cm approximately.

The basic robotic construction, after a meticulous testing and tuning process, was assigned a realistic mission: to perform soil moisture measurements, in a semi-autonomous manner. For this reason the robot was equipped with an electro-mechanical attachment (of one degree of freedom) and a reliable soil conductivity sensor was fixed on it, the ThetaProbe ML2 [53] device, manufactured by the Delta-T company. The ThetaProbe is a popular and reliable model, which is used by the scientists, for years, typically in research works like the one described in [54]. This sensing device was adapted at the end of a wooden arm of "Π"-shape, able to move up and down, assisted by a custom jackscrew lifting mechanism which is capable of rotating grace to a small geared DC motor. The jackscrew mechanism can apply to the sensor a pushing force of up to 35 N, approximately, thus making its nose to penetrate inside the soil so as to provide accurate measurements. Two position switches and a small part of arduino code automate the measuring sequence process, thus leaving the main unit of the robot available for more time-critical or computationally-heavy tasks. The overall implementation enhancement details are shown in Figs. 4 and 5. After experimentation, it was decided to attach the photo

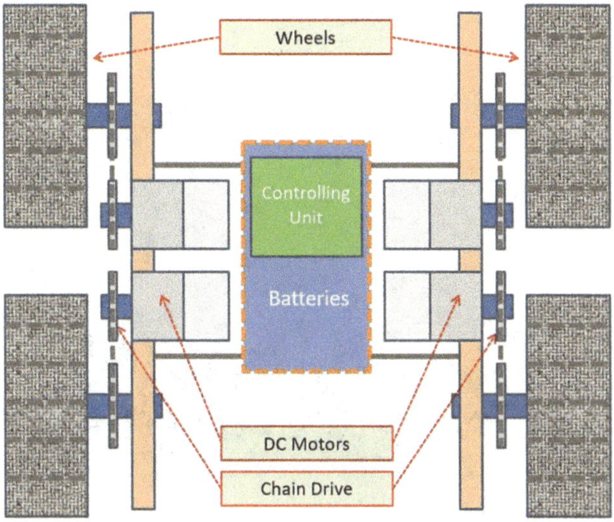

Fig. 3 More advanced robotic vehicle design providing two independent motors per side and all-wheel drive and skid-steering

interrupters to the driving sprockets of the rear wheels, for better traction, and to use separate discs for them, for increased control time granularity and thus faster response.

The second robotic vehicle example is a taller construction which is intended to pass over the young plants, mainly for inspecting and spraying them with fertilizers or herbicides. The frame of this prototype vehicle is made of wooden rectangular rods. Its length, width, and height are approximately 70 cm 75 cm, and 75 cm, respectively. The front part of the robot is equipped with the two traction wheels, while two caster wheels are fixed in the rear part. Two separate motors, one per side, are used for moving the spraying robot, these motors, apart from being larger, are equipped with accurate hall sensors, for better speed control results. The differential steering method is followed by this robot as well, leading to a more compact, simpler, and cheaper solution. For better maneuverability, the front part is slightly wider than the rear one. The rear part is connected to the rest of the frame in a way that provides for rotations by small angles, on a layer vertical to the vehicle's main moving direction, in order to surpass small obstacles in the fields, without any wheel to lose contact with the ground. The front part is designed to carry the spraying reservoirs and it is much heavier than the rear part. As the vehicle is not an all-wheel drive, this arrangement results in not losing traction, as the rear caster wheels have considerably less weight to carry. The front "Π"-shaped side, apart from the electric motors system and the reservoirs, is equipped with the low-pressure spraying components and a set of electric centrifugal pumps, similar to the ones used in the windshield washers of the cars. The necessary electronic components are hosted at the top of the robot, while the two batteries of 7.2 Ah are fixed, one at each side. The

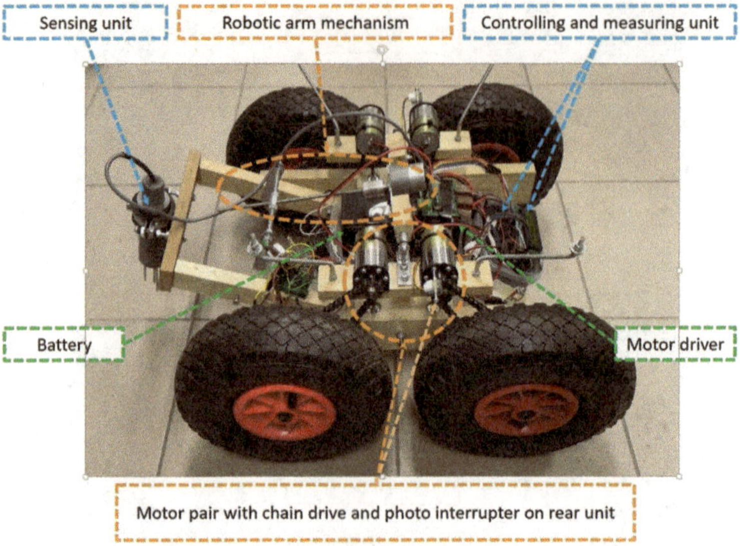

Fig. 4 More advanced robotic vehicle implementation for soil measurements, using two larger DC motors per side, providing all-wheel drive and skid-steering

Fig. 5 Details of the photo interrupter sensor disc, attached to the rear wheel sprocket (left), and the solar panel assisted version of the soil measuring robot (right)

total weight of the construction being described is 16 kg, approximately, or 24 kg, with the two reservoirs that are ready for spraying.

Typically, the "brain" of this electro-mechanical construction was an arduino uno unit, used for the low-level tasks, in cooperation with a raspberry pi unit, used for the high-level tasks and the communication with the user. The presence of the quite powerful raspberry pi unit allows to host and exploit advanced software and hardware tools.

The need for modular system, of satisfactory accuracy and capable of providing more connectivity and monitoring options, yet maintaining configuration simplicity, favored the adoption of a navio2 [55] "hat" board along with the raspberry pi platform. Indeed, the navio2 is a promising alternative to the combination of a 9DOF IMU unit with an external GPS that provides an easy-to-use navigation system and a rich set of functions. The navio2 cooperates with conventional RF controlling systems and provides with many useful telemetry and health status data, via a UDP (User Datagram Protocol) or a serial connection with the hosting raspberry pi unit. Furthermore, the navio2 also provides pins for controlling various servomotors. The navio2 requires a modified raspbian version to be installed on the raspberry pi unit. This linux operating system version supports the popular ArduPilot [56] vehicle monitoring and controlling software that cooperates fluently with pairing tools at the operator's end, like MissionPlanner [57], for desktop computers, or QGroundControl [58], for smart phones/tablets. These tools allow for supervision and control of robotic vehicles of any type as well as for setting way points for the robots to follow. The software of the navio2 system also supports the ROS [59] robot operating system, which is among the standard robot communication protocols.

The presence of a raspberry pi unit, along with the navio2 board, provides the potential to perform a rich set of custom assistive tasks in parallel, like remote communication or log-file storage or even local (i.e., performed by the robot) off-line voice recognition, with the latter option been investigated in [36, 48]. Among the most valuable features that the computational power of the raspberry pi allows for, is the machine vision. More specifically, as the vision-based techniques, combined with IMU and GPS data, comprise the most commonly used method for robot navigation in the fields, the presence of the raspberry pi unit, at a "gluing" role, encourages the hiring of the OpenCV tool [60] for adding machine vision assistance to the robots.

Nevertheless, the OpenCV is not the ideal environment for the less-inexperienced users. For these users the pixy2 smart camera system [61] provides a more attractive component for introduction to the real-time image recognition concept. The pixy2 is an open hardware and open software innovative module that can be easily configured to identify objects according to their color and to report their coordinates, into the active camera frame, via SPI or USB interfaces to an arduino or a raspberry pi unit respectively. This functionality has been used to add a smart spraying functionality to the robot. More specifically the robot intercepts the existence of the plants nearby, and sprays only when it is above them. The fusion of the camera data with the distance data provided by two ultrasonic distance sensors (one vertical and one horizontal) delivered even better results. The pixy2 data have also been used for small corrections in the moving direction of the robot, in order for the latter to be kept aligned with the plantation line.

Finally, two small solar panel units can be fixed on the top of the robot to increase its autonomy. These 15 W solar panels are able to deliver at about 0.8 A at 18 V under good sunlight conditions, increasing the vehicle's autonomy by 25%. The

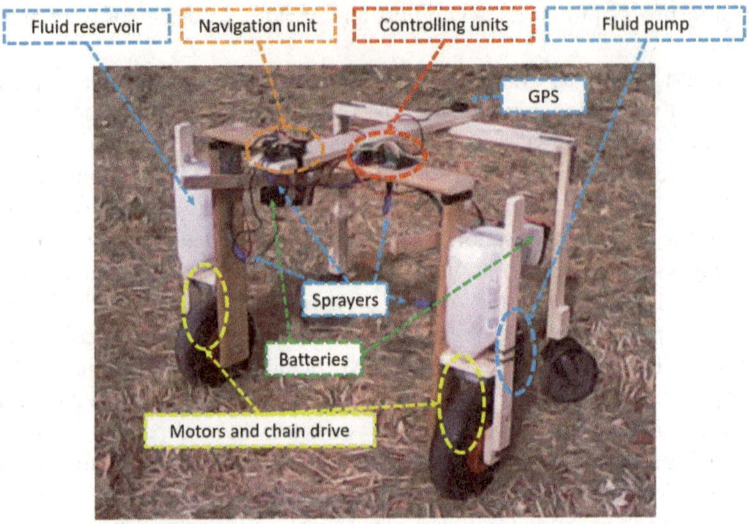

Fig. 6 The spraying robot variant, using one DC motor per side, differential drive and special navigation and monitoring equipment

solar panels add slightly more than 2.5 kg to the total weight. The overall implementation enhancement details for this robotic variant are shown in Figs. 6 and 7.

As with the robotic implementations previously described, in the spraying robot case, the low-level controlling tasks are addressed by an arduino unit while the high-level and more complicated tasks are left to a raspberry pi unit. As already discussed, the WeMos board is offering an attractive and more compact alternative to the raspberry pi—arduino uno cooperation schema, for comparatively lightweight tasks, and thus the WeMos controlled space and energy—saving variants proved their practicality and flexibility in many of our experiments. Furthermore, the WeMos also has a much faster boot-up time that the raspberry pi unit, which is an important factor during experiments using a large number of consecutive tests.

For all implementations, referring to the low-level tasks, the arduino mega board was offering a safe alternative to the arduino uno board, with its increased number of GPIO pins and memory. Similarly, for the WeMos board case, a good upgrade was the ESP-32 board. Table 1 summarizes the characteristics of the proposed robotic vehicles, compared with the initial implementation. The detailed information that Table 1 contains is further highlighted in Sect. 4 of this chapter.

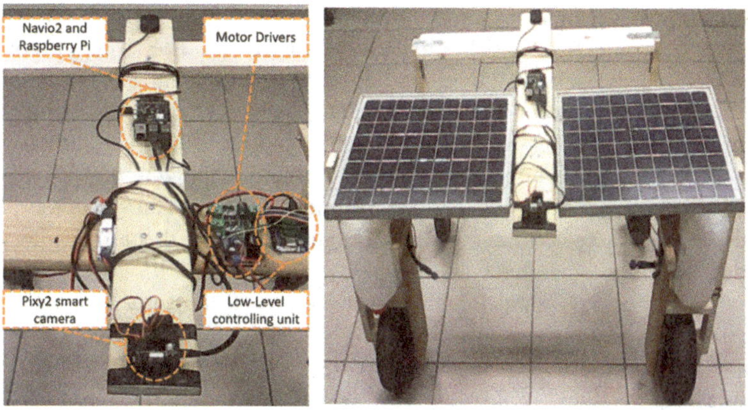

Fig. 7 Further details of the electronic components of the spraying robot variant (left) and its solar panel assisted version (right)

Table 1 Characteristics of the robotic vehicles

Robot type	Initial model	For soil measuring	For spraying
Transmission system	4 × 4, chain drive	4 × 4, chain drive	4 × 2, chain drive
Motor number	2 (electric)	4 (electric)	2 (electric)
Dimensions (cm) (L × W × H)	40 × 40 × 20	50 × 50 × 30	70 × 75 × 75
Weight (net) (kg)	5	10	16
Weight (gross) (kg)	5	11	24
Steering method	Skid steering	Skid steering	Differential steering
End effectors	None	ThetaProbe ML2	Sprayers
Max speed (ms^{-1})	0.70	0.70	0.65
Power consumption (A)	2.5 A	4.2 A	4.6 A
Power source	Acid lead batteries	Acid lead batteries	Acid lead batteries
Solar assistance	No	Yes (1 × 15 W)	Yes (2 × 15 W)
Indicative cost (bare) (€)	125	250	300
Indicative cost (total) (€)	445	570	620

4 Lessons Learned, Results and Discussion

During the construction of the proposed vehicles, a very challenging issue was the selection and the gluing of all the necessary components that had to be collected from a very diverse set of sources, varying from recycled materials and local hardware stores to official electronic suppliers. The robots being presented have low power consumption and while they are far larger than the typical STEM (Science-Technology-Engineering-Mathematics) approaches, they are easily deployable, cost effective and modular enough to allow for students' practice and for further variants implementation. The modular nature of the experimental robotic layout allows for a meticulous study of numerous combinations involving cutting-edge electronics and one or more popular microcontrollers. The hiring of more than one "core" units for each robot does not appear to be a problem, as long as the total power consumption is kept low and there are interconnections options available between the different components. In addition, a good practice being followed was to use a separate power source (e.g., a power bank) for supplying the "logic" of the controlling circuit, as motors have rather a "crude" behavior that sometimes results in sudden voltage drops on the main battery units supplying them. Use of fuses for protecting the controlling elements is also suggested.

Most of the remote interaction scenarios, even with the earlier robot versions, have been carried out through Wi-Fi interfaces, while some of them involved LoRa interfaces, to extend the effective controlling distance. In the latter case, the adoption of a LoRa dragino shield [62] component on a secondary arduino uno unit at both communication ends performing commands bridging, via its USB port, was the most preferred solution. The LoRa protocol was capable of establishing up to a few kilometers connectivity, at a few kbps rate, with a small power consumption of the order of 100 mA. Nevertheless, this radio is not intended to serve for continuous telemetry and command data delivery, similarly with its Wi-Fi counterpart, but rather to carry scarce supervision commands, with respect to the 1% fair policy rule suggested for LoRa communicating devices [63].

In order to provide the necessary flexibility and customization options for efficiently supporting the various energy consumption metric scenarios, for a considerable amount of amperage (i.e., up to 8 A), without shifting to more expensive equipment, efforts were made to make our own measuring circuit using low-cost discrete elements and one of the available arduino units. This circuit contains an electric current divider, which is made by two 0.56 Ω resistors of 5 W each connected in parallel, and a capacitor, in parallel, at the output of the circuit. The input of this system drains energy from the battery while its output supplies the motors and the controlling units of the robot. The A/D converter of the arduino, through a voltage divider system, is used to provide differential measurements of the voltage drop the two-resistor system. From the latter difference, the current amperage that the robot consumes is calculated. The battery voltage of the robot is also reported. These data are periodically delivered, as a time trace sequence, towards a smart phone or tablet device (which is usually the monitoring unit at the human's end) for further processing and permanent storage.

Figure 8 depicts the indicative energy consumption trace (motors only), acquired by the custom energy metering mechanism, for the soil moisture measuring robot, while performing a simple move sequence on smooth terrain, i.e., of going forward, stopping, turning left, turning right and going back and stop. The peaks in power consumption correspond to the turning actions, which are quite greedy for this type of vehicle which is using skid-steering.

The actual speed being exhibited by the first robotic implementation varied from 0.25 to 0.70 m s^{-1}, and thus, they were similar to the ones of a slow walking man, as it was initially planned. The dragging force of the robot initially was nearly 4 kg and its weight about 5 kg, while its cost was below 125€. The speed specifications of the

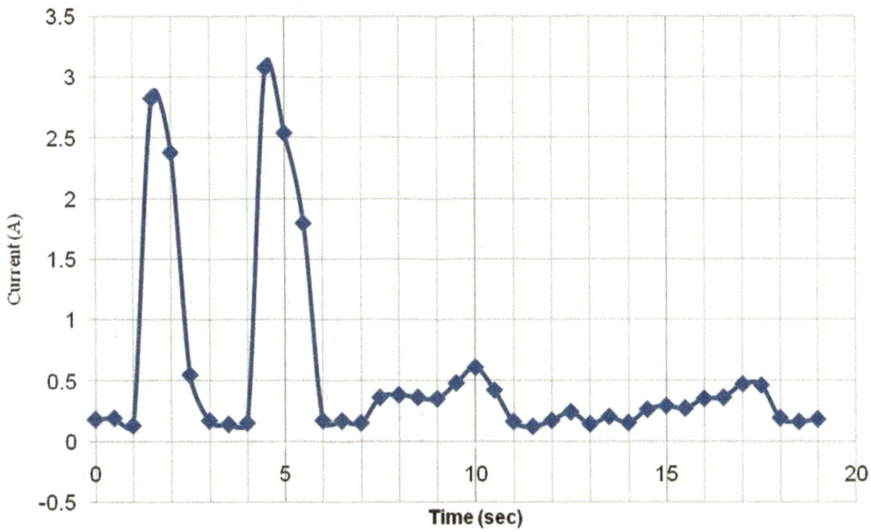

Fig. 8 Indicative energy consumption trace (motors only), for the soil moisture measuring robot, while performing a simple move sequence on smooth terrain

derived layout implementations were kept almost unchanged while the dragging force was increased to 10 kg, per side, approximately, and the cost of the electromechanical and driving units almost doubled. The weight of the new robotic constructions shifted to 11 kg, for the soil measuring version, and to 16 kg, for the spraying version, respectively. The flexible chassis layout capabilities of the soil-measurement vehicle allow for graceful obstacle passing as depicted in Fig. 9. Up to 10 cm height obstacles are successfully surpassed without any wheel to lose contact with the ground.

The spraying robot has a 0.75 m by 0.75 m layout and a max speed of 0.65 ms^{-1} (0.70 ms^{-1} for the soil measuring version). Its electronics (WeMos implementation) consume 100 mA approximately. Its net weight is 16 Kg, while, with both reservoirs loaded, the weight is reaching the 24 Kg. The presence of two photovoltaic panels of 15 W each adds an extra 2.5 Kg. The water supply of each water pump is 0.2 L min^{-1}, while the pump motor consumes 1 A, approximately. Considering the 4 L reservoir capacity corresponding to each pump unit, the spraying process can last up to 40 min at 0.2 L min^{-1} or up to 20 min, if both pumps are used simultaneously. This type of vehicle, without load on flat horizontal surfaces, consumes 1.4 A, approximately, while, at full load, it consumes 1.5 A, approximately. These consumption values increase to 3.2 A and 4.6 A, respectively, when the robot is immobilized by holding its rear end while its wheels are spinning.

In Fig. 10, the navigation and monitoring equipment details for the spraying robot at the operator's end are depicted, involving a smart phone, custom software, developed using the MIT App Inventor tool, and the QGroundControl application. The accuracy of the navigation process depends heavily on the accuracy of the GPS

Fig. 9 The flexible chassis layout capabilities of the soil-measurement vehicle allow for graceful obstacle passing

device being used, which for our experiments was of the order of 1 m. Things became better after the fusion with the data provided by the pixy2 camera, giving error near to ±0.1 m, for identifying the young plants, and thus, aligning with the plantation line. It must be noted that the pixy2 camera is running a quite simplified machine vision algorithm and cannot be used for a wide set of cases.

In both variants, the presence of a RPi3 unit (that costs 45€) plus a navio2 shield is adding an amount of 650 mA approximately at 5 V which incurs a total rise of 350 mA in consumption, at the 12 V battery-supplying level, provided that an efficient switching regulator unit is used. The use of sophisticated sensors like the navio2 or the pixy2 camera are increasing the cost by 200€ and 75€, respectively (total cost line in Table 1).

The findings of this research indicate that the gap between DIY and commercial implementations can become smaller and the robots' manufacturing cost barrier can be even lowered, i.e., from the 10,000 to 15,000€ range to drop below the 1000€ border, for comparatively smaller and lightweight, but yet efficient constructions of reduced accuracy. At this cost level, the educational role of the proposed robotic implementations seems to be served fluently. Indeed, according to interviews, there are promising indications that the students assisting in the design, implementation and testing stages of the discussed robots, have reinforced their technological background and acquired soft skills that might be beneficial for their careers as engineers. This feeling, which is in accordance with the findings of other similar research works [36, 37], should be further assessed and studied in both quantitative and qualitative manners. Apparently, from technical aspect, still much of work should be done to provide more accurate solutions exhibiting increased autonomy and robustness during their operation. Indeed, more sophisticated methods can be investigated, including fusion with more precise sensors and especially with more

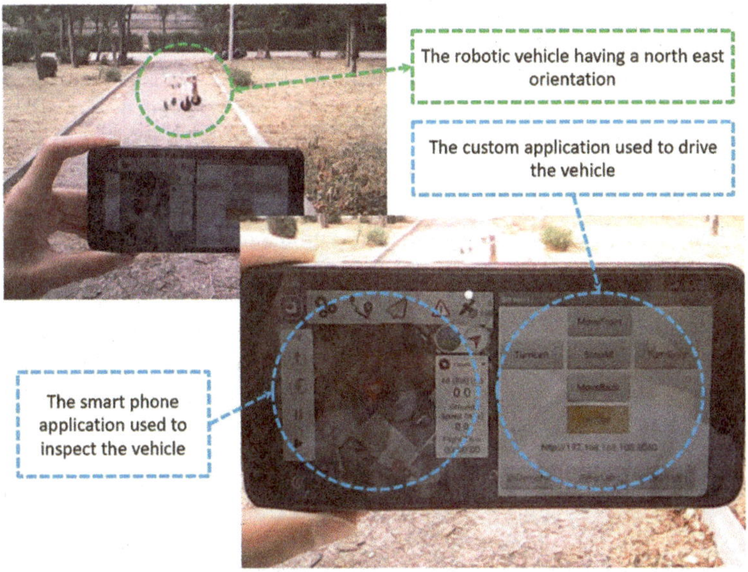

Fig. 10 Navigation and monitoring equipment details at the operator's end, involving smart phone, custom software and the QGroundControl application

advanced AI techniques, but the continuous flourishing of the relevant research and industry production assists that.

5 Conclusions and Future Work

The work being presented reports on the efforts of our team to bridge the gap between educational-laboratory level and commercial-industrial level robotic vehicle implementations, in a cost-effective manner, with additional pedagogical impact. This goal is served by highlighting the trials for building DIY agricultural robots in non-negligible size and providing the necessary functionality and intelligence to support simple real-world agricultural tasks. The recent rapid technological advances resulted in a plethora of low-cost electronic components and credit card-sized systems that greatly encourage similar advanced project trials. Two basic forms of experimental robotic layouts have been implemented and tested, one for spraying over the plants and the other one for performing all-terrain soil moisture measurements. Simple electromechanical parts along with popular microcontroller units, tablet or smart phone devices and innovative sensor boards have been combined effectively. Both visual and textual programming environments were used to properly program the robots. The remote interaction with the prototype robots, in most cases, was performed using Wi-Fi interfaces while, a step beyond that, the experimentation with LoRa interfaces provided a promising solution for supervising the robots, at increased distances. Mechanisms for measuring the energy consumption of the vehicles were also implemented and solar panel assistance solutions were also taken into account. Our findings indicate that the gap between DIY and commercial implementations can become smaller, but still much of work should be done to provide more accurate and more autonomous solutions.

Plans for the feature include a deeper experimentation with the robotic vehicles, mainly in terms of working accuracy, autonomy and energy consumption. Both layouts being presented can be seen as "vanilla" platforms that can be easily customized to support a rich set of use cases by picking a diverse set of sensors and algorithms. Beyond that, the investigation of further variants having an even larger and more robust electro-mechanical layout, is an apparent priority as well. Finally, an ambitious goal is the construction of a large number of vehicles and then the application and study of various swarm management techniques, towards increasing the productivity and minimizing the relevant costs.

Acknowledgments We would like to thank the personnel of the Farm Machinery Laboratory and the students of the Dept. of Natural Resources Management and Agricultural Engineering of the Agricultural University of Athens, Greece, for their assistance in tools and labor, during the construction stages of the robotic vehicles. We would also like to thank the CEO and staff of Hellas Digital company for their great assistance in replying to our numerous questions and offering an abundant set of high-end electronic equipment for supporting the experiments.

References

1. FAO (2018) 'Climate-Smart Agriculture Sourcebook. Second Edition', online at: http://www.fao.org/climate-smart-agriculture-sourcebook/en/. Accessed on 20 September 2019.
2. Symeonaki, E.G., Arvanitis, K.G. and Piromalis, D.D. (2019), 'Cloud computing for IoT applications in climate-smart agriculture: A review on the trends and challenges towards sustainability'. In Theodoridis, A., Ragkos, A. and Salampasis, M. (Eds.), Innovative Approaches and Applications for Sustainable Rural Development, HAICTA 2017, Springer, Cham, Earth System Sciences Series, Vol. 29, Chapter 9, pp. 147–167, 2019. https://doi.org/10.1007/978-3-030-02312-6_9.
3. Bochtis, D.D., Sørensen, C.G.C. and Busato P. (2014). "Advances in agricultural machinery management: a review", Biosyst Eng, Vol. 126, October 2014, pp. 69–81, https://doi.org/10.1016/j.biosystemseng.2014.07.012
4. O'Grady, M. and O'Hare, G., (2017) "Modelling the smart farm" in Information Processing in Agriculture, Vol. 4, Issue 3, May 2017, https://doi.org/10.1016/j.inpa.2017.05.001.
5. Bechar, A. and Vigneault, C. (2016) 'Agricultural robots for field operations: Concepts and components', Biosystems Engineering, 149, pp. 94–111. https://doi.org/10.1016/j.biosystemseng.2016.06.014
6. Bechar, A. and Vigneault, C. (2017) 'Agricultural robots for field operations. Part 2: Operations and systems', Biosystems Engineering, 153, pp. 110–128. https://doi.org/10.1016/j.biosystemseng.2016.11.004
7. Krishna, K.R., (2016), 'Push button Agriculture: Robotics, drones, satellite-guided soil and crop management', Apple Academic Press, Oakville, Ontario, Canada. ISBN-13: 978-1-77188-305-4 (eBook - PDF).

8. UK-RAS Network: Robotics & Autonomous Systems (2018), "Agricultural robotics: The future of agricultural robots", UK-RAS White Papers, ISSN: 2398-4414. Online at: https://arxiv.org/ftp/arxiv/papers/1806/1806.06762.pdf.
9. Roldan JJ, del Cerro J, Garzon-Ramos D, Garcia-Aunon P, Garzon M, de Leon J, Barrientos A. (2017) Robots in agriculture: state of art and practical experiences. IntechOpen: Service Robots; 2017.
10. Pedersen, S.M., S. Fountas, and S. Blackmore, (2008) Agricultural robots—Applications and economic perspectives, in 740 Service robot applications. 2008, IntechOpen.
11. Shockley, J.M. and C.R. Dillon. (2018) An economic feasibility assessment for adoption of autonomous field machinery in row crop production. in 2018 international conference on precision agriculture. 2018. Montreal, QC, Canada: ICPA.
12. Thomasson JA, Baillie CP, Antille DL, Lobsey CR, McCarthy CL. (2019) Autonomous technologies in agricultural equipment: a review of the state of the art. American Society of Agricultural and Biological Engineers; 2019.
13. Gonzalez-de-Santos P, Ribeiro A, Fernandez-Quintanilla C, Lopez-Granados F, Brandstoetter M, Tomic S, Pedrazzi S, Peruzzi A, Pajares G, Kaplanis G, et al., (2017) "Fleets of robots for environmentally-safe pest control in agriculture". Precis Agric 2017; 18(4):574–614.
14. Gaus, C.-C., et al., (2017) Economics of mechanical weeding by a swarm of small field robots, in 57th Annual Conference 742 of German Association of Agricultural Economists (GEWISOLA). 2017, German Association of Agricultural 743 Economists (GEWISOLA): Weihenstephan, Germany.
15. Shamshiri, R., Weltzien, C., Hameed, I., Yule, I., Grift, T., Balasundram, S., Pitonakova, L., Ahmad, D., and Chowdhary, G. (2018). "Research and development in agricultural robotics: A perspective of digital farming", International Journal of Agricultural and Biological Engineering, 11. 1–14. https://doi.org/10.25165/j.ijabe.20181104.4278.
16. Bochtis, D., Griepentrog, H.W., Vougioukas, S., Busato, P., Berruto, R., Zhou, K. (2015) Route planning for orchard operations. Comput. Electron. Agric. 2015, 113, 51–60.
17. Reina, G.; Milella, A.; Galati, R. (2017) Terrain assessment for precision agriculture using vehicle dynamic modelling. Biosyst. Eng. 2017, 162, 124–139.
18. Yang, L., Noguchi, N. (2012) Human detection for a robot tractor using omni-directional stereo vision. Comput. Electron. Agric. 2012, 89, 116–125.
19. Emmi, L., Gonzalez-de-Soto, M., Pajares, G. and Gonzalez-de-Santos, P. (2014). New Trends in Robotics for Agriculture: Integration and Assessment of a Real Fleet of Robots. The Scientific World Journal. 2014. Article ID 404059. https://doi.org/10.1155/2014/404059.
20. Blackmore, B.S., Fountas, S., Gemtos, T.A., Griepentrog, H.W. (2009) A specification for an autonomous crop production mechanization system. In Proceedings of the International Symposium on Application of Precision Agriculture for Fruits and Vegetables, Orlando, FL, USA, 1 April 2009, Volume 824, pp. 201–216.
21. Utstumo, T., Urdal, F., Brevik, A., Dørum, J., Netland, J., Overskeid, Ø., Berge, T.W., Gravdahl, J.T. (2018) Robotic in-row weed control in vegetables. Comput. Electron. Agric. 2018, 154, 36–45.
22. Katupitiya, J. (2014) An autonomous seeder for broad acre crops. In Proceedings of the American Society of Agricultural and Biological Engineers Annual International Meeting 2014, St Joseph, MI, USA, 13–16 July 2014, Volume 1, pp. 169–176.
23. Schor, N., Bechar, A., Ignat, T., Dombrovsky, A., Elad, Y., Berman, S. (2016) Robotic disease detection in greenhouses: Combined detection of powdery mildew and tomato spotted wilt virus. IEEE Robot. Autom. Lett. 2016, 1, 354–360.
24. Mueller-Sim, T.; Jenkins, M.; Abel, J.; Kantor, G. (2017) The Robotanist: A ground-based agricultural robot for high-throughput crop phenotyping. In Proceedings of the IEEE International Conference on Robotics and Automation, Singapore, 29 May–3 July 2017; pp. 3634–3639.

25. Adamides, George. (2017). Agricultural Robots in Targeted Spraying: A mini State-of-the-Art review. Robotics & Automation Engineering Journal. 2. https://doi.org/10.19080/RAEJ.2018.02.555581
26. Feng, Q.; Zheng, W.; Qiu, Q.; Jiang, K.; Guo, R. (2012) Study on strawberry robotic harvesting system. In Proceedings of the CSAE 2012 IEEE International Conference on Computer Science and Automation Engineering, Zhangjiajie, China, 25–27 May 2012; Volume 1, pp. 320–324.
27. Zahid, A., He, L., Zeng, L. (2019) Development of a robotic end effector for apple tree pruning. In Proceedings of the 2019 ASABE Annual International Meeting, Boston, MA, USA, 7–10 July 2019.
28. Grimstad, L., and From, P.J. (2017) "The Thorvald II agricultural robotic system". MDPI, Robotics, 2017, Vol. 6, Issue 4, https://doi.org/10.3390/robotics6040024
29. Fountas S, Mylonas N, Malounas I, Rodias E, Hellmann Santos C, Pekkeriet E. (2020) Agricultural Robotics for Field Operations. Sensors. 2020; 20(9):2672.
30. Doran, M. V., and Clark, G. W. (2018) 'Enhancing Robotic Experiences throughout the Computing Curriculum', SIGCSE'18, February 21–24, Baltimore, MD, USA (pp. 368–371).
31. Alimisis D., Alimisi R., Loukatos D., Zoulias E., (2019) Introducing Maker Movement in Educational Robotics: Beyond Prefabricated Robots and "Black Boxes" In: Daniela L. (ed.) Smart Learning with Educational Robotics, Springer, June 2019, ISBN 978-3-030-19913-.5, https://doi.org/10.1007/978-3-030-19913-5_4, pp. 93–115.
32. Alimisis D., Loukatos D., Zoulias E., Alimisi R. (2020) The Role of Education for the Social Uptake of Robotics: The Case of the eCraft2Learn Project. In: Pons J. (eds) Inclusive Robotics for a Better Society. INBOTS 2018. Biosystems & Biorobotics, vol 25. Springer, Cham. https://doi.org/10.1007/978-3-030-24074-5_30
33. Anwar, S., Bascou, N., Menekse, M. and Kardgar, A. (2019). A Systematic Review of Studies on Educational Robotics. Journal of Pre-College Engineering Education Research (J-PEER). 9. https://doi.org/10.7771/2157-9288.1223.
34. Rbr (2017). "The Ultimate Guide to Agricultural Robotics". Retrieved on 20 September 2019 from: https://www.roboticsbusinessreview.com/agriculture/the_ultimate_guide_to_agricultural_robotics/
35. Tan, J. T. C., Iocchi, L., Eguchi, A. and Okada, H. (2019) "Bridging Robotics Education between High School and University: RoboCup@Home Education", 2019 IEEE AFRICON, Accra, Ghana, 2019, pp. 1–4, https://doi.org/10.1109/AFRICON46755.2019.9133791.
36. Loukatos D., Arvanitis K.G. (2019) "Extending Smart Phone Based Techniques to Provide AI Flavored Interaction with DIY Robots, over Wi-Fi and LoRa interfaces", MDPI – Education Sciences, August 2019, vol. 9, issue 3, pp.224–241, https://doi.org/10.3390/educsci9030224.
37. Phan M-H, Ngo HQT. (2020) A Multidisciplinary Mechatronics Program: From Project-Based Learning to a Community-Based Approach on an Open Platform. Electronics. 2020; 9(6):954.
38. Loukatos D., Tzaninis G, Arvanitis K.G., Armonis N. (2019) "Investigating Ways to Develop and Control a MultiPurpose and Low Cost Agricultural Robotic Vehicle, under Scale", Proceedings of XXXVIII CIOSTA & CIGR V International Conference (CIOSTA2019), Rhodes, Greece, June 2019.
39. Markham, T. (2011) Project Based Learning. Teach. Libr. 2011, 39, 38–42.
40. Arduino Uno (2019). Arduino Uno board description on the official Arduino site. Retrieved on 15 April 2019 from the site: https://store.arduino.cc/arduino-uno-re
41. Raspberry (2019). Raspberry Pi 3 Model B board description on the official Raspberry site. Retrieved on 20 April of 2019 from the site: https://www.raspberrypi.org/products/raspberry-pi-3-model-b/
42. WeMos (2019). The WeMos D1 R2 board. Retrieved on 20 April 2019 from: https://wiki.wemos.cc/products:d1:d1
43. Wi-Fi (2019). The IEEE 802.11 Standard. Retrieved on 20 April 2019 from the site: http://www.ieee802.org/11/
44. LoRa (2019). LoRa protocol description on Wikipedia. Retrieved on 20 April 2019 from the site: https://en. wikipedia.org/wiki/LoRa

45. SPI (2020). Serial Peripheral Interface description on Wikipedia. Retrieved on 18 October 2020 from the site: https://en.wikipedia.org/wiki/Serial_Peripheral_Interface
46. I2C (2020). Inter-Integrated Circuit description on Wikipedia. Retrieved on 18 October 2020 from the site: https://en.wikipedia.org/wiki/I^{2}C
47. TTL (2020). Transistor–transistor logic description on Wikipedia. Retrieved on 18 October 2020 from the site: https://en.wikipedia.org/wiki/Transistor-transistor_logic
48. Loukatos D., Arvanitis K.G., Armonis N., (2020a) "Investigating Educationally Fruitful Speech-Based Methods to Assist People with Special Needs to Care Potted Plants", In: Human Interaction and Emerging Technologies. IHIET 2019. Advances in Intelligent Systems and Computing, vol 1018. Springer, Cham Human Interaction and Emerging Technologies, https://doi.org/10.1007/978-3-030-25629-6_25.
49. Loukatos D., Kahn K., Alimisis D. (2020b) Flexible Techniques for Fast Developing and Remotely Controlling DIY Robots, with AI Flavor. In: Moro M., Alimisis D., Iocchi L. (eds) Educational Robotics in the Context of the Maker Movement. Edurobotics 2018. Advances in Intelligent Systems and Computing, vol 946. Springer, Cham. https://doi.org/10.1007/978-3-030-18141-3_14
50. Ardublock (2019). Official site of the ArduBlock programming tool. Retrieved on 15 April 2019 from the site: http://blog.ardublock.com/
51. App Inventor (2019), The MIT App Inventor programming environment. Retrieved on 20 April 2019 from the site: http://appinventor.mit.edu/explore/
52. LSM9DS1 (2019). Description of the SparkFun 9DoF IMU Breakout - LSM9DS1. Retrieved on 20 September 2019 from the site: https://www.sparkfun.com/products/13284
53. ThetaProbe (2019). Official description of the ThetaProbe ML2 soil moisture sensor. Retrieved on 20 September 2019 from the site: https://www.delta-t.co.uk/wp-content/uploads/2016/11/ML2-Thetaprobe-UM.pdf
54. Kargas, G. and Kerkides, P. (2009). "Performance of the theta probe ML2 in the presence of nonuniform soil water profiles", Soil & Tillage Research - SOIL TILL RES. 103. 425–432. https://doi.org/10.1016/j.still.2008.12.007.
55. Navio2 (2019). "Navio2: Autopilot HAT for Raspberry Pi", official description of the Emlid Navio2 board. Retrieved on 20 September 2019 from: https://emlid.com/navio/
56. ArduPilot (2019). Official description of the ArduPilot, an open source vehicle-controlling software. Retrieved on 20 April 2019 from the site: https://ardupilot.org/ardupilot/
57. MissionPlanner (2019). Official description of the Mission Planner ground station application. Retrieved on 20 September 2019 from: https://ardupilot.org/planner/
58. QGroundControl (2019). Description of the QGroundControl (QGC) ground station controlling software. Retrieved on 20 September 2019 from: http://qgroundcontrol.com/
59. ROS (2019). Official description of the Robot Operating System (ROS) framework. Retrieved on 20 September 2019 from: https://www.ros.org/about-ros/
60. OpenCV (2019). Description of the OpenCV, an open source computer vision and machine learning software. Retrieved on 20 September 2019 from the site: https://opencv.org/about/
61. Pixy2 (2019). Description of the Pixy2 AI-assisted robot vision camera. Retrieved on 20 September 2019 from the site: https://pixycam.com/pixy2/
62. Dragino (2019), The LoRa Dragino shield for arduino. Retrieved on 20 April 2019 from the site: http://www.dragino.com/products/module/item/102-lora-shield.html
63. ETSI (2012) 'ETSI EN 300 220-1: Electromagnetic compatibility and radio spectrum matters (ERM); short range devices (SRD); radio equipment to be used in the 25 MHz to 1000 MHz frequency range with power levels ranging up to 500 mw; part 1 (2012)', online at: https://www.etsi.org/deliver/etsi_en/300200_300299/30022001/02.04.01_40/en_30022001v020401o.pdf. Accessed on 20 September 2019.
64. Eaton E. (2017): Eaton, E. (2017). Teaching Integrated AI through Interdisciplinary Project-Driven Courses. AI Magazine, 38(2), 13–21. https://doi.org/10.1609/aimag.v38i2.2730

Evaluation of Spray Coverage and Other Spraying Characteristics from Ground and Aerial Sprayers (Drones: UAVs) Used in a High-Density Planting Olive Grove in Greece

Athanasios Gertsis and Leonidas Karampekos

1 Introduction

New olive systems for olive oil production are expanding worldwide, using high and super high planting densities, ranging from 1000 to 2200 trees per hectare (ha) and are fully adaptable to mechanical harvesting with straddle type harvesters used to harvest grapes [1]. In Greece, conventional olive production systems have a planting density ranging from 150 to 300 trees per hectare for table olives or olive oil production, rainfed or irrigated. The new linear and high density groves were first planted in 2007 and are expanding mainly in Central and North Greece [2] due to the available area abandoned or replaced by other crop species in these areas and the climate change resulting to the temperature increases and rainfall distribution. However, their expansion rate is limited mainly due to legislative reasons, limited management expertise, since the establishment of new olive grove is not supported financially by the National Agricultural legislation and Farmer support system, proportionally to the conventional density plantations.

Also, the new systems are so far mainly established by agricultural corporations focusing on achieving higher Return on Investment (ROI) rates as compared to traditional systems, due to much lower harvesting cost by the machines. In general, most tree production systems are globally changing from conventional and low to medium planting densities into linear systems of higher densities, to facilitate use for both manual and mechanical systems, mainly concerning harvesting. A main issue of all these systems is spraying with ground machinery and equipment, varying in size and capacity in order to be to adaptable to the high variable Greek farm sizes and

A. Gertsis (✉) · L. Karampekos
Perrotis College-American Farm School, Thessaloniki, Greece
e-mail: agerts@afs.edu.gr

D. D. Bochtis et al. (eds.), *Information and Communication Technologies for Agriculture—Theme IV: Actions*, Springer Optimization and Its Applications 185,
https://doi.org/10.1007/978-3-030-84156-0_13

location relief where olives are grown. The average olive grove in Greece is in small size (<2 ha) and most olive oil production varieties are grown under the rain fed (not irrigated) conditions and usually in field with high slope (>5–10%).

Due to tree characteristics and depending on the cultivation method, agrochemicals are frequently applied uniformly over the entire orchard area, without taking into account the spatial variability of the target tree profile and often using higher dosages than needed. This phenomenon increases spray losses associated with deposition on the ground, off-target application and drift which in turn decreases application efficiency. One of the most important off-target application phenomena is pesticide runoff caused either by overdosing, as a consequence of not having an appropriate dosing system, or by performing low uniformity treatments as a result of the inadequate use and poor maintenance of application equipment [3]. Olive farmers often apply products to the point of runoff to guarantee a high biological efficacy, contributing this way to environmental pollution [4]. The advantages of new prototype nozzle in spaying olives was reported in olives by Miranda-Fuentes et al. [17]. They developed new airblast sprayers to optimise application efficiency and overcome the limitations of conventional sprayers used in traditional and intensive olive orchards. This approach resembles the way used in centrifugal nozzle used in some drones. Issues of economic aqssessement are outlined by Martinez-Guanter et al. [19], indicating strong evidence of economic advantage as a basis for a debate about the current legislation.

Main issues regarding plant protection quality in olives are the uncertainty about optimal application parameters and the influence of operational parameters on the application homogeneity and efficacy [3]. Moreover, olive grove characteristics and the improper spraying equipment, especially in new systems such as super high-density olive groves, can negatively influence the application efficiency [4]. Unmanned aerial vehicles (UAV) are precision agriculture solutions for developing spraying processes that are probably safer and more precise than the solutions of manned agricultural aircraft and traditional ground sprayers. Compared to ground technology, aerial systems allow faster spraying without the need for traversing the crop field [5], even in sloping groves.

The question of which systems are more efficient in terms of spaying coverage (Percent Coverage, PC %), has not been adequately evaluated among the various systems used [6, 7]. In addition, the most recent developments in the UAVs business, has generated new technological tools in agriculture and nozzle type for application of agrochemicals and other production inputs and concepts (fertilizing, seeding, monitoring and crop scouting, etc., [8]).

The present study was undertaken to evaluate the long-term effects of spraying by the most common ground spraying systems used in Greece and to add information from aerial spraying by using two different nozzle technologies, namely: electrostatic vs. conventional. It is a subsequent study from the first datasets collected in the reported study from the same olive grove [6]; however, additional treatments including a new UAV with two nozzle types and ground sprayers using the electrostatic nozzle technologies, were added to provide more aspects of droplet distribution in spraying.

In a comprehensive review of electrostatic system parameters, many aspects and principles of this technology, as it is applied to superpose charges to spraying droplets for improving abaxial and adaxial surface deposition, retention, and coverage, were reported by Appah et al. [9]. They mentioned that the electrostatic nozzles can be placed in any spraying system-ground or aerial- and they improve droplet deposition, impinging, rebound, and drift inefficiencies since they are better attracted to the substrate leaves, branches, and fruits. They also identify future trends to improve pesticide application for crop protection.

Patel et al. [10] reported a new system which appears to be a contribution to the development of a new and more efficient nozzle with increase in liquid deposition and better uniformity. The subsequent environmental benefits from such systems are apparent, and mainly refer to minimizing drifting, reducing the volume of pesticides sprayed, and providing a better coverage of crops.

A major issue with drone spraying remains the current legislation in EU. The machinery directive 2006/42/EC was published on 9 June 2006 and became applicable on 29 December 2009 [18]. It was amended by Directive 2009/127/EC of the European Parliament and of the Council of 21 October 2009, with regard to machinery for pesticide application, and by Regulation (EU) No 167/2013 of the European Parliament and of the Council of 5 February 2013, on the approval and market surveillance of agricultural and forestry vehicles, among others [15].[1]

Outside EU members but still within Europe, Switzerland has become the first country in Europe to develop a process to authorize spraying drones. According to the Swiss government, drones are well suited as replacements for helicopters in spraying pesticides [16].[2] One question for EU officials commonly asked by farmers and other spraying drone users, is the following: "*The prohibition of aerial spraying in the EU since the SUD (Sustainable Use Directive) also precludes precise spraying with drones. How do you address this issue*?". A common answer is: "*Regulations may have to be re-considered and re-appraised as new technologies and new evidence accumulate to support changes in legislation*".

The Directive 2009/128/EC of the European Parliament and of the Council of 21 October 2009 establishing a framework for Community action to achieve the sustainable use of pesticides provides more on the aerial spraying in EU member countries. The Chapter IV SPECIFIC PRACTICES AND USES—Article 9 refers to Aerial spraying and the first paragraph is:1. Member States shall ensure that aerial spraying.[3]

[1]Read the consolidated text of the directive: https://ec.europa.eu/growth/sectors/mechanical-engineering/machinery_en

[2]https://www.electricvehiclesresearch.com/articles/17812/switzerland-authorises-crop-spraying-drones

[3]For the entire directive visit: https://eur-lex.europa.eu/legal-content/EN/TXT/PDF/?uri=CELEX:02009L0128-20091125&from=EN

Fig. 1 The Educational Research Demonstration high density olive grove in Perrotis College, American Farm School of Thessaloniki, Greece

2 Methodology

2.1 Olive Grove

This long-term study is located in a high-density olive grove (Fig. 1) established in 2011 in Perrotis College, American Farm School, Thessaloniki Greece (Lat 40° 34′ 13″ N Long 22° 59′ 12″ E). The olive grove is composed of 12 rows of 50 m each, with between-rows spacing equal to 4 m for all rows, oriented North to South. In-the-row spacing is 1.5 m, 2,5 m and 5 m for the Super High Density (SHD), High Density (HD) and Medium Density (MD) resulting in tree densities of 1670, 1000 and 500 trees ha^{-1}, respectively. The long-term objectives in this olive grove are to evaluate olive production levels, agronomic characteristics and olive oil quality under the new high density, linear systems, adapted for mechanical harvesting. Another specific goal is to evaluate the Input Use Efficiency (IUE) for major agronomic production inputs in the framework of LISA and SOCRATEES© [11] (Table 1).

The experimental variables include the three planting densities mentioned above, two olive oil producing varieties (Arbequina and Koroneiki), two levels of irrigation and fertilization (conventional and deficit irrigation, c.a. 40% less than the conventional and low fertilization, c.a. 50% less than the conventional), along with foliar

Table 1 The various spraying systems used

1	Commercial—Electrostatic boom sprayer –1000 lph used (2000 L tank)
2	Electrostatic boom sprayer –200 lph used (2000 L tank)
3	Electrostatic gas operated blower back pack (16 L tank)
4	Commercial electrostatic-mixed boom sprayer (1500 L tank)
5	Commercial conventional boom sprayer 1500 L tank
6	Back pack sprayer—battery operated (1 hose used-16 L tank)
7	Back pack blower/mist type sprayer—gas operated—slow walk (16 L tank)
8	Back pack blower type sprayer—gas operated—fast walk (16 L tank)
9	UAV with electrostatic nozzles—flight between the rows (16 L tank)
10	UAV with conventional nozzles—flight over the rows (16 L tank)
11	UAV with conventional nozzles—flight between the rows (16 L tank)
12	UAV with electrostatic nozzles—flight over the rows (16 L tank)

Table 2 Water Sensitive Papers (WSP) position numbers in each tree and canopy volume per tree

Tree number in the line	Canopy volume (m^3)	Number of WSPs used per tree
9	3.06	5
30	3.39	6
23	3.61	6
2	3.99	7

spray of a bio-stimulant product. Many yield and agronomic parameters are evaluated since the beginning of olive fruit production, 3 years after planting.

This specific study reports on various spraying systems commonly used in large-medium-small size olive groves, representative of the Greek olive production systems, site relief and average farmer's land. It is a continuation and a further expansion and evaluation of different spraying systems, of a study on this olive grove, conducted in spring-summer 2018 [6]. A number of trials were performed using the ground and an aerial (UAV—Model Joyance JT16L-2N1) vehicle with a 16 L spraying tank, shown in Table 2 and Fig. 5.

Each system sprayed one SHD line (50 m long), selected as a representative of a very thick canopy scenario, around the line's left and right side (West and East orientation), while the UAV was sprayed in two configurations and at a height approximately 1 m above the top canopy, to fly over the row and around the row. A battery operated back-pack sprayer was used twice, in a slow and a fast walking pace in order to simulate average farmer walking modes, since there is a variability among workers in their walking speed, as affected by fatigue initially and towards the end of the spraying process later on.

For the tests performed, four trees were randomly selected from one of the four super-high density lines. In each of the four replicate trees, Water Sensitive Papers (WSP-1 $\times$ 2 in by Sygenta—Figs. 2 and 3) were placed in the same positions to provide an integrated evaluation in all directions of the canopy, shown in Fig. 4, for

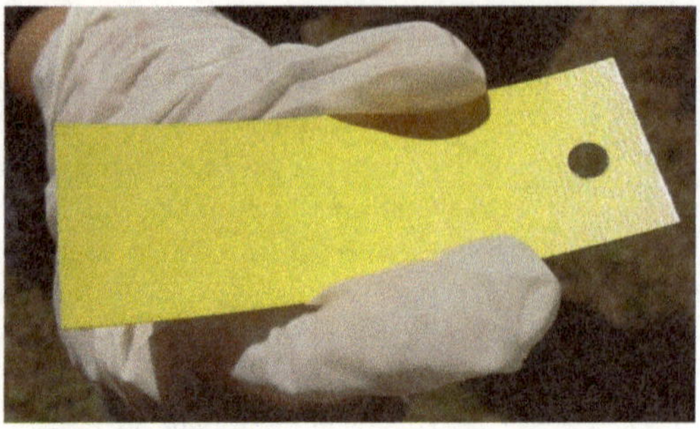

Fig. 2 Water Sensitive Paper (WSP, dimensions 1 in × 2 in used—by Sygenta)

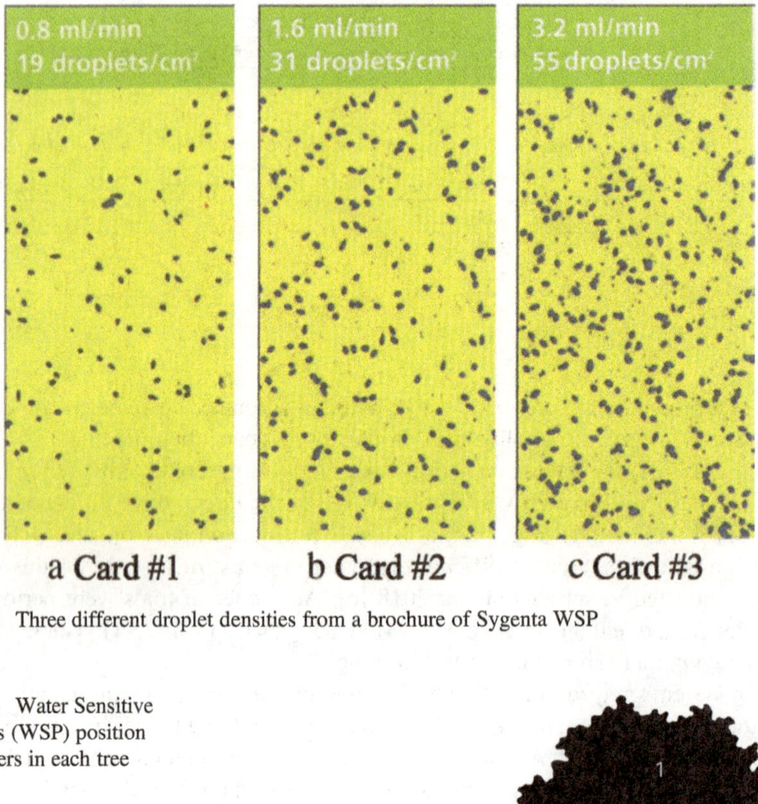

Fig. 3 Three different droplet densities from a brochure of Sygenta WSP

Fig. 4 Water Sensitive Papers (WSP) position numbers in each tree

all trials. The number of WSP per tree was 5, 6 or 7 and were determined based on tree measured volume (Fig. 4 and Table 2). The trees and WSP positions remained the same for all systems and trials used. A light pruning (mainly the suckers were removed) was performed to maintain the leaf canopy in approximately the same volume per tree, as the initial one.

The WSP were collected after 40–60 min period left to be dried out and then scanned with two software: DepositScan[4] and SnapCard[5] to determine the Percent Coverage (PC%) and other droplet features, such as number of droplets, and size distribution.

Statistical analysis was performed using the statistical software JMP SE version 14[6] for means comparisons using the four replicated trees. Also, to determine if any effects were evident among the four trees and among each WSP position, similar means comparisons were done using Student's *t*-test (Fig. 5).

3 Results and Discussion

Analysis of data will be presented from the DepositScan software, because the correlation was very strong between the two scanning software used and since DepositScan also reports on the size and number of droplets, while SnapCard reports only the PC%. Both software recorded the percent coverage (PC %) in a very similar mode as shown in Fig. 6. Their linear correlation was highly significant ($r = 0.978$) and the two-line equations are:

$$SnapCard\% = 1.2360056 + 0.6008362 * DepositScan\%$$

or

$$DepositScan = -4.533 + 1.4291 * SnapCard$$

Results of PC% from the DepositScan are shown in Table 3. In general, all spraying systems using the electrostatic nozzles performed better and in most case the PC% was significantly higher than the conventional sprayers nozzles. Also, the uniformity of droplets per area unit was higher from the electrostatic nozzles. The UAV showed the lowest PC% although not significant different from some of the large commercial boom sprayers. It must be mentioned that the measurements with the initial UAVs took place under non-optimal (relatively high) wind speed conditions and they will be repeated for further evaluation. Chen et al. [7] introduced

[4] https://www.ars.usda.gov/midwest-area/wooster-oh/application-technology-research/docs/depositscan/ and Zhu et al. [12]

[5] https://www.agric.wa.gov.au/grains/snapcard-spray-app and Nansen et al. [13]

[6] www.jmp.com

Fig. 5 The various spraying systems used. (Top left: the electrostatic and conventional back pack gas operated blower. Top right: the two commercial electrostatic and conventional large size blower sprayers-2000 L tank. Middle left: the spraying UAV 16 L tank. Middle right: the 16 L tank back-pack electrostatic. Bottom left: the 16 L battery operated back pack and Bottom right: the 1000 L tank conventional mixed spraying system)

an overspray index and defined it "*as any situation with spray coverage greater than 30%, which was based on WSP samples*". Calculation of this index was done by normalizing the PC% using the following equation:

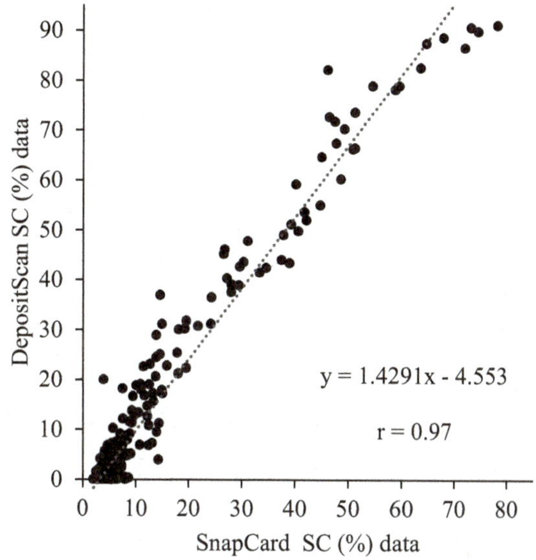

Fig. 6 Comparison between coverage (%) data obtained by DepositScan and SnapCard. A strong correlation (r = 0.978) exists between the two data sets

Table 3 The mean Percent Coverage (PC%) and statistical differences of each spraying system and means comparisons using Student's t test

Spraying system						PC%
Commercial—electrostatic boom sprayer –1000 lph used	A					28.1
Electrostatic gas operated blower—back pack	A	B				26.4
Commercial—electrostatic—mixed boom sprayer (air blast)	A	B				26.3
Commercial—Conventional boom (air blast) sprayer		B				20.1
Back pack sprayer—battery operated (1 hose used)			C			10.1
Back pack blower type sprayer—gas operated—slow walk			C	D		8.7
Commercial electrostatic boom sprayer (air blast) –200 lph used			C	D	E	7.5
Back pack blower type sprayer—gas operated—fast walk			C	D	E	6.1
UAV with electrostatic nozzles—flight between the rows				D	E	2.46
UAV with electrostatic nozzles—flight over the row				D	E	1.83
UAV with conventional nozzles—flight between the rows					E	1.39
UAV with conventional nozzles—flight over the row					E	0.69

Note: Spaying systems not connected by same letter are significantly different (Student's t-test, a = 0.95)

$$I_o = (C - 30)/(100 - 30),$$

where I_o is the overspray index and C is the spray coverage on WSP (%). The values of I_o range from 0 to 1, with 0 representing no coverage and 1 representing saturated spray (100%) coverage. However, this index has not been substantiated by any field evidence data and therefore it may represent their own "personal proposal". Also, the calculations on the above formula are not valid for any spray percentage less than

Table 4 The amount of spraying and time of operation for the various spraying systems used

Spraying system	Water amount (L)	Operation time (s)
1. Commercial—electrostatic boom sprayer –1000 lph used (2000 L tank)	36.5	68 s
2. Electrostatic boom sprayer –200 lph used (2000 L tank)	8.6	62 s
3. Electrostatic gas operated blower back-pack (16 L tank)	3.5	205 s
4. Commercial electrostatic—mixed—boom sprayer (1500 L tank)	40.2	137 ss
5. Commercial conventional—mixed—boom sprayer 1500 L tank	32.7	64 s
6. Back-pack sprayer—battery operated (1 hose used-16 L tank)	2.6	272 s
7. Back-pack blower/mist type sprayer—gas operated—slow walk (16 L tank)	4.8	312 s
8. Back-pack blower type sprayer—gas operated—fast walk (16 L tank)	2.6	190 s
9. UAV with electrostatic nozzles—flight between the rows (16 L tank)	2.15	78 s
10.UAV with electrostatic nozzles—flight over the row (16 L tank)	1.0	51 s
11. UAV with conventional nozzles—flight over the row (16 L tank)	1.1	55 s
12. UAV with conventional nozzles—flight between the rows (16 L tank)	1.9	105 s

30%. It can be logically argued that the "30% coverage" or practically any percent coverage can be differently translated for different plant species (such as field crops or tree production systems), planting or seeding densities and, most importantly, for different type of chemicals used, i.e., contact type vs. systemic pesticides for which the mode of action significantly differs. In our study, no system exceeded the above "overspray index", since the highest percentage was 28.2%.

The amount of water used by the different systems (Table 4) ranged from c.a. 32–40 L for the large commercial blower (blat air) sprayers, c.a. 3–5 L used by the back-pack systems to less than 2 L used by the UAV system, while the time of operations are between c.a. 1–6 min among the various systems. This information should be also considered when evaluating the efficiency of spraying systems along with the higher flexibility of UAV flight [5]. The advantages of spraying UAVs are expected to be higher in farm specific situations where the large tractor operated systems may not be able to enter in the field, due to excessive soil moisture, sloppy parts of the field, or other unusual soil-weather conditions. The significant amount of water used by the large ground systems does not result necessarily in significantly higher coverage.

Comparison of an independent trial for the two UAVs with conventional and electrostatic nozzles and in two directions (Table 5) has shown that the electrostatic nozzle drones had significantly higher PC% over the conventional nozzles. This finding is also in line with the ground systems equipped with electrostatic nozzles and can be used by farmers as a tool to further increase their pesticide application

Table 5 Comparison of two systems in the whole olive grove (0.3 ha)

Spraying system	Solution used (L)	Operation time (min)
Commercial conventional tractor boom sprayer	165	14.5
UAV with electrostatic nozzles—flight between the rows	10	5.5

Table 6 Comparison of an independent trial for the two drones with conventional and electrostatic nozzles and in two directions

Type of UAV and flying pattern			Mean PC%
UAV with electrostatic nozzle (between the rows)		A	2.46
UAV with electrostatic nozzle (over the row)		A	1.83
UAV with conventional (between the rows)		B	1.39
UAV with conventional (over the row)		B	0.69

Table 7 Results for the effect of WSP position in all trees for the Table 5

Position of WSP in the trees		Mean PC%
2	A	0.61
5	A	0.60
1	A	0.58
6	A	0.45
4	A	0.44
3	A	0.36
7	A	0.28

Table 8 Results for the effect of replicated trees for the Table 5

Tree number		Mean PC%
2	A	0.56
30	A	0.54
9	A	0.53
23	A	0.42

efficiency and reduce environmental issues related to agrochemical use and application methodologies. The cost of centrifugal-electrostatic nozzles in drones is not prohibiting. However, there is a significant difference in the cost of round equipment using electrostatic sprayers. It is expected that new system, developed or under development, will reduce this cost [10].

The possible effects on PC% from the WSP position and tree Number (possible tree characteristics) was assessed and are presented in Tables 6 and 7 for the new trials of drones only and Tables 8 and 9 from all the systems evaluated. In the case of the two nozzle type drones, there was no difference either in Tree no or in WSP position, indicating a very uniform distribution of spraying droplets, in all directions

Table 9 Results for the effect of WSP position in all trees for the Table 2

WSP position in the trees				Mean PC%
5	A			18.85
4	A	B		16.59
6	A	B	C	15.28
7	A	B	C	14.86
2	A	B	C	13.34
3		B	C	10.87
1			C	7.82

Table 10 Results for the effect of replicated trees for the Table 2

Tree no.		Mean
9	A	15.79
2	A	14.59
30	A	13.26
23	A	11.60

in the olive canopy. In the case of all systems, there was no difference among the trees, but significant differences were shown among the positions. Position #1 (Fig. 4) has shown the lowest PC% while position # 5 had significantly higher PC % from #1. These finding suggest a variable spraying coverage by the ground systems mainly, resulting to not so uniform droplet distribution all around the tree. Therefore, these findings support the hypothesis that drones result in more uniform all around coverage then the ground systems. Also, the superiority of electrostatic nozzles over the conventional ones is evident.

The results from the possible effects of the trees (Table 10) indicated no significant differences; therefore, we accepted that the four replicated trees were uniform in the distribution of applied liquid and had not effect on the percent coverage comprehensive results.

The extrapolation of spraying results from one line in the whole olive grove—15 lines—outlined in Table 5, indicated that the amount of chemical solution used by the major ground system was significantly higher than the amount used by the drone, and it was about 17.5 times higher. Extrapolated to cost reduction for the farmer and effect to environment it can be the major factor in the future for decision making in term of equipment used for spraying and ROI. This difference provides concrete evidence of the higher efficiency of the drone over the most common ground systems in medium to large groves. However, the efficacy of the spraying over insects and disease control remains to be evaluated in separate studies, which are under progress for the coming growing season 2020. A preliminary study in Greece has provided initial and very encouraging results using conventional and drone spraying systems for the control of the olive fruit fly (*Bactroceraoleae, Dacusoleae*), representing the most important problem in olive production systems [14]. Their preliminary results showed that using bait spaying for this insect was proven very effective as compared to backpack sprayers used in nearby located sloppy olive groves.

4 Conclusions

Spraying with various agrochemicals and other substances represent a major cost for the farmer and impact the entire eco-agro system and the environment. Spraying time and timing are crucially important for efficient application of agrochemicals. The efficiency and efficacy of each spraying is not well understood and has not been evaluated, due to miscellaneous factors, including very specific conditions of crops, machinery and type of agrochemical used. Therefore, this study contributes significantly to this specific area and fills some gaps, to provide information available for further research and for farmers use. A first approach study reported here, aiming to evaluate typical ground sprayers used by farmers and aerial spraying by drones, for a high density olive grove, indicated a variety of PC% and mainly a large discrepancies of amount of water solution used, which can lead to over-spraying with all consequent effects to the environment.

The quantities used by the conventional ground systems were up to 45 times more than the drone's, while the operation time for ground systems was up to 6 times more than the drone's. In addition to ground sprayers, the first effort to evaluate aerial means in the form of drone (UAV) was presented with satisfactory results and quite comparable in their PC% with most of the ground systems used. In addition, the potential for spraying systems equipped with electrostatic nozzles was shown to be superior in coverage and uniformity/droplet dispersion than the conventional nozzles used. This study will be continued in the next years with new UAS models, to further evaluate other aspects of spraying characteristics and chermical use efficacy. An important issue, however, remains the need for an update of the EU Directive 2006/42/EC. Any legislative adjustment to facilitate use of drones for spaying, under controlled conditions and with all necessary prerequisites, will open a new world in spraying systems and provide more and efficient tools to EU farmers, environmentalists and all other users in small and large scale applications.

Acknowledgments The authors express their sincere gratitude to two companies for the kind collaboration and provision of the spraying UAV (3D S.A.—General Aviation Applications, Thessaloniki, Greece www.3dsa.gr/en) and the electrostatic and other boom sprayers (G. XIGKAS—Agricultural Machinery, Thessaloniki, Greece www.higas.gr).

References

1. Castro-Garcia, S and L. Ferguson. 2017. Mechanical harvesting of olives. *IN* Kiritsakis, A. And F. Shahidi. 2017. Olives and olive oil as functional foods: Bioactivity, chemistry and processing. John Wiley & Sons Ltd.
2. Gertsis, A., D. Fountas, I. Arpasanu and M. Michaloudis. 2013. Precision Agriculture applications in a high density olive grove adapted for mechanical harvesting in Greece. Procedia Technology 8(2013) 152-156.

3. Miranda-Fuentes, A., Llorens, J., Rodríguez-Lizana, A., Cuenca, A., Gil, E., Blanco-Roldán, G. L. et al. 2016. Assessing the optimal liquid volume to be sprayed on isolated olive trees according to their canopy volumes. Science of the Total Environment 568 296–305.
4. Miranda-Fuentes, A., Rodríguez-Lizana, A., Gil, E., Agüera-Vega, J. and Gil-Ribes, J. A. 2015. Influence of liquid-volume and airflow rates on spray application quality and homogeneity in super-intensive olive tree canopies. Science of the Total Environment 537 250–259.
5. Faiçal, B. S., Freitas, H., Gomes, P. H., Mano, L. Y., Pessin, G., de Carvalho C.P.L.F., Bhaskar Krishnamacharid, J. Óueyama. 2017. An adaptive approach for UAV-based pesticide spraying in dynamic environments. Computers and Electronics in Agriculture 138 210–223.
6. Aru, F., A. Gertsis, G. Vellidis, F. Morari. 2019. Investigation of spraying efficiency of an aerial system in a super-high density olive grove in Greece. Proceedings of European Conference on Precision Agriculture (ECPA), Montpellier, France (July 8–11, 2019).
7. Chen, Y., Ozkan, H. E., Zhu, H., Derksen, R. C. and Krause, C. R. 2013.Spray deposition inside tree canopies from a newly developed variable-rate air-assisted sprayer. Transaction of the ASABE-American Society of Agricultural and Biological Engineers 56 1263-1272.
8. Patel, M. K., Sahoo, H. K., Nayak, M. K., Kumar, A., Ghanshyam, C. and Kumar, A. 2015. Electrostatic nozzle: new trends in agricultural pesticides spraying. Seventh Sense Research Group-International Journal of Electrical and Electronics Engineering 6–11.
9. Appah, S, P. Wang, M. Qu, C. Gong and W. Jia. 2019. Review of electrostatic system parameters, charged droplets characteristics and substrate impact behaviour from pesticide spraying. Int. J. Agric& Biol. Eng., Vol. 12(2):1–9, [Open access www.ijabe.org].
10. Patel, M.K., B. Praveen, H. K. Sahoo, B. Patel, A. Kumar, M. Singn, M.K. Nayak and P. Rajan. 2017. An advance air-induced air–assisted electrostatic nozzle with enhanced performance. Computers and electronics in Agriculture,135 (2017) 280-288.
11. Gertsis, A.C. and C. Vasilikiotis, (2018). The LISA and SOCRATEES© Approach for Sustainable Crop and Soil Management. *In* Sengar, S. (Ed.) Eco-friendly Agro-biological Techniques for Enhancing Crop Productivity (pp. 89–110). Singapore, Singapore: Springer Nature Singapore.
12. Zhu, H., Salyani, M. and Fox, R. D. 2011.A portable scanning system for evaluation of spray deposit distribution. Computers and Electronics in Agriculture 7638–43.
13. Nansen, C., Ferguson, J.C., Moore, J., Groves, L., Emery, R., Garel, N. et al. 2015. Optimizing pesticide spray coverage using a novel web and smartphone tool, SnapCard. Agronomy for Sustainable Development 35:1075-1085.
14. Tsagarakis, A., I. Trilivas, E. Kontogiannis, V. Sideris and G. Lagoutaris. 2019. Application of bait spray from UAS for protection of olive fruit fly (*Bactosperaoleae*). Agriculture-Crop and Animal Husbandry, Issue No 12, December 2019 (In Greek).
15. https://ec.europa.eu/growth/sectors/mechanical-engineering/machinery_en.
16. https://www.electricvehiclesresearch.com/articles/17812/switzerland-authorises-crop-spraying-drones.
17. Miranda-Fuentes, A., Rodríguez-Lizana, A., Cuenca, A., Gonzàlez-Sànchez, E. J., Blanco-Roldàn, G. L., Gil-Ribes, J. A. 2017. Improving plant protection product applications in traditional and intensive olive orchards through the development of new prototype air-assisted sprayers. Crop Protection 94 44–58.
18. Zwetsloot, Hannah M., Nikol, Lisette, and Jansen, Kees, 2018. The General Ban on Aerial Spraying of Pesticides of the European Union: The Policy-Making Process between 1993-2009. Wageningen, Rural Sociology. 34 pp.; 1 tab.; 62 ref.; https://doi.org/10.18174/442443
19. Martinez-Guanter, J, Pablo Agüera, J. A., and Pérez-Ruiz, M. 2019. Spray and economics assessment of a UAV-based ultra-lowvolume application in olive and citrus orchards. Precision Agriculture. https://doi.org/10.1007/s11119-019-09665-7

Predictive Model for Estimating the Impact of Technical Issues on Consumers Interaction in Agri-Logistics Websites

Damianos P. Sakas and Dimitrios P. Reklitis

1 Introduction

The modern digital revolution has altered the competitive structure of the commercial industry and increasingly requires marketers to switch their efforts from the real market to the digital one. This digital transition has occurred in light of the growing consumer focus on digital services. Several scientific works contribute to the examination of this phenomenon by analyzing evolving digital marketing tactics. Existing literature on the topic has extensively examined the logistics industry and social analysis data sets with the view to inform marketer's decision-making processes [1]. Indeed, existing literature has examined and proposed various approaches which can be used as marketing strategies to increase consumer interaction with a company's digital platform; ultimately providing a company with a competitive digital edge [2].

Despite the existing body of literature, there is limited research which examines the current limitations of marketing tactics employed by the niche agri-logistics companies which are attempting to adapt to the increasing digital market. In turn, this has led to the emergence of a problem within agricultural websites, as they contain little consumer interaction. Similarly to the existing literature which focuses on the wider logistics sector, extensive research regarding digital marketing optimization strategies must be carried within the agri-logistics industry. This research should seek to begin filling the existing gap by prognosticating digital marketing strategies which can be utilized within the agri-logistics sector to increase digital consumer interaction.

D. P. Sakas (✉) · D. P. Reklitis
Department of Agribusiness and Supply Chain Management, School of Applied Economics and Social Sciences, Agricultural University of Athens, Athens, Greece
e-mail: d.sakas@aua.gr

D. D. Bochtis et al. (eds.), *Information and Communication Technologies for Agriculture—Theme IV: Actions*, Springer Optimization and Its Applications 185,
https://doi.org/10.1007/978-3-030-84156-0_14

By taking into account existing theories on logistic website's digital marketing strategies, it is anticipated that the various variables of a website interact with each other in a manner which either positively or negatively impacts the efficiency of an agri-logistic digital marketing strategy. For instance, it is suspected that the time it takes for a web page to load impacts the web page's ranking on digital servers.

This Chapter seeks to identify the various correlations which exist between the variables which affect the efficiency of the digital marketing strategy. Based on existing literature, the present work sets as a hypothesis that the existing correlations between different web-variables have a direct impact on the efficiency of an agri-logistic digital marketing strategy. Within the scope of this research, this hypothesis is considered with a view to, firstly, prognosticate the most efficient digital marketing strategies that can be employed by agri-logistic websites and to, secondly, enable the long-term forecast of digital marketing within the agri-logistic sector.

In order to address the above question, namely, how to increase consumer interaction with agri-logistics websites, this Chapter adopts the following structure. The first section carries out a literature review on existing digital marketing theory, search engine optimization (SEO) theory, Big Data theory, and web analytics theory. The second section sets out the predicted hypotheses, while the third section sets out the adopted three-stage methodology process. The fourth section carries out a verification of the output in terms of the statistical analysis. The fifth section utilizes the results of the statistical analysis to generate a Fuzzy Cognitive Map (FCM) approach in order to assist in the creation of a predictive model. Finally, the creation of the predictive agent-based simulation model will inform the later discussion on strategies which can be employed by agri-logistic websites to optimize consumer interaction. The Chapter concludes with a wider consideration of existing literature and opportunities for further research in the field.

2 Background

2.1 Digital Marketing

The most crucial piece of information which a company must acquire is a competitive advantage [3]. Nowadays, a competitive edge is not only crucial for the development of the company but for its survival in this free jungle market. This is especially true for logistics companies which operate in a highly competitive market [4]. Information Technology has great impact on business advantage [5, 36]. According to Lai et al. [5], a competitive advantage could be achieved by focusing on IT. IT capability is achieved through the connection of IT strategy with IT management skills. Third Party Logistics (3PL) managers can acquire advantages from their investments in IT only if the execution of IT strategy is higher than that of their competitors. This is the reason why Digital Marketing is crucial. The results of Digital Marketing are more measurable than the results of controversial marketing [6, 7, 38]. According to Chaffey and Ellis-Chadwick [8], the definition of "Digital

Table 1 The advantages of digital Marketing, as opposed to controversial Marketing

Interactivity	With digital marketing there is a possibility of higher interaction between the corporation and the customer which leads to higher sales and close rates [10]
Intelligence	Digital Marketing cost is significantly lower than the Controversial cost. Researchers can gather great amounts of information by sending to customers for example Google Forms [8]
Individualization	With digital marketing the information gathering is easier and cheaper. With this gathered information, which are gathered using techniques and procedures that are aligned with the rules of General Data Protection Regulation (GDPR) [39] Regulation of the European Parliament and the European Council (2016), digital marketers can suggest the right product to the right customer to fulfill the required need [11]
Integration	Digital marketing covers both the outbound and inbound Internet-based communications. The effective and beneficial communication from the company to the customer and vice versa. Therefore, there is a need to identify and analyze the web analytics that will be explained later on [8]
Industry restructuring	One of the key elements of restructuring is the reintermediation that should be examined by any organization developing their Digital Marketing strategy [12]
Independence of location	The organization can supply products to the global market without having a local store [13]

Marketing" is *"achieving marketing objectives through applying digital technologies and media"*. Digital Marketing is different than the controversial and obsolete Marketing because it has introduced new ways of interaction. According to McDonald and Wilson [9], the benefits of Digital Marketing as opposed to controversial Marketing are listed in Table 1.

2.2 Search Engine Optimization

According to Chaffey and Ellis-Chadwick [8], search engine optimization (SEO) is the technique which is used in order to maximize the quantity and quality of traffic in a website. Businesses have realized that twenty-first century marketing is being indexed on the Internet. More specifically, the goal of a business is to appear at the top of search engines. In order to achieve this, technical variables must be taken into account in order to optimize the ranking of these machines. The so-called optimization process and the techniques and strategies used to raise the page at the top of the rankings are called SEO [14].

2.3 Big Data

According to Thucydides [41] (404 b.c.) information leads to power. In ancient Greek, the word King refers to the one that holds the information. Today, the need to acquire a relative advantage over competitors can be achieved by using and structuring Big Data [15]. According to SAS Institute Inc. [16], "*Big Data refers to the ever-increasing volume, velocity, variety, variability, and complexity of information*". The term "Big Data" was introduced in order to describe the vast and ever-increasing amount of data that corporations store and analyze in light of their enormous data resources. The volume of data is expanding as it takes in a greater range of sources, of which the vast majority are in an unstructured form. Companies need to extract value from the extraction of the data but also need to employ the results to resolve their security problems [17, 40]. According to Chaffey and Ellis-Chadwick [8], "Big Data" is a term used to describe the analysis techniques and systems that "can be used to make use of the large volumes of data that are available for marketers". According to Soubra [18], the opportunity which is now available to marketers can be classified in three dimensions (Table 2):

All businesses operating in this sector strive to improve, distinguish and refine their websites as these represent a key marketing tool for a business. The enhancement of a website is achieved by utilizing all accessible metrics through web analytics. The management of Big Data is a crucial element for the achievement of this goal. Properly handled, it enables the adoption of precise, easy and efficient decision-making strategies.

By treating a website, not only as an online store, but also as a core tool which strategically advertises a business' aims, the need for its continuous improvement becomes apparent. A business' ulterior motive in relying on websites is to not only confront existing industry competitors but to also create a competitive advantage for the business. To achieve this, two actions must be carried out simultaneously. Namely, all the information which exists by virtue of the business' physical presence must be gathered and harmonized with the information arising from the web analytics and the Big Data [19, 20].

Table 2 Big data properties

Data volume	Relates to the increasing size of the data which is disposable for online interactions between Digital media platforms. In the past the creation of the data was up to employees, nowadays everyone generate data for the company including employees, clients, other companies and even machines
Data velocity	The data are streaming into the company's computers in real time so, the marketing professionals have the opportunity to take it in consideration
Data variety	Represents the potential that provide the unstructured data as texts, photos, videos, sensors, CRM etc. This also provides the possibility of unifying alternative sources of data to get more client insight

This research gathers data from GTMetrix and presents the extracted data from the leading agri-logistic companies. For the purpose of this Chapter, the selected leading companies will be taken to represent the entire industry.

2.4 Web Analytics and Metrics

One of the core benefits that Digital Marketing provides to organizations, is the ability to fully or partially understand consumer behavior through the utilization of web data analytics, while simultaneously placing a plurality of quantitative metrics [21]. According to Chaffey and Ellis-Chadwick [8], web analytics (WA) are techniques used to gather data, evaluate and optimize the contribution of Digital Marketing to a company, including clickstreams, online reach data customer satisfaction surveys, leads and sales. In light of this research, the analytics of multiple logistics websites were gathered through the use of the descriptive web mining analytics tool GTMetrix. GTMetrix is a website that scores a website's performance and speed. GTMetrix examines the Page Speed Score, YSlow score, Fully Loaded Time, Total Page Size and the number of the Requests of the website that are provided by the URL. GTMetrix proposes solutions on how to make the given website more efficient and faster. This website provides the complete structure of the elements that contribute to the total size of the site [22]. WA is used by more than 66% of all websites globally [23].

In spite of the fact that there is a relatively good adoption rate, the available research on WA is insufficient. Furthermore, according to Järvinen et al. [24], the data extracted from WA are only used to solve specific problems and are not adopted as a part of the company's wider strategy; such use could allow a business to acquire a competitive advantage. In addition, according to Germann et al. [25], many marketing managers are reluctant to take into consideration extracted data from WA platforms in their decision-making processes and prefer to make decisions according to their experience and intuition. Also, the SEO analysis tool can be used for the recognition rectifications that need to be carried out to enable the growth of the website's visibility [26, 27].

The research of Järvinen and Karjaluoto [21] emphasized the managerial incapability to understand, define and use WA metrics. In light of these findings, this paper goes on to consider the description and explanation of the five-web analytics metrics; namely, the Fully Loaded Time, Page Speed Score, Requests, Total Page Size and YSlow Score, as described in Table 3 [22]. This will benefit digital marketing managers and practitioners by informing their decision-making processes through WA with the purpose of understanding these WA metrics and their effect. A further benefit of these findings is setting a strategic *modus operandi* of gathering objective web analytics data of users' behavior in websites and to understand the interrelationships and interconnections that these web analytic metrics might have. As a final benefit, these findings may serve to improve managers' and practitioners' analytics skills by utilizing the gathered web behavioral data. In turn, these can be used to

Table 3 Description of the examined Web Analytics Metrics

Web analytics metrics	Description of the WA metrics
Page Speed Score	According to Google Developers [37], "Page Speed Score reports on the performance of a page on both mobile and desktop devices, and provides suggestions on how that page may be improved"
YSlow Score	According to Duran [28], the tool: "Grades web page based on one of three predefined ruleset or a user-defined ruleset; it offers suggestions for improving the page's performance; Summarizes the page's components; Displays statistics about the page; Provides tools for performance analysis"
Fully Loaded Time	Fully Loaded Time is the point after the Onload event fires and there has been no network activity for 2 s [22]
Total Page Size	"The term page size in the SEO world refers to the downloaded file size of a given web page" [29]
Requests	"Decreasing the number of components on a page reduces the number of HTTP requests required to render the page, resulting in faster page loads" [22]

design, implement, evaluate and potentially optimize a performance measurement system [21].

The investigation of the probable interrelationships and interconnections between the five Web Analytic metrics, establishes a significant step for marketers to realize how the Fully Loaded Time, YSlow Score and the Total Page Size of their website influences the average time users spend on the marketer's website. For example, the Fully Loaded Time could be a critical component affecting the time spent on the site and page views per visitor. The lower the Fully Loaded Time is, the higher the time that is spent and the page views that are recorded on the websites. Contrastingly, the lower the Fully Loaded Time, the lesser the time users spend and the pages they view within the site. It is still unknown if and how the level of Fully Loaded Time impacts the overall traffic of the website. In other words, it is unknown whether the lower the Fully Loaded Time of a website is the higher the website traffic will be, or if, the higher the Fully Loaded Time, the fewer the number of visitors that will access the website. The following section presents the research hypotheses of the study and the implementation of a descriptive modeling technique as a precondition of the construction of the predictive agent-based model. Finally, the Simulation Modeling process is used to represent and estimate all the scenarios in a graphical way [30].

3 Descriptive Modeling Process and Research Hypotheses

First, the initial method of generating a prediction model illustrates the creation of a descriptive model that defines the relationships between the variables [31]. This book chapter shows the possible relationships among the WA metric of the examined sites within the transport sector. To create descriptive modeling, this research applies the Fuzzy Cognitive Mapping (FCM) approach to illustrate the relationships between WA metrics. FCM is used to describe the relationships between factors and also, to demonstrate the importance of each problem.

For this reason, in this study, WA metrics are recorded through a comprehensible representation of these correlations, while defining numerically weighted weights $[0, 1]$ or $[+1, -1]$ for each relation of the described descriptive model [32, 33]. A benefit of relying on the adopted prediction model is that it illustrates, in a visual manner, the correlation and causal relationship between the various entities [34]. Its main advantages are the flexibility of visualization of a model and the ease of understanding even by those who do not know [35]. Figures 1 and 2 show the FCM and then the Research Cases.

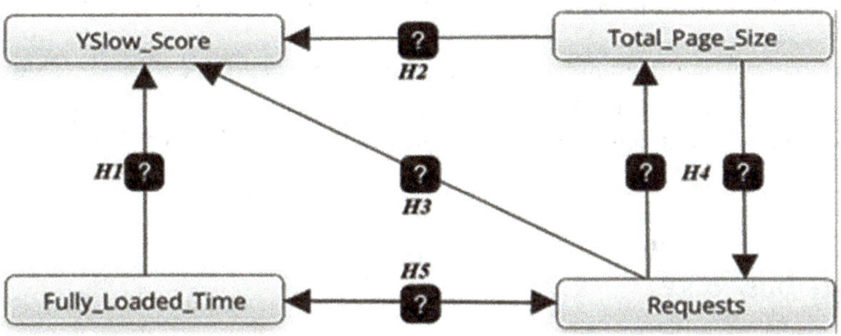

Fig. 1 Fuzzy cognitive representation of possible relationships between the WA metrics via Mental Modeler software as a FCM builder

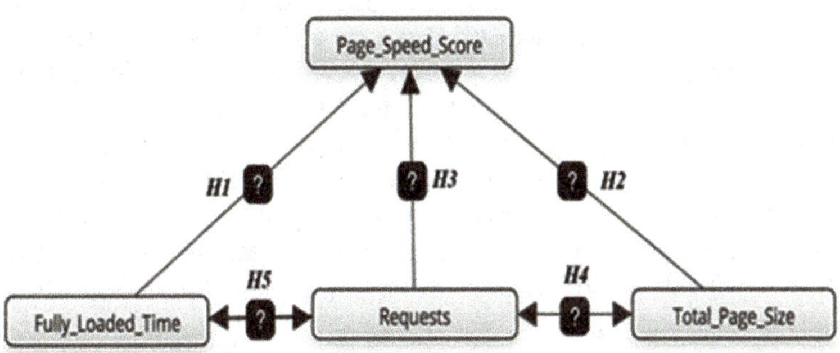

Fig. 2 Fuzzy cognitive representation of possible relationships between the WA metrics via Mental Modeler software as a FCM builder

Based on Figs. 1 and 2, the possible relationships between these WA metrics were evaluated. Figures 1 and 2 were designed on the online editor "Mental Modeler". The formulation of the research questions was completed by taking into account two parameters. On the one hand, the potential benefit that agri-logistics companies will obtain from the answers to these questions. On the other hand, the function they have with each other and the function they have with SEO. With that being said, as far as all research questions are concerned, similar research has been conducted in other research fields. For example, it is widely known that there is a function between Total Page Size and Rankings. Is this function also available in agri-logistics businesses? Hence, all the research questions posed took into account which variables may or may not be related to the rankings and the SEO and even between them.

The research hypotheses set out in the present book chapter, serve to clarify and analyze the factors which are necessary for the improvement of corporate websites within the agri-logistics sector. These research hypotheses are formulated in the following manner under five hypotheses:

- H1: The Fully Loaded Time of the page affects the score rendered by YSlow Score and Page Speed Score Rakings;
- H2: The Total Page Size affects the score rendered by YSlow Score and Page Speed Score Rakings;
- H3: Requests affect the score rendered by YSlow Score and Page Speed Score Rakings;
- H4: Is there a correlation between Requests and Total Page Size for YSlow and Page Speed Score Rakings?
- H5: Is there a correlation between Requests with Fully Loaded Time?

4 Methodology

SEMRush APIs were used to collect website usage data for measurements over a period of 100 days from seven different websites of leading companies in the agri-logistics industry. The process of mining and collecting data on the Internet of transport websites for Page Speed Score, Fully Loaded Time, Total Page Size, Requests, and YSlow Score was carried out on a daily basis in order to have a deeper understanding of the range of variations that occurred in each website throughout the 100-day period. The Spearman factor was used to estimate the possible relationships based on the five hypotheses. This method allows for the calculation of r the possible linear correlations between the WA data over a time period of 100 days. The closer the r is to 1 or -1, the greater the correlation between the measurements. The positive linear r correlation suggests that when a variable yis increased, the variable y also increases. The negative linear correlation r shows that when the variable yy changes, the variable x increases.

5 Hypotheses Testing and Results

Table 4 represents the Spearman correlations ρ in relation to the examined logistics websites. Table 4 further presents descriptive statistics that could be useful for the ABM modeling process (Table 5).

For the H1, the Fully Loaded Time impacts positively the Page Speed Score with $\rho = 0.204$ and the Fully Loaded Time impacts negatively to the YSlow Score with $\rho = -0.267$. This practically means to the analysts that when the Fully Loaded Time *level* of a website is increased, the Page Speed Score is increased as well, and the YSlow Score is decreased.

Table 4 Spearman coefficient of correlations between the WA metrics of logistics website. Abbreviations PSS, YS, FLT, TPS, CC and SIG correspond to Page_speed_score, Yslow_score, Fully_Loaded_Time, Total_page_size, Correlation Coefficient and Sig (two-tailed) respectively

			Correlations				
			Page Speed Score	Yslow Score	Fully Loaded Time	Total Page Size	Requests
Spearman's rho	Page Speed Score	Correlation coefficient	1000[a]	.488[a]	.204[a]	−.165[a]	−.661[a]
		Sig. (two-tailed)	.	.000	.000	.000	.000
		N	840	840	840	840	840
	Yslow Score	Correlation coefficient	.488[a]	1000	−.267[a]	−.223[a]	−.683[a]
		Sig. (two-tailed)	.000	.	.000	.000	.000
		N	840	840	840	840	840
	Fully Loaded Time	Correlation coefficient	.204[a]	−.267[a]	1000	−.025	−.132[a]
		Sig. (two-tailed)	.000	.000	.	.000	.000
		N	840	840	840	840	840
	Total Page Size	Correlation coefficient	−.165[a]	−.223[a]	−.025	1000	.527[a]
		Sig. (two-tailed)	.000	.000	.000	.	.000
		N	840	840	840	840	840
	Requests	Correlation coefficient	−.661[a]	−.683[a]	−.132[a]	.527[a]	1000
		Sig. (two-tailed)	.000	.000	.000	.000	.
		N	840	840	840	840	840

[a]Correlation is significant at the 0.01 level (two-tailed)

Table 5 Spearman coefficient of correlations between the WA metrics of logistics websites

	Mean	Mode	Std. Deviation	Median
Page Speed Score	64.35	36	14.621	70.00
YSlow Score	61.20	58	6.059	60.00
Fully Loaded Time	4.256	4.2	3.1848	4.000
Total Page Size	1.85855	1.810	1.066318	1.71000
Requests	89.90	153	35.774	75.00
N = 100 days	Correlation is significant at the 0.01 level (two-tailed)			

In H2, there is a negative correlation between the Total Page Size with the Page Speed Score and the YSlow Score with $\rho = -0.165$ and $\rho = -0.223$. This means that when the Total Page Size is increased, the Page Speed Score and the YSlow Score are increased as well.

For the H3, there is a negative correlation between the Requests with the Page Speed Score and YSlow Score, with $\rho = -0.661$ and $\rho = -0.683$. This practically means that when the Requests are increased, the Page Speed Score and the YSlow Score are decreased.

In H4, the results showed that there is a positive correlation between the Requests and the Total Page Size that the websites receive with $\rho = 0.527$. That is, the more the Requests on the examined websites, the better the Total Page Size will be.

Lastly, for the H5, there is a negative correlation between the Requests and the Fully Loaded Time with $\rho = -0.132$. This means that when the Requests are increased the Fully Loaded Time is decreased.

5.1 *Predictive Modeling*

The latest research approach is the use of Anylogic in order to simulate the process to make a prediction of the system. Unlike the Anylogic system, the use of statistics represents a mere juncture of time without capturing the entire image of the issue. The present findings confirm the statistical analysis. Following the presentation of the model, this work analyzes the fluctuations occurring over a period of 100 days. Agent based model (ABM) enables marketers to perceive and use all the data they have from their customers and also enables them to better understand their business by analyzing this data for the benefit of both businesses and customers. This will allow researchers to anticipate all those variables that affect the SEO up to a number of important elements of the website. Finally, in contrast to previous studies which used social analysis data sets, the present aggregate quantitative behavioral data was suitable for visualizing real situations and, as such, ABM was adopted [1]. To start, the user (visitor) is encouraged to enter the website through affiliate marketing. This process, which is portrayed as an abscess in an unconditional form because it has not been studied in the present work, which applied communication strategy persuaded him to enter the website (accidental click, information, order). Requests are generated as soon as the visitor enters the page. The model confirms that there is a negative correlation between Requests and Fully Loaded Time. Figure 3 below shows the exact values. Secondly, the Total Page Size, which is made up of the data gathered from the variable triangular model, is presented. Finally, the final rankings are produced through the Fully Loaded Time in the variable rate with triangular model, as shown in Fig. 3 below.

During the 100-day period in which web analytics data was gathered, a plurality of fluctuations in the agents' populations, of the seven (7) agri-logistics websites, was noticed (Fig. 4). As the agents' distributions were examined, it was observed that specific dates showed higher or lower values than others. These are the discrepancies of highest and lowest values in the graph composed by the highest values and lowest values of Requests, Total Page Size, Fully Loaded Time, Page Speed Score and YSlow Score. Tables 6 and 7 indicate the lowest and the highest values of Page Speed Score and YSlow Score because there are the final scoring metrics. The above

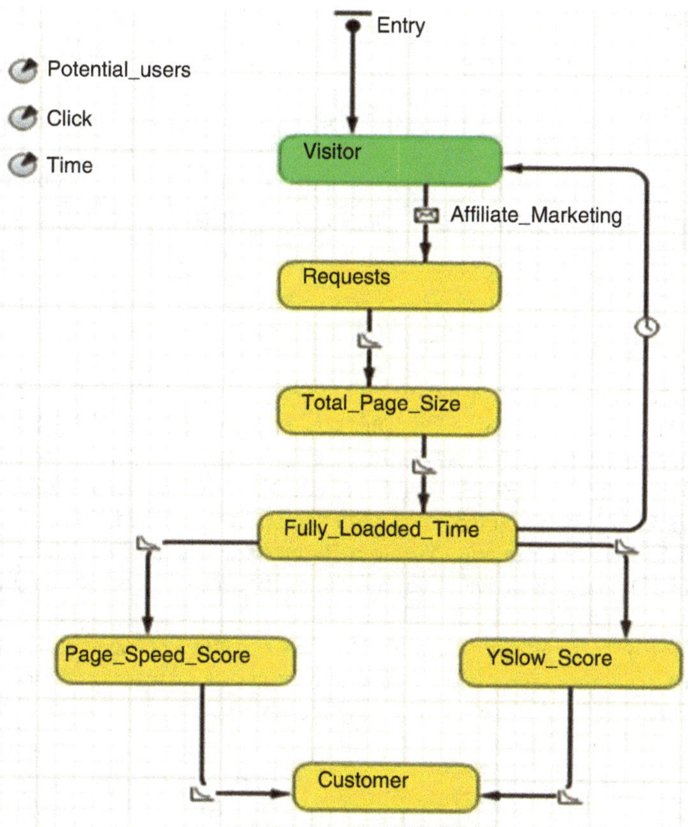

Fig. 3 Agent based modeling elements

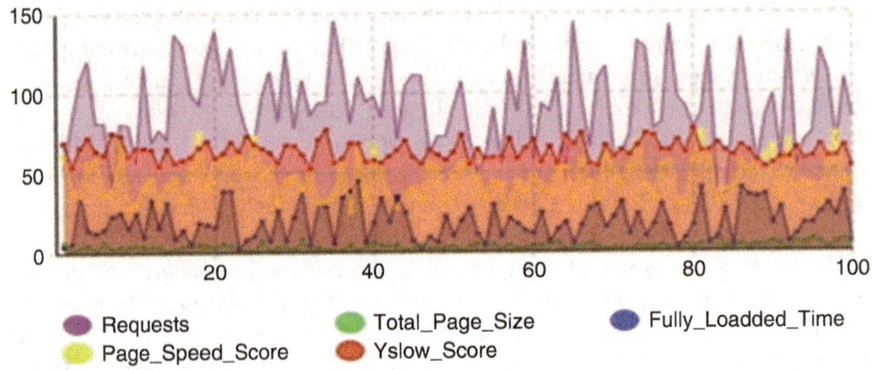

Fig. 4 Range of metrics

Table 6 Specific days of the overall data range and the lowest and highest values of Page Speed Score

Specific days	Lowest values of Page Speed Score	Specific days	Highest values of Page Speed Score
37	25	18	71
58	24	81	71
97	24	98	70

Table 7 Specific days of the overall data range and the lowest and highest values of YSlow Score

Specific days	Lowest value of YSlow Score	Specific days	Highest value of YSlow Score
32	51	7	74
87	52	34	76
100	52	80	76

statistical analysis is confirmed in all variables. More specifically, as for the H1, Fully Loaded Time has a positive effect on Page Speed Score and Fully Loaded Time has a negative effect on YSlow Score. In H2, there is a negative correlation between Total Page Size with Page Speed Score and YSlow Score. As for H3, there is a negative correlation between Requests and Page Speed Score and YSlow Score. In H4, there is a positive correlation between Requests and Total Page Size. And finally, for the H5, there is a negative correlation between Requests and Fully Loaded Time.

In Tables 6 and 7, these days show the highest and lowest prices of the two metrics that provide the score of the websites. In practice, developers and marketers can have a competitive advantage by analyzing these variables.

6 Discussion

This chapter attempted to highlight the potential influences and interactions of Requests, Total Page Size, Fully Loaded Time, Page Speed Score and YSlow Score of agri-logistics websites. Both statistical correlations and modeling and simulation results produced useful suggestions specifically for marketers and developers who manage agri-logistics websites. In this case, statistical analysis and the simulation model had similar results. In addition, developers have the ability to compare all the prices produced as well as the application and their effects. This paper is useful as it combines two fields of research that have not yet been explored.

A difficulty that was often encountered within previous research was an abundance of WA which in turn posed a difficulty in determining which of the presented information was more important [21]. In light of this obstacle, the present work has presented a methodology that first explains and analyzes the metrics of agri-logistics companies in order to reduce misunderstandings about their usefulness. In addition, it has listed a descriptive modeling approach through the Fuzzy Cognitive Mapping in order to understand the dynamics and influences that metrics have on each other. Furthermore, the adopted approach allows developers and marketers to further expand their mapping needs [32].

As a next step, the statistical analysis of web analytics measurements was carried out using the Spearman method. By relying on this method, marketers can better understand which variables are really affecting their system and what the correlations are between these variables.

Finally, previous studies had indicated that ABM was suitable for the social analysis of data sets [1]. In light of these findings, this paper adopted ABM as its aggregate quantitative behavioral data was suitable for showing the real situation. By relying on ABM, it is possible to combine previous experience and knowledge using useful information from the generated data.

This research contributes to the modeling construction based on agent-based model and more specifically to the quantitative approach using the metrics drawn. It emphasized the production of correlations between the variables that contributed to the behavior of agents [2]. Finally, when the model was tested, the above statistical analysis was confirmed in all variables. Furthermore, this paper sets out the largest and lowest prices of metrics that provide a score in order to assists marketers and developers within their decision-making process. By relying on these types of agent-based models, marketers and developers are able to better analyze obscure and complex data.

This research, through the extraction and management of consumer behavior data, is part of the research project in digital marketing and optimization strategy. The branch of application is in agri-logistics companies. Further research is needed in the field of agri-logistics with the help of FCM based models because it will lead to the long-term forecasts for the studied industry.

The authors of this study have already begun examining the entire range of web analytics in agri-logistics companies, including measurements of paid advertising, social media loyalty measurements and measurements that may affect the presentation of websites and the ranking of search engines.

References

1. Kavak, H., Padilla, J. J., Lynch, C. J., & Diallo, S. Y. (2018, April). Big data, agents, and machine learning: towards a data-driven agent-based modeling approach. In *Proceedings of the Annual Simulation Symposium* (p. 12). Society for Computer Simulation International.
2. An, L. (2012). Modeling human decisions in coupled human and natural systems: Review of agent-based models. *Ecological Modelling*, *229*, 25–36.
3. Porter, M. (2001) Strategy and the Internet, Harvard Business Review (March), 62–78.
4. Yew Wong, C. and Karia, N. (2010). Explaining the competitive advantage of logistics service providers: A resource-based view approach. *International Journal of Production Economics*, 128(1), pp. 51–67.e
5. Lai, F., Zhao, X. and Wang, Q. (2006). The impact of information technology on the competitive advantage of logistics firms in China. *Industrial Management & Data Systems*, 106(9), pp. 1249–1271.
6. Hennig-Thurau, T., Malthouse, E., Friege, C., Gensler, S., Lobschat, L., Rangaswamy, A., & Skiera, B. (2010). The impact of new media on customer relationships. Journal of Service Research, 13(3), 311–330.
7. Wilson, R. D. (2010). Using clickstream data to enhance business-to-business web site performance. Journal of Business & Industrial Marketing, 25(3), 177–187.
8. Chaffey, D. and Ellis-Chadwick, F. (2016). *Digital marketing*. Harlow: Pearson.
9. McDonald, M. and Wilson, H., (1999). *E-Marketing*. Harlow: Financial Times Prentice Hall [for] Cranfield University School of Management.
10. Levy, S. and Gvili, Y., 2015. How Credible is E-Word of Mouth Across Digital-Marketing Channels?. *Journal of Advertising Research*, 55(1), pp. 95–109.
11. Smith, K., 2010. Digital Marketing Strategies that Millennials Find Appealing, Motivating, or Just Annoying. *SSRN Electronic Journal*.
12. Sheth, J. and Sharma, A., 2005. International e-marketing: opportunities and issues. *International Marketing Review*, 22(6), pp. 611–622.
13. Agarwal, A., Hosanagar, K. and Smith, M., 2008. Location, Location, Location: An Analysis of Profitability of Position in Online Advertising Markets. *Journal of Marketing Research*, 48.
14. Yalçın N., & Köse U., (2010) "What is search engine optimization: SEO?", World Conference on Learning, Teaching and Administration Papers Procedia - Social and Behavioral Sciences, Vol. 9, pp. 487–493.
15. Barham, H., 2017. Achieving Competitive Advantage Through Big Data: A Literature Review. *2017 Portland International Conference on Management of Engineering and Technology (PICMET)*.
16. SAS Institute Inc. (2015) THE POWER TO KNOW. Available at: https://www.sas.com/en_us/insights/big-data/big-data-marketing.html. [Accessed 1 September 2019]
17. Tankard, C. (2012). Big data security. *Network Security*, 2012(7), pp. 5–8.
18. Soubra, D. (2012) The 3Vs that define Big Data. Blog post published 5 July 2012, www.datasciencecentral.com/forum/topics/the-3vs-that-define-big-data.
19. Richey, R. Morgan, T., Lindsey-Hall, K. and Adams, F. (2016), "A global exploration of Big Data in the supply chain," *International Journal of Physical Distribution & Logistics Management*, 46 (8), 710–739.

20. Wu, L., Xiaohang Y, Jin, A. and Yen, D. (2016) "Smart supply chain management: a review and implications for future research", *The International Journal of Logistics Management*, 27 (2), 395–417
21. Järvinen, J. and Karjaluoto, H., (2015). The use of Web analytics for digital marketing performance measurement. Industrial Marketing Management, 50, pp. 117–127.
22. Gtmetrix.com. (2019). *GTmetrix | Website Speed and Performance Optimization*. [online] Available at: https://gtmetrix.com/ [Accessed 6 Aug. 2019].
23. Web Technology Surveys. (2019). Usage of traffic analysis tools for websites. August 2019. Available at: http://w3techs.com/technologies/overview/traffic_analysis/all [Accessed 30 August 2019].
24. Järvinen, J., Töllinen, A., Karjaluoto, H., & Jayawardhena, C. (2012). Digital and social media marketing usage in B2B industrial section. Marketing Management Journal, 22(2), 102–117.
25. Germann, F., Lilien, G., & Rangaswamy, A. (2013). Performance implications of deploying marketing analytics. International Journal of Research in Marketing, 30(2), 114–128.
26. Drivas, I., Sakas, D. and Reklitis, P. (2019). Improving Website Usability and Traffic Based on Users Perceptions and Suggestions—A User-Centered Digital Marketing Approach. *Strategic Innovative Marketing*, pp. 255–266.
27. Sarlis, A., Drivas, I. and Varveris, A. (2017b). The Cooperative Role of Marketer and Programmer on SEO Strategies in Scientific Journals. *Strategic Innovative Marketing*, pp. 429–435.
28. Duran, M. (2020). *YSlow - Official Open Source Project Website*. [online] Yslow.org. Available at: http://yslow.org/ [Accessed 11 Jan. 2020].
29. SEOptimer: SEO Audit & Reporting Tool. White Label Option. (2020). *The Importance Of Page Size And How It Affects Your Seo | SEOptimer*. [online] Available at: https://www.seoptimer.com/blog/webpage-size/ [Accessed 11 Jan. 2020].
30. Sarlis, A., Drivas, I. and Sakas, D. (2017a). Implementation and Dynamic Simulation Modeling of Search Engine Optimization Processes. Improvement of Website Ranking. *Strategic Innovative Marketing*, pp. 437–443.
31. Hair, J., 2007. Knowledge creation in marketing: the role of predictive analytics. European Business Review, 19(4), pp. 303–315.
32. Glykas, M.(ed.) (2010). Fuzzy Cognitive Maps. Springer
33. Salmeron, Jose L.; Papageorgiou, E. (2012). "A Fuzzy Grey Cognitive Maps-based Decision Support System for Radiotherapy Treatment Planning". Knowledge-Based Systems. 30 (1): 151–160.
34. Kosko, B., (1993). *Fuzzy Thinking*. New York: Hyperion.
35. Vergini, E. & Groumpos, P., (2017). New concerns on fuzzy cognitive maps equation and sigmoid function. In Control and Automation (MED), 2017 25th Mediterranean Conference on (pp. 1113–1118). IEEE.
36. Fernie, J., Sparks, L. and Mckinnon, A.C. (2010) Retail logistics in the UK: past, present and future, International Journal of Retail and Distribution Management, 38(11–12), 894–914.
37. Google Developers. (2020). *About PageSpeedInsights Google Developers*. [online] Available at: https://developers.google.com/speed/docs/insights/v5/about [Accessed 11 Jan. 2020].
38. Mussomeli, A. and Gish, D. (2016), *"The rise of the digital supply network"*, Research Report Deloitte University Press.
39. *REGULATION (EU) 2016/679 OF THE EUROPEAN PARLIAMENT AND OF THE COUNCIL of 27 April 2016 on the protection of natural persons with regard to the processing of personal data and on the free movement of such data and repealing Directive 95/46/EC (General Data Protection Regulation)* (2016). [online] Available at: https://eur-lex.europa.eu/legal-content/EN/TXT/HTML/?uri=CELEX:32016R0679&from=EN [Accessed 23 Aug. 2019].
40. Sanders, N. (2016), "How to Use Big Data to Drive Your Supply Chain," *California Management Review, 58*(3), 26–48.
41. Thucydides, Warner (TRANSLATOR), R. and Finley (EDITOR), M. (1954). *History of the peloponnesian war*. Harmondsworth, Middx., Engl.: Penguin Books.